PESTICIDE FORMULATIONS

PESTICIDE FORMULATIONS

edited by **WADE VAN VALKENBURG** (cd)

3M Company
Central Research Laboratories
St. Paul, Minnesota

1973

MARCEL DEKKER, INC.   New York

PREFACE

The cost of discovering and developing a new pesticide runs between five and ten million dollars. The time lag between discovery and marketing averages 77 months: 7430 compounds are screened for every marketable toxicant. The number of man years of effort to clear a new pesticide for the U.S. market averages 65.* Economic and manpower commitments of this magnitude have required that increased attention be directed towards greater understanding of all aspects of pesticidal formulations.

This is not a "how to" formulations recipe book. Rather, its objectives are to foster a greater understanding of fundamental principles involved in research on pesticidal formulations.

Once a candidate pesticide is mixed with anything (a slurry of the compound in water qualifies) the composition may be considered a pesticidal formulation. Hence, any research on a combination of an active compound and a second material is research on a pesticidal formulation. Any instance where the added material will affect the chemical, physical, and biological properties of an active biocide is fair game for this book.

Let us briefly survey some of the principles touched upon in this volume. The biological activity of a compound can be mathematically correlated with experimentally determined physical properties of that compound. Two parameters which correlate very well are (1) a solubility coefficient denoting the ability of a compound to move to an active site and (2) the Hammett sigma constant of one or more substituents which can be related to the reactivity of a compound at the active site. An inert ingredient is capable of changing the apparent solubility coefficient of the toxicant. Hence it is logical that the current new technology on biological correlations be presented as a first chapter in this book.

A few additional principles emphasized in the following chapters include (1) the importance of solubility relationships as indicated by the HLB system in selecting an emulsifier, (2) the effect of the phase inversion temperature on emulsion stability, (3) the dehydrating effect of fertilizer salts on emulsifiers and the resultant stability or instability of the emulsion,

*Chemical Week, July 26, 1972, page 18.

(4) the catalytic effect of clays and other carriers on the degradation of a pesticide, (5) the all-important flowability property of a mixture as it relates to the successful manufacture of a dry pesticidal formulation, and (6) the physical properties of formulations and how they affect the size of the particles in a spray and the movement of the particles in air.

The physical properties of a formulation and its resultant spray solution as indicated by particle size and surface properties (surface tension, contact angles, spreading coefficient) will dictate whether or not a spray droplet will adhere to a plant surface. Once a pesticide is on a target site it is faced with a lipid barrier which must be penetrated to affect biological activity. An understanding of the physical and chemical properties of this barrier can aid the formulator in designing his composition for optimum penetration and efficacy. And finally, when the formulation finds its way to the soil, the adsorption characteristics of the ingredients of the composition on soil colloids will dictate how far the pesticide will move with possible concomitant danger of contamination of soils and ground waters far removed from the site of application.

It is obvious that one should not routinely add inerts to pesticides just to get the right handling properties. One must consider the effect of the additives on the biological activity of the toxicant, on processing, and on the movement on or in the host and its environment. Through a better understanding of all the principles involved, one hopes that the formulator will be able to optimize his formulations in terms of efficacy and reduction of deleterious side effects.

The scope of this book is very broad, as is the whole subject of the evaluation of pesticidal formulations. It could not have been written without the expert cooperation of all the contributors to this volume. They are all recognized experts in their fields and I express my sincere gratitude to them all for participating in this endeavor.

St. Paul, Minnesota   Wade Van Valkenburg
October, 1972

CONTRIBUTORS TO THIS VOLUME

NORMAN B. AKESSON, University of California, Davis, California

D. E. BAYER, Department of Botany, University of California, Davis, California

PAUL BECHER, ICI America Inc., Wilmington, Delaware

V. H. FREED, Department of Agricultural Chemistry and Environmental Health Science Center, Oregon State University, Corvallis, Oregon

R. HAQUE, Department of Agricultural Chemistry and Environmental Health Science Center, Oregon State University, Corvallis, Oregon

D. R. JOHNSTONE, Ministry of Overseas Development, Tropical Pesticides Research Unit, Porton Down, Salisbury, Wilshire, England [*]

PAUL LINDNER, Witco Chemical Company, Chicago, Illinois

J. M. LUMB, Department of Botany, University of California, Davis, California

JAMES A. POLON, Industrial Products Research, Minerals & Chemicals Division, Engelhard Minerals & Chemicals Corporation, Menlo Park, Edison, New Jersey

WADE VAN VALKENBURG, The 3M Company, St. Paul, Minnesota

J. F. VINE, C. Eng. M. I. Mech. E., Chigwell, Essex, England

C. F. WILKINSON, Department of Entomology and Limnology, Cornell University, Ithaca, New York

WESLEY E. YATES, University of California, Davis, California

[*]Present affiliation: Overseas Development Administration, Centre for Overseas Pest Research, Division of Chemical Control.

CONTENTS

PESTICIDE FORMULATIONS

Chapter 1

CORRELATION OF BIOLOGICAL ACTIVITY WITH CHEMICAL STRUCTURE AND PHYSICAL PROPERTIES

C. F. Wilkinson

Department of Entomology and Limnology
Cornell University
Ithaca, New York

I. INTRODUCTION

Throughout the course of the past century and particularly during the last two decades, we have witnessed a remarkable "chemical revolution" by which man has achieved a great measure of success in the development and utilization of synthetic organic chemicals. Although these have been, and continue to be, of great benefit to almost all aspects of modern life, there is little doubt that the most remarkable advances have been made in the chemical control of those organisms which are either directly or indirectly opposed to man's well-being. Thus the widespread use of a vast number of synthetic agricultural chemicals which include insecticides, herbicides, fungicides, and veterinary drugs has dramatically increased both the quality and quantity of the food and fiber derived from our crops and domestic animals. Similarly our suffering has been greatly alleviated and life expectancy enhanced by the successful development of drugs to control diseases associated with pathogenic organisms and of chemical correctives to counterbalance our natural deficiencies. Because the one common property of all these chemicals resides in their ability to interact with living systems they can all be said to possess some form of biological activity. The fact that many are biocidal indicates that this activity has lethal consequences to some forms of life.

The first organic materials employed by man for various medicinal and agricultural purposes were generally of natural origin and unknown structure. As the chemical nature of the active components of these materials was recognized and the compounds themselves were isolated, characterized, and subsequently synthesized in pure form, considerable interest was shown in possible relationships which might exist between their chemical structure and biological effect. Interest in this area has continued to increase and today investigations involving the correlation of chemical structure with biological activity have far-reaching ramifications in many different fields. For the chemist and biochemist such studies may serve to provide information on possible reaction mechanisms or on the structural nature of the biochemical receptor sites involved in certain processes. For the toxicologist they can often afford an important method of studying the mode of action of toxic agents. Perhaps most important of all, correlation studies have considerable practical application in providing some rationale to aid the designer of new and potentially more effective drugs and agricultural chemicals.

The reader is referred to the following books, reviews, and discussions all of which excellently cover many aspects of this vast field of interest (1-5). Although these are largely concerned with medicinal and pharmaceutical chemistry, all cover important principles of general relevance to any consideration of materials possessing biological activity.

This chapter constitutes a discussion of some of the major factors involved in considerations of structure-activity relationships and of mathematical approaches which attempt to express these factors in quantitative terms. Although the examples given to illustrate specific points will be taken largely from the field of agricultural chemistry and will mainly involve chemicals of a toxic nature, it should not be construed that the importance of structure-activity relationships is limited to those chemicals which interfere with normal living systems. On the contrary, it should be emphasized that the normal functioning of any living organism depends on a multiplicity of highly specific, highly organized relationships between the structure and biological activity of its naturally occurring chemical components.

II. BIOLOGICAL ACTIVITY

A. General

All living organisms are chemically dynamic systems; they behave and function as living entities as a direct result of an amazing complexity of interdependent chemical reactions which although in continuous flux are maintained at any given time in a delicate state of balance. The presence of a "foreign" chemical within a living system can readily upset this balance by enhancing, inhibiting, or otherwise interacting with one or more of the chemical reactions or components on which its integrity depends. Such a chemical can be said to possess some form of biological activity.

Biological activity can take many different forms and may be measured in different ways depending on the level at which the investigation is conducted. When the critical site and mechanism of action of a chemical are known, biological activity can be measured directly in terms, for example, of the degree of inhibition or enhancement of an enzyme system as measured in vitro. More usually, however, biological activity is measured in an indirect manner through in vivo observations of the end result of the chain of events initiated by the interaction of the chemical with some unknown biochemical component. In the case of a pesticide, for instance, it is customary to measure biological activity in terms of the per cent mortality of an organism without necessarily having any knowledge of the mode of action of the material at the molecular level. More strictly defined this should be termed biological effect or response. The observable in vivo response to some biologically active chemical is often difficult to relate to the critical disturbance at the molecular level from which it results. Thus in the case of a mammal poisoned with an organophosphorus insecticide one would not readily connect death through respiratory failure, the observable biological effect, with the inhibition of cholinesterase which constitutes the true biological activity of this particular group of compounds.

B. Factors Determining Biological Effect

Studies involving the correlation of biological activity with chemical structure are extremely complex and until recently have been based almost entirely on empirical observations of a qualitative or semi-quantitative nature. This results largely from the complex series of events which can take place between the initial application of a material to an organism and its arrival at, and subsequent interaction with, a biological receptor (Fig. 1). These factors which effectively compete with the receptor for the chemical include: failure to penetrate and translocate to the site of action, storage in inert tissues, and degradative metabolism and excretion. The relative importance of any one factor depends on the physical and chemical characteristics of the material in question. A chemical, therefore, must not only possess the correct structure to interact with a specific receptor, but must also incorporate structural features which will allow it to successfully circumnavigate these competing factors, each of which constitutes a potential barrier to prevent the material from reaching its site of action. As a result, the interpretation of data obtained in structure-activity investigations usually increases in complexity as the level of the investigation moves from the true in vitro system to that existing in the intact organism.

When structure-activity investigations are carried out in vitro, the material of some specific tissue or organ is homogenized and the enzyme or other cell component under investigation is isolated and perhaps purified to some extent. Under these relatively uncomplicated conditions the chemical under consideration can be placed effectively in direct contact with the target or receptor site, and the results usually serve as a good qualitative indicator of the absolute structural features required to effect a specific type of biological activity. In view of the high degree of complementary character which must exist between a chemical and its biological receptor, it is likely that structure-activity relationships determined in vitro will truly reflect the structure of the receptor surface. Consequently a great deal of our present knowledge regarding the three-dimensional structure of biological receptors has been obtained from in vitro studies on the structural nature of the chemicals with which they interact.

The in vitro activity of any chemical depends primarily on steric factors such as size, shape, and stereochemical configuration. It is these properties which determine the relative position of specific substituent groups

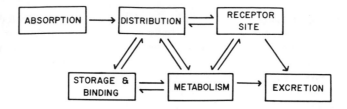

FIG. 1. The interactions of a chemical between its initial application
and its reaction with a target receptor.

through which binding and/or chemical reaction with the receptor takes
place. The actual types of interactions between chemicals and cellular
components are many and varied, ranging from the largely irreversible
formation of covalent bonds to rather loose, usually reversible complexes
resulting from hydrogen bond formation, ionic and dipole interaction,
van der Waals forces, or hydrophobic bonding. Most of these interactions are effective over only relatively short intermolecular distances so
that optimal biological activity is obtained only if the molecular size and
stereochemical configuration of the chemical allows it to come into close
juxtaposition with the relevant receptor surface.

Although still considered to be at the in vitro level, the results of
studies involving the use of intact isolated cells in the form of either tissue slices or discrete organs become increasingly more difficult to evaluate. In this case the chemical must not only possess the properties previously discussed in relation to its activity at the receptor surface but, in
addition, must incorporate properties which allow it to traverse one or
more of the lipophilic membranes or ion impermeable barriers which will
otherwise prevent it from reaching its site of action. Under these conditions it is fairly obvious that the physical properties of the chemical can
have a marked quantitative effect on its biological activity even if the material possesses the necessary structural requirements to interact at the
molecular level. As we shall see such properties as lipid/water partition coefficients and ionic dissociation often play a dominant role in determining biological activity.

In the intact living organism a chemical must meet still further structural criteria before it is able to elicit a biological response. First it
must be capable of reaching its site of action which may be far removed
from its point of application to the organism. This involves penetration
through such tissue as mammalian skin, the highly lipophilic epicuticular
layer of insect cuticle, the polysaccharide phosphoprotein materials of
bacterial membranes, or the outer cuticle and cellulose cell walls of

plant tissues. Having achieved the penetration of this outer protective
sheath the material must move relatively freely through a number of lipo-
protein cell membranes to some site where it must seek out and interact
with a specific receptor. During this translocation process the chemical
is often exposed to the action of strong acids, as in the mammalian gastric
juices, or to alkalis, as in the gut contents of many lepidopterous larvae.
It must in addition be capable of withstanding the potentially degradative
action of a multiplicity of enzyme systems and must avoid being effectively
removed from the system through binding with the large variety of inert
proteinaceous and lipophilic materials with which it comes into contact.

For these reasons considerable caution should always be exercised in
attempts to predict the probable in vivo effect of a material based solely
on information regarding its in vitro performance. Often a chemical which
demonstrates a high degree of biological activity in vitro proves to be entirely inactive when applied to the intact organism. Alternatively metabolic
alteration of the chemical in vivo may produce a compound of much greater
activity than would be suggested by the in vitro activity of the parent compound. The insecticide parathion is inactive as an inhibitor of cholinesterase in vitro; however, it is extremely toxic to many forms of life as a result of its oxidative conversion in vivo to the potent anticholinesterase
paraoxon.

In summary therefore the biological activity of any material is governed
by several major factors which include its ability to successfully penetrate
the organism and to subsequently translocate to its site of action, its ability to avoid binding and storage in inert components and tissues, its ability
to withstand the action of degradative enzymes, and ultimately of course,
its ability to interact with some essential biological receptor. The extent
to which a chemical satisfies any of these requirements is a function of its
chemical structure and physical properties.

III. ABSORPTION AND DISTRIBUTION

A. General

In order to exert its biological effect a chemical must be capable of
penetrating the several barriers frequently interposed between its point
of application to the organism and its biochemical receptor site. When
a drug or pesticide has the necessary structural characteristics for biological activity in vitro, its failure to initiate a response in the intact organism often results from the fact that the material does not possess the
necessary physico-chemical properties which allow it to cross one or

more of these biological barriers. The nature of the barriers likely to be encountered is determined by the route of application, the target site, and the type of organism involved. The barriers may be broadly classified as either external or internal.

The external barriers constitute the outer integuments of living organisms and as such have a primary protective function. They include such materials as insect cuticle, mammalian skin, bacterial membranes, and the outer cuticle of living plants. They are often of great importance in determining the penetration of "foreign compounds."

In addition to these integumental barriers, others surround the tissues of internal organs. Included in this category are membranes such as the gastro-intestinal epithelium and the plasma-cerebrospinal fluid barrier of mammals, and the somewhat analogous mid-gut epithelium and ganglionic nerve sheath of insect species. Not only do these membranes protect delicate tissues from mechanical damage, but they also serve an extremely important role in selectively determining the nature of the compounds to which they will allow ingress. Other membranes surround individual tissue cells and even the intracellular organelles such as mitochondria and nuclei but these seldom present additional penetration problems and will not be given special consideration.

Penetration of both the outer integument and inner tissue membranes of an organism is determined mainly by the physico-chemical properties of a material. Since the major principles involved in the penetration of materials through internal membranes are more clearly understood and are of general applicability these will be discussed prior to considerations of integumental penetration.

B. Membrane Penetration

In view of the largely aqueous cellular environment and the fact that most normal metabolites are highly polar, water-soluble materials, it is perhaps not surprising that order is maintained, and a large degree of protection afforded by internal membranes of a characteristically lipophilic nature. Most biological membranes are considered to have a similar structure (6), comprising an inner bimolecular layer of lipids covered on each side by a layer of protein. The lipid molecules are orientated perpendicularly to the membrane surface and their hydrophilic ends are closely associated with groups on the protein.

There already exist a number of excellent and comprehensive reviews concerning the physico-chemical factors which determine the penetration and transport of drugs and pesticides through biological membranes (7-11). These reviews indicate that the ability of any chemical to penetrate a biological membrane depends largely on its lipid/water partition coefficient and that the lipophilic "foreign compounds," with which we will be chiefly concerned, cross membranes by a process of simple diffusion.

Lipophilic materials will pass from an aqueous environment into the lipid phase of the membrane to an extent determined by their lipid/water partition coefficients. A diffusion gradient is established across the membrane and the material is subsequently transported into the aqueous medium on the other side according to the physical laws governing partition equilibria. Theoretically, movement of a substance across the membrane will continue until at equilibrium the ratio of its concentration on each side of the membrane reaches unity. With those substances of low lipid solubility, however, such a ratio is seldom attained. Passage through the membrane itself takes place at a rate determined by Fick's law of diffusion, which predicts that the rate of penetration will obey simple first-order kinetics. This can be expressed by Eq. (1), where P is the permeability constant

$$P = \frac{1}{t} \log_n \frac{c}{c-x} \qquad (1)$$

(first-order rate constant), c is the initial concentration of the chemical on one side of the membrane, and x is the decrease in this concentration at time t. An important consequence of the fact that first-order kinetics are obeyed is that the time required for the penetration of any definite fraction of a chemical is independent of its initial concentration. Thus the measurement of the half-time of penetration $(t_{1/2})$ affords an important parameter for a material and can be used in comparative investigations as well as in the calculation of P, as shown in Eq. (2).

$$P = \frac{0.693}{t_{1/2}} \qquad (2)$$

Organic compounds which are highly polar and therefore of low lipid solubility will penetrate membranes only with difficulty, and completely ionized materials will effect passage even less readily unless they contain large lipophilic groups (8, 9). In addition to the fully ionized organic electrolytes (such as quaternary ammonium compounds) a number of compounds used as drugs and pesticides are ionizable organic acids or bases whose degree of ionization is a function of their dissociation constants, K_a, and the pH of the surrounding medium. For example, the strength of a base (that is, its ability to accept a proton) is expressed by its pK_a (-log K_a) which is the pH of the medium at which it is 50% ionized (protonated). The stronger a base, the higher its pK_a value. The degree of ionization at any pH can readily be determined by the Henderson equation which for a base can be expressed:

$$pK_a - pH = \log_{10} \frac{c \text{ ionized}}{c \text{ un-ionized}} \tag{3}$$

and for an acid:

$$pK_a - pH = \log_{10} \frac{c \text{ un-ionized}}{c \text{ ionized}} \tag{4}$$

As a result of the fact that only the un-ionized form of the compound shows any appreciable degree of lipophilicity, the ability of an organic acid or base to penetrate biological membranes will depend on its degree of ionization (determined by its pK_a and the pH of the surrounding medium) and the lipid/water partition coefficient of the un-ionized material.

Evidence supporting the importance of these factors is demonstrated by the penetration of the plasma-cerebrospinal fluid (CSF) barrier by certain ionizable drugs (12, 13). This barrier constitutes a typical lipophilic membrane separating the blood plasma from the cerebrospinal fluid of the mammalian central nervous system, and as such appears to be largely impermeable to ionized compounds. Although different drugs are found to penetrate the membrane at vastly different rates, penetration (as measured by the rate constant P) is found to follow the first-order kinetics predicted by Fick's law (12, 13). Two major factors are found to limit the rate of penetration of any particular drug. With those compounds which are mainly ionized at pH 7.4 (the approximate pH of both plasma and cerebrospinal fluid), the rate of penetration is found to vary in a manner determined by the pK_a of the drug and therefore its degree of ionization at this pH value (Table 1). The penetration rate increases in direct relation to the amount of material in the un-ionized form. In the case of those drugs which are mainly un-ionized at pH 7.4, the degree of ionization is no longer the rate limiting step and the rate of penetration is found to correlate well with the lipid solubility of the un-ionized form of the material as determined by its heptane/water partition coefficient (Table 1).

A similar situation exists with regard to the absorption of materials through the gastric mucosa of the mammalian alimentary tract. Thus the efficiency of absorption of orally administered drugs depends primarily on their degree of ionization in the acidic contents (pH 1) of the gastric lumen. In general it is found that weak acids ($pK_a < 2.5$) and very weak bases ($pK_a > 2.5$) which are mostly in their un-ionized form at pH 1 readily enter the plasma through the gastric mucosa (9, 14).

TABLE 1

Penetration of Drugs into Cerebrospinal Fluid of Dogs[a]

Drug	pKa	Per cent un-ionized at pH 7.4	Heptane/water partition coeff un-ionized drug	Permeability const
Drugs mainly ionized at pH 7.4				
5-Sulfosalicylic acid	(strong)	0		0.0001
N-Methylnicotinamide	(strong)	0		0.0005
5-Nitrosalicylic acid	2.3	0.001		0.001
Salicylic acid	3.0	0.004		0.006
Mecamylamine	11.2	0.016		0.021
Quinine	8.4	9.09		0.078
Drugs mainly un-ionized at pH 7.4				
Sulfaguanidine	10.0	99.8	0.001	0.003
N-acetyl-4-amino antipyrine	0.5	99.9	0.001	0.012
Barbitone	7.5	55.7	0.002	0.026
Antipyrene	1.4	99.9	0.005	0.12
Pentobarbitone	8.1	83.4	0.05	0.17
Amidopyrene	5.0	99.6	0.21	0.25
Aniline	4.6	99.8	1.1	0.40
Thiopentone	7.6	61.3	3.3	0.50

[a]Data taken from Brodie et al. (9, 13).

Although physico-chemical properties such as lipid/water partition coefficient and pK_a often provide extremely useful parameters from which the
medicinal chemist or insecticide designer can predict the rate of penetration of a new compound, it should not be inferred that biological membranes
act as absolute barriers to all ionized molecules. There is little doubt that
the penetration rates of polar materials are usually greatly reduced, but
considerable evidence presently exists to indicate that many ionized compounds can penetrate biological membranes quite readily. As a result the
ion barrier effect should be considered relative rather than absolute.

This viewpoint has been emphasized by Eldefrawi, O'Brien and colleagues (15-20) who reported the results of comprehensive investigations
on the penetration of organic ions into insect ganglia. The latter comprise
the central nervous system of insect species and are invested in a nerve
sheath, the neural lamella, which appears to have somewhat similar characteristics to the mammalian plasma-cerebrospinal fluid barrier. It is
now recognized that in insects the acetylcholine-cholinesterase system is
confined entirely within the ganglia of the central nervous system and is not
involved, as it is in mammals, in nerve impulse transmission at the per-
ipheral neuromuscular junction. As a result, anticholinesterases such as
the organophosphates and carbamates must penetrate the ganglionic nerve
sheath in order to promote insecticidal action. Investigations on the
physico-chemical properties of materials to which it will allow access have
therefore attained significant toxicological importance.

Working with a homologous series of $1-^{14}C$ labeled fatty acids, Eldefrawi
and O'Brien (15, 20) found that even at pH 7.0 when the acids are virtually
completely ionized, penetration of the acidic anions into the ganglia of the
American cockroach (Periplaneta americana) occurs at a rate which is directly correlated with their octanol/water partition coefficients (Fig. 2).
As a result of the decrease in penetration rate observed in the presence of
2,4-dinitrophenol it was concluded that metabolism of the acids, effectively
resulting in their removal from within the ganglia through fixation in some
non-diffusible form, tended to increase the rate of influx. This is the probable explanation of the rather high in/out molar ratios observed with the
fatty acids (Fig. 2) with which the exception of acetate considerably exceed
the theoretical ratio of 1.0. The general significance of metabolism was
further substantiated by the finding that the influx into the nerve cord of
acetylcholine itself was markedly decreased in the presence of $10^{-4}M$ eser-
ine, which blocked the cholinesterase responsible for acetylcholine hydrolysis (16, 20).

Similar correlations between penetration rates and octanol/water partition coefficients were obtained in investigations with a homologous series
of aliphatic alcohols (17), but in this case a cut-off point at butanol indicated

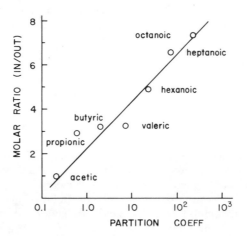

FIG. 2. The relationship between partition coefficient and penetration
into cockroach nerve cord of a homologous series of fatty acids [after
Eldefrawi and O'Brien (15)].

the existence of some limiting factor possibly associated with either molecular size or free diffusion rate.

In addition, the effect of molecular size on the rate of penetration of
materials into cockroach ganglia was investigated with two series of alkyl
substituted quaternary ammonium compounds (16, 20). In one series,
$RN^+(CH_3)_3$, increases in the alkyl chain R were effected without changing
the minimum cross-sectional diameter of the molecule, whereas in the
other series, $R_3N^+C_2H_5$, the minimum diameter increased as the series
ascended. In the first series penetration rate increased with increasing
apolarity with the exception of the rapid rate of penetration observed with
the first member of the series (where R = CH_3). In the second series the
results indicated a decrease in influx rate with increasing minimum molecular diameter from R = CH_3 up to R = C_4H_9. An increased rate of penetration of the compound in which R = C_5H_{11} indicated that the increase in apolarity of this compound more than compensated for the additional increment
in minimum molecular diameter. Minimum molecular diameter tends to
progressively decrease as the series is ascended due to folding of the longer
alkyl chains.

These investigations establish without doubt that organic anions and cations can indeed penetrate lipophilic membranes. Furthermore, it is found
that the first-order rate constants for these materials fall within a rather

similar range to those observed with nonionic materials (20). It is equally clear however that, as previously discussed, membranes do present a relative barrier to ionized materials and the degree to which this is effective can conveniently be assessed by observing the pH dependent penetration rates of ionizable materials. Thus, Eldefrawi and O'Brien demonstrated that the penetration rate of butyric acid into the cockroach nerve cord markedly decreased with increasing pH and that the reverse was observed with the ionizable base butylamine (Fig. 3). In contrast, the penetration rates of both un-ionizable butanol and of the butyltrimethylammonium ion were unaffected by pH, although as expected penetration of the latter was relatively small. From this and related investigations it appears that the ganglionic nerve sheath of insects introduces a barrier effect which slows the penetration of ionized materials 5-15-fold (19, 20). This undoubtedly accounts for the low insect toxicity of many ionized and ionizable anticholinesterases which are highly toxic to mammals where the acetylcholinecholinesterase system is not protected by an ion barrier. Thus the organiphosphate, Amiton [1], a base of pK_a 8.5, is 330-fold more toxic to the mouse than to the housefly (21). This fact correlates with the relatively poor penetration of the ionized form of Amiton into the cockroach nerve cord (22). Similarly, a number of ionized carbamates such as prostigmine

$$(C_2H_5O)_2 \ P {\overset{\displaystyle \nearrow O}{\diagdown}} \ SCH_2CH_2 \overset{+}{N}H(C_2H_5)_2$$

[1] Protonated

$$(CH_3)_2 N \overset{\displaystyle O}{\overset{\|}{C}} O - \langle \ \rangle - \overset{+}{N}(CH_3)_3$$

[2]

$$\langle \ \rangle \overset{\underset{\displaystyle N+}{|}}{\underset{(CH_3)_3}{}} O \overset{\displaystyle O}{\overset{\|}{C}} NHCH_3$$

[3]

[2] and the quaternary analogs of m-dimethylaminophenyl N-methylcarbamate [3] are of low toxicity to insects despite their high anticholinesterase activity as determined in vitro (23, 24). On the other hand, structure-activity investigations with a large number of nicotinoids (25) have established that insecticidal activity is associated with those compounds having pK_a values between 7.4 and 9.0. Basic compounds of this type can be expected to penetrate membrane barriers only with difficulty at physiological pH's (nicotine, pK_a, 7.9, is 88.9% ionized at pH 7) and the relatively high pK_a requirement strongly suggests that nicotine and its analogs act in a monocationic form at the receptor site (Sec. V.B.1).

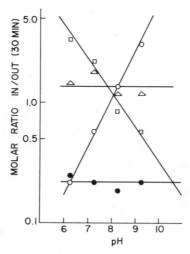

FIG. 3. Dependence on pH of penetration into cockroach nerve cord of butanol (△), butyltrimethylammonium (●), butyrate (□), and butylamine (○) [after O'Brien (20)].

C. Integumental Penetration

When the biological effect of a material is evaluated by its cutaneous or topical application to the integument of an organism the material must clearly be capable of penetrating this outer protective barrier. Although at first sight the nature of the integument appears to vary widely in different organisms, closer investigation reveals a rather general structural pattern, comprising an outer lipophilic layer (often nonliving) within which lie progressively more hydrophilic polar layers. Thus in actual physico-chemical terms the epidermal outer layer of mammalian skin (cellular) is rather similar to the epicuticular layer of insects (noncellular) and the waxy cuticle of the higher plants. Similarly, the mammalian dermal tissue is not dissimilar to the insect's exo- and endocuticle in that both constitute fairly permeable tissues with some degree of polarity. Although a good deal is known about integumental structure, surprisingly little unequivocal information exists regarding the physico-chemical properties required for efficient penetration through this barrier. The literature in this area abounds with inconsistencies and by far the most rational review of this tangled problem is that of O'Brien (11). Only a very brief discussion will be included here.

As pointed out by O'Brien (11), variation in the mode of application of the material to the integument probably represents the major source of error from which arises most of the apparently conflicting data on penetration. In in vivo investigations of internal membrane penetration (Sec. III. B) the form in which a material is ultimately presented to the membrane is determined by the constant conditions of the physiological environment (e.g., in the blood) over which the investigator can exert little control. However, when the material is administered through the integument, the mode of application, particularly the solvent vehicle employed, is under complete control of the investigator and is usually determined by either convenience or personal preference. Widely different techniques have been employed (11).

Early investigations with insects indicated that materials such as DDT and other highly lipophilic chlorinated hydrocarbons were rapidly adsorbed through the cuticle and that the low toxicity to mammals of cutaneously administered DDT (compared with that observed following injection) resulted from its failure to penetrate mammalian skin. More recently however, it has been established that the half-penetration time of DDT (about 26 hours) into rats (26) is essentially identical to that found in the American cockroach (27) and as a result O'Brien concludes that the observed differences in toxicity probably arise from variations in the rate of metabolism of DDT in each of the two species. A general assumption has been fostered that rate of penetration through insect cuticle usually increases with increasing lipophilicity of the material concerned. However, recent investigations (27) have shown that highly polar inorganic materials such as H_3PO_4 and K_2PO_4 can penetrate the integument of the American cockroach at rates considerably greater than those observed with DDT, if they are applied to the epicuticular surface in a volatile solvent (acetone). Thus the half-penetration time for H_3PO_4 was only 16 minutes compared with 26.4 hours for DDT (27). It was suggested that this initially surprising phenomenon results from the fact that the materials are introduced directly into the least polar phase of the cuticle and that subsequent diffusion into the more polar inner layers is aided by the hydrophilic nature of H_3PO_4. If H_3PO_4 is applied in a highly polar solvent such as water, the rate limiting step becomes the partitioning from the solvent into the epicuticular waxes, in which case penetration occurs very slowly or not at all until the water evaporates. Highly lipophilic materials such as DDT tend to penetrate the lipophilic epicuticle extremely rapidly, but in this case the rate limiting step becomes the penetration of the more polar exo- and endocuticular layers within. Similar results have been obtained with a number of other organisms (28).

It appears therefore, that for effective integumental penetration a material must possess the correct hydrophile-lipophile balance. A highly

polar material will not pass into the outer lipophilic phase unless it is
"artificially" incorporated into the layer by means of some convenient
solvent, and similarly a highly apolar material will fail to gain access
to the organism because it will not readily partition out of the lipophilic
phase into the more polar phases within. The whole problem of integumental penetration requires reevaluation and should prove a fruitful field
of research.

D. Storage and Binding

It is widely recognized that many lipophilic materials, particularly the
chlorinated hydrocarbon insecticides DDT, DDE, lindane (γ-hexachloro-
cyclohexane), and the cyclodienes, partition from the aqueous body fluids
into the adipose tissues of the organism (29-32). Storage in adipose tissues can therefore represent an important site of loss for highly lipophilic
substances and effectively decreases the concentration for initiation of
toxicological response. It should however, be remembered that storage
is not a passive, irreversible phenomenon. The partitioning process itself constitutes a reversible equilibrium, and in addition, the fat depots
of the organism are in a state of dynamic flux, being continuously broken
down and reformed. Thus the concentration of dieldrin in the adipose tissues of man has been correlated with its concentration in whole blood (31)
and the latter has also been shown to be inversely related to the mass of
the individual's body fat (32).

In addition to storage in fat depots, biologically active materials may
bind to plasma proteins, a process which can be of considerable significance in determining both the degree and duration of their biological effect (33, 34). The binding of a material to proteinaceous binding sites is
a reversible process, so that the bound and unbound forms of the material
are in equilibrium at all times. Serum albumin is the most abundant of
the plasma proteins and is undoubtedly of major significance in binding polar materials. Binding may involve any of the intermolecular forces discussed later (Sec. V. B). Highly lipophilic materials may bind to the α and
β lipoproteins also present in the plasma and it is likely in this case that
binding is effected by a partitioning process similar to that involved in
storage in adipose tissues.

IV. STRUCTURALLY NONSPECIFIC NARCOTICS

Most biologically active chemicals are structurally specific, i.e.,
their activity ultimately depends on the fact that they possess the necessary

structural configuration and chemical reactivity to interact with some specific receptor. As previously discussed (Sec. III. B) the intrinsic biological activity of these materials is often considerably modified by physicochemical properties which determine their ability to penetrate the several membranous barriers en route to their site of action.

In other cases however, the physical properties of a chemical are found to have a much more direct relationship to its intrinsic biological activity and the presence or absence of specific structural moieties is of secondary importance. Such chemicals which include the alkanes, alcohols, ethers, phenols, and alkyl benzenes comprise the so-called "physical poisons." They are structurally nonspecific and their biological activity, usually narcosis and toxicity, appears to result solely from their physical presence in some lipophilic phase of the cellular environment. Consequently their biological activity is determined by the physical properties which allow them to penetrate and accumulate in cell tissues.

As a result of early investigations it was proposed by Overton (35) and Meyer (36) that the narcotic potential of a drug was directly related to its lipid/water partition coefficient. Narcosis was considered to result from accumulation of the drug in some critical lipophilic component of the cell termed the biophase.

Almost forty years later Meyer and Hemmi (37) suggested that narcosis occurs when the molar concentration of drug in the biophase attains a critical threshold level. Evidence to support this view was obtained from measurements of the concentration of a number of materials required to cause narcosis in tadpoles (Table 2). Thus at levels effecting narcosis the lipid concentration of a number of materials was found to remain remarkably constant in spite of large variations in concentration of the external aqueous phase. The Meyer-Hemmi concept can be expressed in simple form by Eq. (5), where C is the external concentration of a material

$$PC = \text{constant} \tag{5}$$

required to elicit a given response and P its lipid/water partition coefficient. Unfortunately, the veracity of calculations depends on the rather tenuous assumption that the lipid biophase resembles the olive oil in which the lipid/water partition coefficients were determined, although subsequent work (38) has demonstrated a relationship between the partition coefficients of a material in similar solvent pairs. Furthermore, as pointed out by O'Brien (11), the Meyer-Hemmi hypothesis is only likely to be true if the volume of the external phase is very large compared with that of the organism so that the choice of tadpoles may have played a fortuitous role in determining the excellent correlations observed.

TABLE 2

Narcotic Action in Tadpoles[a]

Compound	Aqueous conc for narcosis (M)	Oil/water partition coefficient	Lipid conc for narcosis (M)
	C	P	PC
Ethanol	0.33	0.1	0.033
n-Propanol	0.11	0.35	0.380
n-Butanol	0.03	0.65	0.02
Valeramide	0.07	0.30	0.021
Antipyrine	0.07	0.30	0.021
Aminopyrine	0.03	1.30	0.039
Barbital	0.03	1.38	0.041
Diallylbarbituric acid	0.01	2.4	0.024
Benzamide	0.013	2.5	0.033
Salicylamide	0.0033	5.9	0.021
Phenobarbital	0.008	59	0.048
o-Nitroaniline	0.025	14	0.035
Thymol	0.000047	950	0.041

[a]Data from Meyer and Hemmi (37).

Earlier results (39) indicated that, in ascending a homologous series, the toxicity or narcotic response of each successive member increased according to a geometric progression (1 : 3 : 3^2 : 3^3 etc.). Similar geometric progressions were also recognized in relation to changes in such physical properties as vapor pressure and partition coefficients. Ferguson (40), noting that the latter were all expressions of some heterogenous phase distribution at equilibrium, suggested by analogy that the concentration of a narcotic in the extracellular phase was in equilibrium with that in the biophase. Under equilibrium conditions the thermodynamic activity or chemical potential of a compound in each phase is the same. Ferguson (40) therefore proposed that equal degrees of biological activity

(narcosis or toxicity) result, not when different substances reach the same
molar concentration in the biophase, but when they attain the same thermodynamic activity. Ferguson's concept is of particular importance because
at equilibrium the thermodynamic potential in the biophase can be readily
determined by its measurement in the extracellular phase, without re-
course to any assumptions regarding the nature of the biophase itself.

Thermodynamic activity (a) can be expressed as shown in Eq. (6) in
terms of the partial molal free energy of a compound (F) compared with
that in a standard state (F^O) where R is the gas constant and T the absolute temperature. A much simpler way of approximating the thermodynamic

$$F - F^O = RT \ \ln a \tag{6}$$

activity of a compound is afforded by measurement of its relative degree of
saturation. Thus the activity of a volatile gas can be expressed as p/p_0
where p is the partial pressure of the gas and p_0 its saturated water pressure. Similarly c/c_0 expresses the activity of a solute, the concentration
of which is c and solubility c_0.

Ferguson and Pirie (41) presented considerable amounts of experimental data to support the proposal that substances were equitoxic when they
were in the biophase at the same proportional saturation. Thus, the toxicity towards the grain weevil (Calandra granaria) of a large number of
volatile hydrocarbons, chlorinated hydrocarbons, and alcohols, correlated
extremely well with the proportional saturation of their vapors (p/p_0)
(Table 3). It is clear that with methyl and ethylene bromide, toxicity is
far greater than would be predicted from proportional saturation, a fact
indicating that the toxicity of these compounds results from factors other
than physical action.

In ascending a homologous series, the biological activity of each successive member does not increase indefinitely. Eventually a point is
reached, the so-called "cut-off point" after which any further increase
in chain length results in a marked decrease in activity. This phenomenon
can be attributed to the limiting effects of water solubility on the geometric
progression associated with increasing biological activity as the series is
ascended. In Fig. 4 water solubility is represented by curve A, whereas
curve B shows the decreasing concentration of the same material required
to produce a given biological response. It can be seen that curves A and B
converge and eventually intersect, at which point (the "cut-off point") the
concentration required to produce a given biological response corresponds
to a saturated solution of that particular compound in the external medium.
In Fig. 4 this is shown to occur with the compound containing five carbon

TABLE 3

Toxicity of Several Compounds to the Grain Weevil (Calandra granaria)[a]

Compound	Vapor pressure at 25°C (mm) (p_o)	LD_{50} (mg/liter)	p/p_o
Ethyl chloride	1170	1124	0.28
Propyl chloride	339	428	0.30
Butyl chloride	107	200	0.38
Amyl chloride	32	73	0.40
Isopropyl chloride	521	740	0.33
Methyl chloride	429	380	0.19
Carbon tetrachloride	114.5	275	0.29
Pentane	511	897	0.45
Hexane	151	353	0.50
Heptane	45.6	137	0.56
Decane	1.6	12	1.00
Methyl bromide	1580	3.3	0.0004
Ethylene bromide	11.0	0.66	0.06

[a]Data from Ferguson and Pirie (41). p = vapor pressure at LD_{50}.

atoms, but the "cut-off point" will vary according to both the organism involved and the homologous series under consideration. Obviously any further increase in chain length results in a marked decrease in the biological response, as the concentration required for the latter exceeds that allowed by aqueous solubility. The convergent nature of curves A and B reflects the steady rise in the proportional concentration required to produce a given response as illustrated with the alkane series in Table 3. When p/p_o or c/c_o reaches 1.0, the point of intersection of curves A and B in Fig. 4 has been reached and successive members of the homologous series will show a decrease in biological activity.

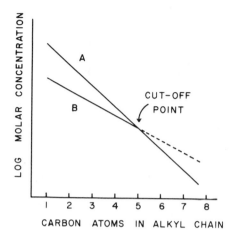

FIG. 4. Possible explanation of the "cut-off point" phenomenon in ascending a homologous series.

It is now evident that there are many substances which can be classified as structurally nonspecific toxicants. Ferguson's hypothesis is of importance, not as mechanistic explanation of narcotic action (this still remains unknown), but by providing a relatively simple method for distinguishing between structurally nonspecific and structurally specific biological activity.

A number of attempts have subsequently been made to modify Ferguson's hypothesis. McGowan, in a series of papers (42-46), has developed equations relating the toxicity of nonspecific substances with energy changes involved in their solubilization in the biophase. According to the "hole" theory of solubility, two types of energy changes are involved in the solubilization process. One of these accounts for the work done in creating a cavity with the appropriate dimensions in the solvent and the other relates to the energy change (E_A) resulting from solvent-solute interactions such as hydrogen bonding. McGowan therefore formulated the expression in Eq. (7), where C_t and C_b are the concentrations of a material outside the

$$-\log_{10}C_t = -\log_{10}C_b + k[P] + E_A \qquad (7)$$

cell and in the biophase respectively, and P represents the parachor. The latter, which is a measure of molecular volume, is employed as a result of its relationship to the energy involved in creating the necessary cavity in the biophase. Parachor is primarily an additive property and parachor equivalents for elements and structural factors such as double bonds etc. have been determined. The constant k in the above equation was found to remain of a similar value (0.012) in many different systems indicating an overall similarity in the nature of the biophase in different organisms. Because parachor is in some cases proportional to the partition coefficient of a material, Eq. (7) is found to correlate quite well with nonspecific narcosis. However, the quantitative aspects of the equation which theoretically allow the calculation of the actual concentration of material in the biophase are questionable and the concept has never enjoyed a great measure of support.

Mullins (47) taking the volume concept even further has suggested that nonspecific materials exert equal degrees of narcosis or toxicity when they attain a constant volume fraction in some nonaqueous phase in the cell. If the narcotic behaves ideally in this nonaqueous phase the product of the thermodynamic activity and molal volume will be constant. Mullins suggests that the observed increase in the thermodynamic activity required for narcosis in ascending a homologous series results from a deviation from ideality that is associated with the increased difficulty in inserting larger molecules into the pores in the biophase. This is associated with an increase in the activity coefficient of larger molecules.

V. CHEMICAL-RECEPTOR INTERACTIONS

A. General

This discussion has covered the major physico-chemical properties which determine the ability of a material to successfully reach its site of action (Sec. III.B). Consideration will now be given to those properties more directly involved in the interaction of a chemical with its biological receptor.

Before chemical reaction can occur, the compound in question must usually be capable of coming into close juxtaposition with certain specific areas of the receptor surface. In many cases these constitute the functional centers of enzyme proteins. The primary structure of a protein consists of a long polypeptide chain containing a unique sequence of amino acids. As a result of intramolecular hydrogen and hydrophobic bonding, folding of the primary chain takes place and determines the characteristic secondary and tertiary protein structure. Proteins, therefore, have a highly specific three-dimensional molecular geometry. Consequently, in order for a small molecule to come into close juxtaposition with an enzyme it must possess structural characteristics which are complementary to that portion of the enzyme surface at which interaction occurs.

Although absolute inflexibility of the enzyme surface should not be inferred (48), the "fit" of a substrate to its enzyme can, for the purpose of this discussion, be conveniently represented by the classical template theory illustrated in Fig. 5. This clearly indicates that as a result of having the necessary molecular complementarity, a substrate is bound and specifically orientated on the enzyme surface. Consequently some critical portion of the substrate (X) (e.g., a bond to be broken) is brought into close contact with a functional group (Y) of the enzyme and chemical reaction subsequently occurs. The template model is useful not only in discussions of enzyme-substrate interactions, but also serves in a much broader sense to indicate those physical and chemical properties likely to determine the ability of a material to interact with any biological receptor.

Fig. 5 indicates that the properties of a material which will allow it to come into close juxtaposition with a receptor include size, shape and stereochemical configuration. In addition it must usually possess the correct electronic distribution and chemical groupings which enable it to successfully bind and in many cases react with specific groups on the receptor surface. Most of these properties are closely interrelated and their separation is of necessity somewhat arbitrary. It is effected in the following discussion purely for the sake of convenience and clarity.

B. Binding Forces

A number of different attractive and repulsive forces are undoubtedly involved in any particular interaction and play an important role in determining the specific orientation of the chemical relative to the functional moieties of the receptor.

1. Ionic Forces

The amino acid chains which constitute the primary structure of enzyme proteins contain a number of ionizable groups which are to some extent dissociated at physiological pH values. Coulombic attraction to these sites on the protein of oppositely charged moieties associated with small molecules is therefore possible and has been recognized to be of importance for some time in substrate-enzyme binding. A good example of this is found in the hydrolysis of acetylcholine by the enzyme cholinesterase. In addition to its functional esteratic site, cholinesterase is considered to have an anionic site which is responsible for binding with the quaternary nitrogen atom of acetylcholine (49, 50). Evidence for the presence of an anionic site on cholinesterase is found in variations in the rate of enzymatic hydrolysis of ionized and un-ionized substrates (49) as well as in the pH dependent inhibition by tetraalkylammonium compounds which decreases markedly below pH 7.0 (51).

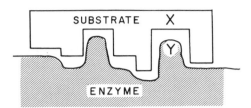

FIG. 5. Template model indicating the complementarity of a substrate for an enzyme surface.

The mechanism whereby cholinesterase hydrolyses acetylcholine is shown in Fig. 6. The cationic group of acetylcholine is coulombically bound to the anionic site of the enzyme and the electrophilic carbon atom of the ester group is similarly bound to a nucleophilic moiety at the active center, subsequently forming a covalent bond. Hydrogen bonding between the ethereal oxygen of acetylcholine and the acidic group of the esteratic site is followed by bond cleavage and rearrangement. Choline leaves the anionic site and the acetylated esteratic site of the enzyme subsequently returns to its original state following nucleophilic attack by a hydroxyl ion.

It is clear from Fig. 6 that the anionic site of cholinesterase does not merely bind the substrate to the enzyme. It is so located that its intramolecular distance from the esteratic site corresponds approximately to the distance between the N atom and ester group of acetylcholine. The latter is therefore brought into close proximity with the functional esteratic group of the enzyme. The intramolecular distance between certain structural features of a chemical relative to that between complementary moieties on a receptor surface is often of great importance in determining biological activity. It is interesting that in addition to acetylcholine, a variety of pharmacodynamic agents incorporate the group $-X-CH_2-CH_2-N-$ (where $X=0$ or N) (3). At maximum separation this group represents a distance of approximately 4-5 $\overset{\circ}{A}$ which may be of some fundamental significance in protein structure.

Employing this rationale it would be predicted that the effectiveness of cholinesterase inhibitors could be enhanced by incorporating in their structure a cationic moiety at the correct distance from the group binding with the esteratic site. This has in fact been demonstrated to some extent with both the organophosphates (52) and the carbamates (53) although a recent critical evaluation of the nature of the active site of cholinesterase (54) suggests that the importance of coulombic binding by the anionic site may have been somewhat overemphasized.

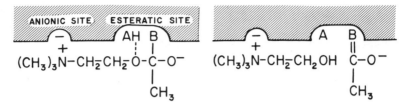

FIG. 6. The cleavage of acetylcholine by cholinesterase.

It is possible that the biological activity of nicotine [4] and related compounds depends to a large extent on their structural similarity with

[4]

[5]

acetylcholine [5] and consequently their ability to bind coulombically with some anionic site of the acetylcholine receptor (55). The fact that all insecticidal analogs of nicotine have pK_a values between 7.4 and 9.0 (25) suggests that they act largely in their monocationic form at the target site, as nicotine itself (pK_a of 7.9) is almost 90% ionized at physiological pH values. An additional requirement for nicotine toxicity is that the cationic nitrogen atom must be separated from that on the pyridine ring by a definite distance.

A further example of the importance of strategically located ionic charges in enhancing biological activity is provided by some of the oxime containing compounds which accelerate the recovery of cholinesterase following its phosphorylation by organophosphorus compounds (52, 56). One of the best materials of this type is 2-PAM (pyridine-2-aldoxime methiodide) [6]. Its cationic group undoubtedly enhances binding to the anionic

[6]

site of the enzyme and at the same time orients the nucleophilic oxime group for attack on the phosphorylated esteratic site.

2. Hydrophobic Bonding and van der Waals Forces

The forces of attraction between nonpolar lipophilic groups such as hydrocarbon chains have been traditionally attributed solely to van der Waals forces resulting from the interaction of mutually induced oscillating dipoles in adjacent molecules. These forces are most powerful when the interacting groups are in very close juxtaposition and maximum bonding energy can attain about 700 cal/mole per methylene group (57). However, their strength is inversely proportional to the seventh power of the distance separating the interacting species and it is now clear that they do not play such an important role as has previously been supposed.

The concept of hydrophobic bonding has recently received increasing attention with regard to its importance in the interaction of small molecules with biological receptors and probably accounts for much of the binding formerly attributed to van der Waals forces. Hydrophobic bonds result from the exclusion of water molecules between two adjacent hydrophobic groups. They are now thought to play a dominant role in many intermolecular interactions between substrates (or inhibitors) and enzymes. They are also considered to be largely responsible for determining the intramolecular folding of polypeptide chains which results in the characteristic tertiary structure of the protein molecule. Hydrophobic bonding has a maximum energy release of about 700 cal/mole per methylene group, almost identical to that associated with van der Waals bonding (57). It is probable that both types of forces are involved in binding the hydrophobic groups of small molecules with similar groups associated with certain amino acids such as leucine or valine on the enzyme surface.

In early work on the inhibition of cholinesterase by N-alkyltrimethylammonium compounds, inhibitory activity was found to increase in a manner which paralleled the increasing length of the alkyl chain (58). This provided good evidence that the combination of certain compounds with cholinesterase could be effected by binding to parts of the enzyme surface other than those directly associated with the anionic or esteratic sites.

Recent work has demonstrated that in several series of phosphate and phosphorothiolate analogs of general structure [7], where X = O or S, and

a : X = O, R = p-nitrophenyl

b (protonated) : X = S,

R = $CH_2CH_2\overset{+}{N}H(C_2H_5)_3$

R = branched or unbranched alkyl chains, anticholinesterase activity increases gradually with increasing chain length R (59, 60). Activity reaches a maximum when R constitutes a six carbon chain and thereafter remains relatively constant with further increases in chain length. These data are in excellent agreement with the results of earlier investigations employing similar alkyl chain analogs of two series of ethylmethylphosphonates (61). It is also probable that similar factors are responsible for the linear relationship between anticholinesterase activity (expressed log I_{50}) and log partition coefficient of a series of N-alkyl-substituted amides (62) as well as for the unexpectedly high inhibitory activity of 3-tertiary butylphenyl and 3-dimethylaminophenyl diethyl phosphates reported in earlier structure-activity studies (63).

The anticholinesterase activity of organophosphorus compounds is normally associated with a high degree of electrophilicity at the phosphorus atom (Sec. V. E). This property, which determines the ability of these compounds to combine with the nucleophilic moiety at the active center, is enhanced by the presence of some electron drawing group such as the p-nitrophenyl moiety of paraoxon [7a]. As mentioned earlier, it is a widely held view that this requirement can be obviated if, as in Amiton [7b], R comprises a cationic chain capable of binding to the anionic site of the enzyme.

The high anticholinesterase activity of compounds in which R is represented by an alkyl chain, and consequently satisfies neither of these requirements, appears at first sight to be anomalous. Bracha and O'Brien (59, 60) conclude that the major contribution to the high inhibitory activity of the alkyl phosphates and phosphorothioates they investigated results from powerful hydrophobic binding to the enzyme surface. This is considered to take place at a hydrophobic patch on the enzyme which is of limited size and is in close proximity (about 4 Å) to the active site. As a result of the activity shown by the carbon analog of Amiton, it was calculated that with Amiton itself, coulombic attraction to the anionic site of the enzyme contributes only about 18% towards its total binding capacity (59, 54). This calculation is based on the assumption that both analogs bind to the same site on the enzyme surface, a fact which still remains to be established. However, the results of others, indicate that increasing anticholinesterase activity with increasing chain length is not a general phenomenon. Fukuto and Metcalf working with a series of ethyl p-nitrophenyl alkylphosphonates [8] (64, 65), as well as similar series of alkyl p-nitrophenyl ethylphosphonates [9] and phosphinates (66, 67), found that increasing the length of R was normally associated with a steady decrease in both anticholinesterase activity and toxicity to houseflies. As a result of ρ-σ-π analysis (Sec. VI. C), Hansch and Deutsch (68) conclude that

$$NO_2 - \bigcirc - \overset{\overset{O}{\|}}{P} \overset{OC_2H_5}{\underset{R}{\diagdown}}$$

$$NO_2 - \bigcirc - \overset{\overset{O}{\|}}{P} \overset{O-R}{\underset{C_2H_5}{\diagdown}}$$

[8]   [9]

hydrophobic bonding plays an insignificant role in the activity of the materials. It was suggested that the decrease in anticholinesterase activity with increasing chain length (R) results mainly from steric hindrance by the bulky alkyl chain and consequently an inability of the phosphoryl group to make close contact with the active site.

Becker et al. (69), employing two series of phosphonates of the same general structure [8] where R = alkyl and phenylalkyl respectively, found that cholinesterase inhibition decreased from C_3 to C_5 in the alkyl phosphonate series and thereafter remained fairly constant until at C_{10} a further decrease in activity was observed. The phenylalkylphosphonates, which were relatively poor anticholinesterases, showed an increase in inhibitory activity up to the phenylbutyl analog in general agreement with the results of Bracha and O'Brien. In contrast, antichymotrypsin activity was found to reach an optimum with the heptyl and phenylpropylphosphonates respectively and a sharp decrease in activity was observed in each series with further increase in chain length.

It is perhaps significant that those compounds which show an increase in anticholinesterase activity with increasing size of the alkyl chain substituent are all relatively poor phosphorylating agents. In contrast, the p-nitrophenyl alkylphosphonates are extremely potent anticholinesterases and are highly toxic in vivo. The electrophilic properties of the p-nitrophenyl moiety are clearly of overwhelming importance in determining the ability of these compounds to phosphorylate the active site and it is perhaps not surprising that hydrophobic binding cannot further enhance this activity. It is indeed possible that with these compounds the observed decrease in inhibitory activity associated with increasing chain length results from hydrophobic bonding which favors a sub-optimal orientation at the active site. This possibility is further substantiated by the negative coefficient of π in the equation of Hansch and Deutsch (68).

It is now apparent that anticholinesterase activity of the phenyl N-methylcarbamates is closely associated with their molecular complementarity to the active site (53, 24). It is equally clear that by enhancing affinity for the active center, hydrophobic interaction between certain phenyl substituents and some lipophilic patch on the enzyme surface can often be of dominant importance in determining their inhibitory activity. This is best illustrated with reference to the activity of the alkylphenyl N-methyl-carbamates of general structure [10]. The anticholinesterase activity and

$$R \diagdown \text{—} O\overset{\text{O}}{\underset{\text{||}}{C}}NHCH_3$$

[10]

enzyme affinity of these compounds is found to increase markedly with increasing size of the alkyl substituent R so that sec-butyl > isopropyl = tert-butyl > ethyl > methyl > unsubstituted (24) (Table 4). In all cases substitution in the meta position of the ring is optimal indicating that binding probably occurs in an area situated approximately 5 Å from the active center. It is possible that this represents the same site which is responsible for binding the alkyl phosphates and which Bracha and O'Brien (59) estimated to be about 4 Å from the esteratic center. However, Metcalf et al. (24) assume that the alkyl binding site is analogous or closely associated with the anionic center of the enzyme and that binding results from van der Waals interactions in this area. Whichever the case, the hydrophobic patch on the enzyme must be quite large, since carbamate affinity is often greatly enhanced if alkyl substitution is effected at both meta positions of the phenyl ring (24).

Enzyme affinity for substrates or inhibitors does not usually increase indefinitely with increasing size of the hydrophobic substituent. The hydrophobic "patches" on receptor surfaces usually occupy a limited area and consequently can only accommodate groups up to a certain critical size. With increasing size of the hydrophobic substituent, a point is reached at which the patch can be considered "saturated" and can exert no further increase in binding efficiency. Compounds containing hydrophobic groups which exceed the size required for saturation will remain of constant activity or will show a marked decrease in potency. When the latter is observed, it probably indicates steric hindrance of the critical receptor interaction as a result of the bulky nature of the hydrophobic group. A good example of the latter is provided by the synergistic activity with carbaryl (against houseflies) in a series of benzoates of α-alkyl substituted piperonyl alcohol (70) (Fig. 7). Replacement of one of the α-hydrogen atoms of piperonyl benzoate by a methyl group results in a remarkable increase in synergistic activity. High activity is maintained when R = C_2H_5 but slowly declines with further increases in R to C_9, thus suggesting that the increasing size of R progressively interferes with the interaction of the 1, 3-benzodioxole ring at the site of action.

In brief summary of this section, it is clear that the presence of hydrophobic substituents can often markedly enhance the ability of compounds

C. F. WILKINSON

TABLE 4

Biological Activity of Alkyl N-methylcarbamates[a]

Substituent	I_{50} cholinesterase	Affinity
Unsubstituted	2.0×10^{-4}	1.0
o-CH_3	1.4×10^{-5}	1.4
o-C_2H_5	1.3×10^{-6}	15
o-iso-C_3H_7	6.0×10^{-6}	33
o-sec-C_4H_9	1.1×10^{-6}	180
m-CH_3	1.4×10^{-5}	14
m-C_2H_5	4.8×10^{-6}	42
m-iso-C_3H_7	3.4×10^{-7}	590
m-sec-C_4H_9	1.6×10^{-7}	1250
p-CH_3	1.0×10^{-4}	2.0
p-C_2H_5	3.8×10^{-5}	5.3
p-iso-C_3H_7	7.0×10^{-5}	2.9
p-sec-C_4H_9	1.8×10^{-6}	110
3,5-di-CH_3	6.0×10^{-6}	33
3,5-di-iso-C_3H_7	3.3×10^{-8}	6060

[a]Data from Metcalf and Fukuto (24).

to bind and subsequently interact with biological receptors through a combination of hydrophobic and van der Waals forces. In some cases, as with
the alkyl-substituted phosphates, this enhanced affinity can adequately compensate for poor reactivity of the functional group. On the other hand, the
often large hydrophobic substituent must be so located in relation to the
overall molecular topography of both the molecule and the receptor surface
that it does not sterically interfere with the critical interaction on which
biological activity depends.

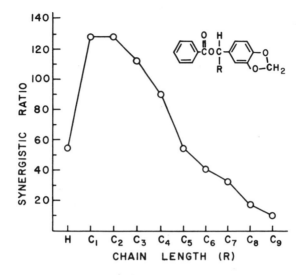

FIG. 7. The effect of chain length (R) on the synergistic activity with carbaryl of a series of benzoates of α-substituted piperonyl alcohols [after Wilkinson et al. (70)].

3. Dipole-Dipole Interactions

In addition to the coulombic attraction between molecules and receptors bearing a complete and opposite unit charge, electrostatic attraction can also take place through the alignment of oppositely charged dipoles which arise from the presence of electron-rich or electron-poor centers on the receptor and the interacting chemical.

Perhaps the best-known and most important of these dipole interactions is that constituting the hydrogen bond. The majority of biologically active materials incorporate groups which can act as either acceptors or donors in hydrogen bond formation. These groups are probably of particular significance in interactions which involve protein receptors. Here both the amino acid side chains and the polypeptide linkages combine to provide a multitude of possible sites at which hydrogen bonding can occur. In view of their potential strength (up to 5 kcal/mole), hydrogen bonds might be expected to play a dominant role in chemical-protein interactions. However, in aqueous solution the degree of hydrogen bonding between a chemical and its receptor will probably be greatly reduced due to the fact that the potential binding sites on both will be to a large extent already associated with water. Thus, although hydrogen bonds are undoubtedly of major significance in the binding and orientation of chemicals with receptor surfaces, their actual contribution to any particular interaction is usually difficult to evaluate with any degree of accuracy.

In a more general sense, electrostatic attraction can occur between any materials containing areas of opposite partial charge and need not necessarily involve the formation of a hydrogen bond. These molecular associations between electron donors and electron acceptors are termed charge-transfer complexes (57).

Charge-transfer complexes have recently been implicated in the mechanism of action of some of the chlorinated hydrocarbon insecticides which have long been known to interfere with nervous function in insects and mammals. Evidence from both spectral data and gel filtration chromatography supports the thesis that DDT can bind with some component of the cockroach (Periplaneta americana) nerve cord (71, 72). Additional studies indicate that dieldrin resistance in the German cockroach (Blattella germanica) may be associated with a decrease of dieldrin binding capacity by nerve components of the resistant insect (73, 74). However, the binding of these highly polar materials to components of the nerve cord may represent only one example of a much broader nonspecific binding capacity involving a large number of proteins and other cellular components (75). Whether this can satisfactorily account for the mechanism of action of the chlorinated hydrocarbon insecticides is by no means clear at the present time.

4. Covalent Bonds

Covalent bonds are those normally encountered in organic chemistry and are produced by two atoms sharing a pair of electrons. The strength of the covalent bond, approximately 100 kcal/mole, is thus considerably more powerful than the other attractive forces so far discussed.

Particularly in the area of medicinal chemistry it is often desirable that some degree of reversibility exists between a chemical and its receptor. As a result, receptor interactions involving the formation of covalent bonds are rather uncommon in pharmaceutical agents. On the other hand, the formation of extremely powerful covalent bonds is often beneficial in interactions designed to promote toxicity in pest organisms. Indeed it is usually found that materials which are capable of interacting with receptors through covalent bonding are of a highly toxic nature to many forms of life.

Good examples of such materials are the alkylating agents which include the nitrogen and sulfur mustards, alkyl methane sulfonates and ethyleneimine compounds. Many of these are used in cancer chemotherapy and are also being investigated for use as insect chemosterilants. Other examples include the interaction of arsenicals with sulfhydryl groups and the inhibition of cholinesterase by the organophosphate and carbamate insecticides (Sec. V. E).

C. Molecular Shape and Size

Structure-activity investigations often indicate that biological activity appears to be associated with the overall shape and size of a compound. This suggests that in order to initiate a biological response the entire molecule must successfully fit onto a receptor surface or into some specific receptor space along the lines proposed by Mullins (47, 76).

In some cases biological activity appears to be associated with planar aromatic rings, a possible indication of the importance of $\pi - \pi$ interactions in binding to the receptor surface. Thus the anticholinesterase activity of 2-isopropylcyclohexyl N-methylcarbamate [11] is approximately 1000 times less than that of the corresponding phenyl N-methylcarbamate [12] (53). In addition, the size and position of the aromatic substituents of

[11]   [12]

the phenyl N-methylcarbamates has a marked effect on their anticholinesterase activity and insecticidal potency. Thus in a series of halogen-substituted phenyl N-methylcarbamates inhibitory activity and toxicity were correlated with the increasing van der Waals radii of the halogen atom (24). This was most marked with ortho substitution (Table 5), and substantiates the importance to the carbamates of molecular complementarity with the active site of cholinesterase.

Molecular planarity also appears to be an important structural requirement for carbamate synergists related to 1,3-benzodioxole [13] (70). The

[13]   [14]

TABLE 5

Activity of Ortho–Halogen–Substituted Phenyl N–Methylcarbamates[a]

Substituent	I_{50} (M) Cholinesterase	Affinity	LD_{50} Houseflies $\mu g/g$
Unsubstituted	2.0×10^{-4}	1	500
o–F	1.6×10^{-5}	12	250
o–Cl	5.0×10^{-6}	40	75
o–Br	2.2×10^{-6}	91	60
o–I	8.0×10^{-7}	200	90

[a]Data from Metcalf and Fukuto (24).

synergistic activity of these compounds depends on the integrity of the methylenic hydrogen atoms and on the presence of the two oxygen atoms in the five–membered ring, although the latter requirement can be satisfied by sulfur with only a slight decrease in potency. However, derivatives of 1, 4–benzodioxane [14] are inactive as carbamate synergists (70). Molecular models of these materials clearly show the rigid planarity of the 1, 3–benzodioxole nucleus compared with the nonplanar structure associated with the 6–membered dioxane ring.

A good example of the importance of molecular shape and size is provided by the cyclodiene insecticides. Comprehensive structure–activity investigations (77) have indicated that the reactive feature of the insecticidal cyclodienes resides in the presence of two electronegative centers. As indicated with reference to aldrin, one of these is associated with the hexachloronorbornene nucleus (A in Fig. 8) and the other with either a double bond (B in Fig. 8) or a suitable electronegative atom such as chlorine, oxygen, nitrogen, or sulfur elsewhere in the molecule. Studies of molecular models have indicated that the two electronegative centers should lie within rather closely defined limits of distance and direction with respect to one another and in addition should be orientated on or across the plane of symmetry of the dichloromethano bridge (77). By outlining the projected shadows of molecular models illuminated from a direction along the bridgehead of the dimethanonaphthalene nucleus (Fig. 8), it is possible to effectively obtain a cross–section of the molecule through the plane of symmetry incorporating

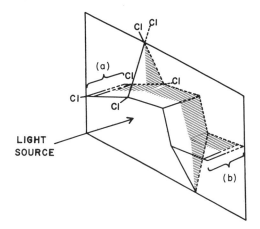

FIG. 8. The three-dimensional structure of aldrin indicating its plane of symmetry and reactive features.

the two electronegative centers (Fig. 8). Using this ingenious method, Soloway has been able to compare and contrast the molecular topography of a number of active and inactive cyclodiene analogs. The topographical outlines of the insecticidally active compounds are virtually superimposable and contrast sharply with the various cross-sectional shapes of the noninsecticidal analogs. The critical requirements of shape and size are exemplified by the fact that the exo-epoxide, dieldrin [15], is about six times more toxic to houseflies than the closely related compound in which the epoxide ring is in the endo position [16] (77).

The wide variation in insecticidal potency of the isomers of hexachlorocyclohexane (HCH) provides another example of the importance of molecular shape and size in determining biological activity. Substitution at each carbon atom of the cyclohexane ring [17] may occur through either an equatorial (e) bond, approximately in the plane of the ring, or through an axial (a) bond positioned perpendicularly to the plane of the ring. Several conformational isomers of HCH are therefore possible depending on the relative positions of the six chlorine atoms in the ring. Of the several possible forms of HCH, only the γ-isomer, in which the chlorine atoms are positioned aaaeee, shows marked insecticidal activity (78).

[17]

As a result of similarities in toxicity, symptomology, and cross-resistance patterns, Soloway (77) attempted to correlate the molecular topography of γ-HCH with that of the insecticidal cyclodienes. A striking similarity with dieldrin was observed and it was suggested that the two electronegative centers of γ-HCH resulted from the correct spatial configuration of the equatorial and axial chlorine atoms in the molecule.

The mode of action of both γ-HCH and the cyclodienes appears to result from their ability to interact with some component in the function of the central nervous system. Although not yet clearly established, it appears likely that the mode of action of these compounds is directly related to their overall size and structural configuration which determine the ability of the two electronegative centers to come into close physical proximity with some receptor site. The lack of chemical reactivity of many compounds within this group suggests that activity may result from some type of physical combination with the receptor, possibly through the formation of charge-transfer complexes as previously discussed (Sec. V. B. 3).

Mullins (76) has attempted to relate the insecticidal activity of γ-HCH and DDT to their ability to "fit" tightly within the pore spaces formed by the macromolecular lattice of the axonic membrane. This hypothesis, which is an extension of that proposed to explain the mode of action of the structurally nonspecific narcotics (47), has received some support but still remains of a purely speculative nature.

Recent studies of structure-activity relationships in a large number of DDT analogs (79) support the view that the insecticidal activity of these compounds is also associated with the size and shape of the entire molecule, possibly through optimal orientation of the phenyl rings and the trichloromethyl group. The insecticidal potency of DDT analogs [18] is influenced markedly by the nature of the substituent groups R_1 to R_6. Although the nature of the substituents R_3 to R_6 may differ considerably and can be various combinations of -H, -F, -Cl, -Br, -CH$_3$, -OCH$_3$, -NO$_2$, and -CN, substituents R_1 and R_2 must be small, relatively nonpolar groups such as -F, -Cl, -Br, -OCH$_3$, -C$_2$H$_5$ and -OC$_2$H$_5$ for maximum activity (79).

[18]

Although it will not be discussed in detail it should be emphasized that molecular shape and size are extremely important in the concept of bioisosterism (80) and in its practical application to the design of antimetabolites (81) which have proved particularly useful in medicinal chemistry. Antimetabolites are compounds which so closely resemble the structure of certain naturally occurring materials such as vitamins or cofactors that they effectively replace the natural materials in the system. Because antimetabolites do not possess the functional activity of the naturally occurring metabolites, their presence usually has a marked biological effect.

The classical example of metabolite antagonism, probably the one which gave the greatest impetus to the development of the entire concept, is found in the antibacterial activity of sulfanilamide. Woods (82) suggested that the activity of sulfanilamide [19] resulted from its competitive action with p-aminobenzoic acid [20]. This was determined to be an essential growth factor in some microorganisms and has since been shown to be incorporated into the essential coenzyme pteroylglutamic (folic) acid.

NH$_2$ NH$_2$

SO$_2$NH$_2$ COOH

[19]   [20]

Many antimetabolites have been developed and investigated as a result of their carcinostatic activity and more recently have been considered as potential agents for use in insect chemosterilization (83). Thus aminopterin [21a] as a result of its structural similarity with folic acid [21b] inhibits the enzyme dihydrofolic reductase (84) which plays a role in DNA synthesis.

Similarly, 5-fluorouracil [22] owes its cytotoxic activity to its ability to inhibit DNA synthesis by blocking the methylation of deoxyuridylic acid to thymidylic acid (85).

a : R = NH$_2$
b : R = OH

D.   Stereochemistry

The stereochemical configuration of any biologically active compound must clearly reflect the stereochemistry of the receptor surface as a direct consequence of the structural complementarity required between a chemical and its target site.   Therefore, when a material can exist in more than one stereochemical configuration, it is frequently observed that only one of the isomeric forms is capable of initiating a particular response. Isomers differ only in the fact that their several substituent groups occupy different relative positions in space, so that investigations of stereospecificity are often of considerable value in elucidating the nature and mechanism of chemical-receptor interactions.

1.   Optical Isomerism

Compounds containing a quadrivalent atom attached to four different groups or substituents occur in two isomeric forms (enantiomorphs), one of which is the nonsuperimposable mirror image of the other. Apart from the ability of each isomer to rotate the plane of polarized light in an opposite direction, other physical and chemical properties are considered identical.

Because of the different relative spatial arrangement of its substituent groups, each optical enantiomorph will present a unique configuration to the receptor surface. If, as illustrated in Fig. 9, the initiation of biological response depends on the superimposition of groups X, Y, and Z of the chemical with functional groups X', Y', and Z' on the receptor surface, only one of the enantiomorphs, A, possesses the necessary structural configuration to interact.   When, however, biological activity results from receptor contact through only two of the substituent groups, it is likely that both isomers will promote a similar response.

Examples of compounds whose enantiomorphic forms show large variations in biological activity are common.   This is particularly true with

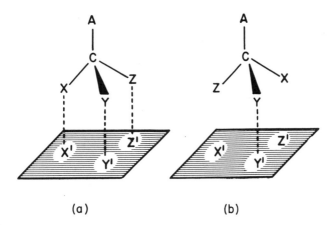

(a)   (b)

FIG. 9. The complementarity of optical enantiomorphs for a hypothetical receptor surface.

naturally occurring materials where it is frequently found that enantiomorph which predominates in nature is the only one capable of initiating a particular response. Thus the naturally occurring l-isomers of the sympathomimetic amines epinephrine and norepinephrine are fifteen and twenty times more active than the corresponding d-forms which do not occur in nature. Similarly, maximum potency of the pyrethroid insecticides is associated with the naturally occurring d-isomers (86) (Sec. V. D. 2).

Nicotine, which in nature occurs as the l-isomer, is somewhat unusual in that variations in biological activity between the d- and l-isomers seem to vary with the particular organism and type of investigation involved (25). Usually, however, the l-form is found to have the greater activity (87).

Optical activity in the organophosphorus compounds may result from either an asymmetrical center in the leaving group moiety, or, in the case of the phosphonates, from the presence of four different groups arranged around the central phosphorus atom. When optical isomerism occurs, differences in both toxicity and anticholinesterase activity are usually found between the enantiomorphic forms. Thus the anticholinesterase activity of the l-isomer of 0-ethyl S-(2-ethylmercaptoethyl) ethylphosphonothiolate is reported to be ten to twenty times that of the d-isomer (88) and in a series of S-alkyl p-nitrophenyl methylphosphonothiolates the d-isomers are found to be more toxic and more effective cholinesterase inhibitors than the l-isomers (89). More recently, investigations with the optical isomers of malathion and malaoxon (90) have demonstrated that maximum mammalian toxicity is associated with the d-isomers, a fact which correlates with the higher bimolecular rate constants (K_i) with cholinesterase.

Investigations on the growth-regulating activity of the enantiomorphs of α-(2, 4-dichlorophenoxy)-propionic [23] and butyric acids indicates that activity is associated almost entirely with the d-isomer of these materials

[23]

(91). As a result, it was concluded that the essential structural features associated with plant growth-regulation in these compounds are the aromatic ring, the carboxylic acid group, and the α-hydrogen atom. Each of these features must therefore occupy a spatially unique position relative to each other, a fact reflecting their necessary alignment with functional groups on the receptor surface.

2. Geometrical Isomerism

In addition to the variation in biological activity associated with different optical enantiomorphs, large differences in activity are often observed between geometrical isomers. Since cis and trans isomers often differ in such physical properties as solubility and partition coefficient, some of the variation in activity may result from different rates of penetration and translocation to the site of action. It is more probable, however, that large variations in activity between cis and trans isomers, particularly when observed in vitro, can be attributed to the fact that the three-dimensional configuration, and consequently functional group alignment of one isomer, exhibits a greater degree of complementarity with certain structural features on the receptor surface. This is illustrated in Fig. 10 which represents the orientation of cis and trans isomers on a hypothetical receptor surface. In the event that biological interaction is dependent on only a two-point attachment of the compound to the receptor surface (e. g., at X and Y), it is probable that both isomers will show a similar biological response, since the relative alignment of X and Y with respect to X' and Y' is the same in each case. It is also possible that with any other combination involving a two-point attachment some activity could result from either isomer. However, if a three-point attachment is required for successful interaction (e. g., at X, Y and Z), it can be seen that only the cis isomer incorporates the necessary functional groups in the correct relative position.

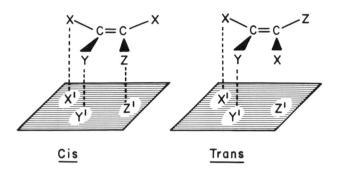

FIG. 10. The complementarity of geometrical isomers for a hypothetical receptor surface.

The organophosphorus compounds Phosdrin [24a] (0, 0-dimethyl-1-carbomethoxy-1-propen-2-yl phosphate) and Bomyl [24b] (the corresponding 1, 3-dicarbomethoxy analog) both occur as cis and trans isomers.

$$
\begin{array}{cc}
(CH_3O)_2\text{-}P & & (CH_3O)_2P
\end{array}
$$

[24]

a : R = H

b : R = COOH$_3$

(cis)   (trans)

Cis Phosdrin is about fifty and twenty times more toxic than the trans isomer to houseflies and mice respectively, and is approximately one hundredfold more effective as an inhibitor of cholinesterase (92). In contrast, the cis and trans isomers of Bomyl are equitoxic and show little difference in anticholinesterase activity.

These data can be rationalized by considering the isomeric variation in intramolecular distance between the phosphoryloxy group and the carbomethoxy moiety (92). On the assumption that the latter is responsible for binding to cholinesterase in the region of the anionic site, optimal alignment of the phosphorus atom to the esteratic site will only occur if the two groups are separated by a distance of approximately 5 Å. With both isomers of Bomyl it was shown that the average distance between the phosphorus atom and at least one of the two carbonyl carbons was 4.8 Å, a value similar to that determined for cis Phosdrin (4.75). However, separation of the two groups in trans Phosdrin is found to be only 3.3 Å, which would consequently tend to preclude interaction of the phosphorus atom at the esteratic center and result in decreased anticholinesterase activity.

In earlier attempts to explain the low anticholinesterase activity of trans Phosdrin, it was suggested (93) that the close proximity of the bulky carbomethoxy group to the phosphorus atom might result in steric interference with the phosphorylation reaction at the esteratic site. However, the high anticholinesterase activity of trans Bomyl is not consistent with this suggestion since, in this case, the position of the carbomethoxy group in relation to the phosphorus atom is analogous with that in trans Phosdrin.

Though considerably more complex, an excellent example of the importance of geometrical isomerism in determining biological activity is provided by the pyrethroid insecticides. The insecticidal components of natural pyrethrum consist primarily of four esters resulting from the condensation of two ketoalcohols, pyrethrolone [25a] and cinerolone [25b], with each of the acids, chrysanthemic [26a] and pyrethric [26b].

a: R = CH_3
b: R = $COOCH_3$

[26]

a: R = $CH_2CH = CHCH = CH_2$
b: R = $CH_2CH = CHCH_3$

[25]

In addition to the existence of optical enantiomorphs resulting from the presence of an asymmetric carbon atom, each alcohol has cis and trans isomers associated with its unsaturated side-chain R. The cyclopropane ring of each of the acid moieties contains two adjacent asymmetric carbon atoms and therefore also has d and l forms. The position of the unsaturated side-chain relative to that of the carboxylic acid group is usually denoted as cis or trans with reference to the rigid plane of the cyclopropane ring. Chrysanthemic acid, therefore, has four forms, d-cis, l-cis, d-trans, and l-trans, with the isomer shown [26a] being one of two trans forms. Pyrethric acid has additional cis and trans isomers as a result of its unsaturated side-chain. The naturally occurring esters always contain d-trans acids and d-cis alcohols and the fact that these combinations are always associated with maximum insecticidal potency has been successfully applied in the development of synthetic insecticidal analogs such as allethrin (86).

E. Chemical Reactivity

Many materials which possess biological activity are to a large extent
chemically unreactive. The effectiveness of such compounds may arise
from either their ability to bind tightly to specific cellular receptors or,
as in the case of the structurally nonspecific narcotics, may result merely
from their physical presence in the biophase. Other materials, however,
are active by virtue of their ability to react chemically with specific functional groups on the receptor surface, a process usually entailing the formation of a covalent bond. The degree of biological activity of these compounds can often be correlated with structural characteristics which determine the chemical reactivity of the specific atom or group through which
the interaction with the receptor occurs.

The inhibition of cholinesterase by the organophosphate insecticides
provides an excellent and thoroughly investigated example of this type of
activity. It is now clearly established that anticholinesterase activity of
the organophosphates results from an electrophilic attack by the phosphorus
atom on some nucleophilic moiety at the active center of cholinesterase
(52, 54, 56, 94, 95). The reaction is shown in Eq. (8) and involves phosphorylation of the enzyme (E) through the covalent bond formation and

$$E : + (RO)_2 P(O)X \longrightarrow E ---- (RO)_2 P(O) ---- X \longrightarrow (RO)_2 P(O)E + X^- \qquad (8)$$

consequent liberation of the so-called leaving group X of the organophosphate. The anticholinesterase activity of organophosphorus compounds
therefore depends to a large extent on the electrophilic character of the
phosphorus atom, which is in turn determined by the nature of the groups
to which it is attached. Because alkaline hydrolysis of the organophosphates occurs through a similar reaction mechanism (nucleophilic attack
by OH^- on the electrophilic phosphorus atom) a linear relationship is
usually observed between hydrolysis constants and the bimolecular rate
constants for reaction with cholinesterase (94-96). If, however, the electrophilicity of the phosphorus atom is too great the material will hydrolize
spontaneously before achieving its inhibitory effect.

Inhibition is not determined solely by the relative facility of the phosphorylation reaction but also depends on the hydrolytic stability of the resulting phosphorylated enzyme. As with the phosphorylation reaction itself, recovery of the phosphorylated enzyme depends on the electrophilic
nature of the phosphorus atom. In this case, however, stability is increased and inhibition consequently enhanced, if the groups attached to
the phosphorus atom are electron-releasing.

The total requirements for an effective organophosphate anticholinesterase [27] are therefore a powerful electron withdrawing group (X) which is

$$R \diagdown P \diagup O(S)$$
$$R \diagup P \diagdown X$$

[27]

readily displaced during the phosphorylation reaction and electron-releasing or weakly electron-withdrawing groups at R and R'. The electrophilic character of the phosphorus atom is further increased by the electronegativity of the oxygen atom in the P=O moiety. The importance of
this group is clearly indicated by the poor anticholinesterase activity of
the corresponding thiono derivatives where the oxygen is replaced by sulfur. It appears doubtful that the large variations in anticholinesterase activity between the phosphates and phosphorothionates can be wholly attributed to differences in the electronegativity of the oxygen and sulfur atoms
(54) and it has been suggested (56) that the O atom may be important in hydrogen bonding.

The nature of R and R' [27] is usually limited to small alkoxy groups
(CH_3O or C_2H_5O) with the phosphates or to alkyl groups in the case of the
phosphonates and phosphinates. The decrease in anticholinesterase activity in compounds with larger alkyl chains appears to be associated with un-
favorable steric effects on the phosphorylation reaction (54, 94), rather
than to differences in the electron releasing ability of R which would give
added stability to the phosphorylated enzyme.

Structural variations of the leaving group X are limited only by the in-
genuity of the organic chemist and many thousands of substituents have been
evaluated. In several series of compounds which differ only in the nature
of X, anticholinesterase activity is often found to be correlated with the
electron-withdrawing power of X. When X is a meta or para substituted
phenyl ring, the electron-withdrawing power can be quantitatively determined by means of the Hammet σ constant of the aromatic substituent (97).
This is obtained from Eq. (9) where k and k_0 are the equilibrium or rate

$$\log \frac{k}{k_o} = \sigma \rho \qquad (9)$$

constants of the substituted and unsubstituted aromatic compounds respectively. The constant σ denotes the electronic character of the substituent
relative to hydrogen (which has a σ value of zero) and ρ is a reaction constant. An aromatic substituent with a positive σ constant possesses an
electron-withdrawing power greater than that of the hydrogen atom, whereas one with a negative σ value shows a tendency to donate electrons to the

system. Good correlations have been obtained between log I_{50} and Hammett's σ constants with several series of organophosphates (98-100). Regression analysis of the anticholinesterase activity of a series of para-substituted diethyl phenyl phosphates (68) indicates that 91% of the variance in the data can be attributed to σ alone, although very unsatisfactory correlations were obtained with a number of meta-substituted derivatives.

When substituents are not directly attached to a phenyl ring, their contribution to the reactivity of the molecule can be obtained by means of Taft's polar substituent constant (σ^*) (101). Thus in a series of methyl 2, 4, 5-trichlorophenyl N-alkylphosphoramidates [28], a linear relationship was found to exist between σ^* and the bimolecular rate constant for cholinesterase inhibition. Regression analysis of these data (68) suggests that although

[28]

activity is promoted by the electron releasing capacity of R and R', the steric effects of these groups are of more importance in determining the observed activity.

F. Free Radicals

In recent years increasing attention has been given to possible involvement of free radicals in some types of biological activity. Free radicals are materials which contain unpaired electrons and because of their high reactivity are usually very short-lived under normal conditions.

However, some compounds, such as the bipyridylium herbicides, paraquot [29] and diquot [30], can be reduced to stable, water-soluble free

[29]   [30]

radicals on addition of one electron. The herbicidal activity of these compounds has been correlated with their ease of reduction to the free radical (103). In addition it has been demonstrated that they are effective only on actively photosynthesizing green tissues (104). No activity occurs under anaerobic conditions in spite of the observed accumulation of the free radical (105) and it appears that herbicidal action may be associated with re-oxidation of the free radicals by atmospheric oxygen (106), possibly through the formation of reactive peroxides during the process. Oxidation of the free radicals is of additional importance as it constitutes an autocatalytic process resulting in the regeneration of the herbicide.

Hansch (107) has recently employed substituent constants and regression analysis (Sec. VI.C) in the studies of structure-activity relationships (108) in a series of carbamate synergists of the 1,3-benzodioxole type. As a result of these investigations Hansch concludes that the activity of the 1,3-benzodioxole synergists may arise from their ability to form homolytic free radicals through removal of one of the methylenic hydrogen atoms of the five-membered ring. Activity is correlated with the nature of the substituent in the phenyl ring and is found to be highest with NO_2 or CH_3O, which are strong promotors of reactions involving homolytic substitution. The radical resulting from homolytic removal of a methylenic hydrogen atom from the 1,3-benzodioxole nucleus is stabilized by delocalization of the odd electron in Eq. (10) and although not yet demonstrated it is possible that the formation of free radicals may play an important role in the biological activity of these compounds.

$$(10)$$

G. Metabolism

Metabolism per se will not be considered in detail since the factors which determine the metabolic interactions of any chemical are identical to those involved in receptor interactions in general. It should be pointed out however, that in some cases metabolism may be the dominant factor in determining the biological effect of a material.

Obviously, degradative enzymatic action can represent an important source of loss of material and may often prevent it from reaching its site of action in a sufficiently high concentration to initiate a biological response. Thus the toxicity of a number of p,p'-disubstituted analogs of DDT to resistant insects has been found to correlate extremely well with the susceptibility of the benzylic hydrogen atom to attack by the enzyme DDT-dehydrochlorinase (79). Variations in degradative enzyme activity between different species can often be beneficially employed, however, in designing toxicants with some degree of selectivity (11, 52, 109). Thus the useful insecticides malathion [31a] and malaoxon [31b] are relatively nontoxic to mammalian species because of the high activity of the enzyme carboxyesterase in the mammal compared with that in most insect species (110).

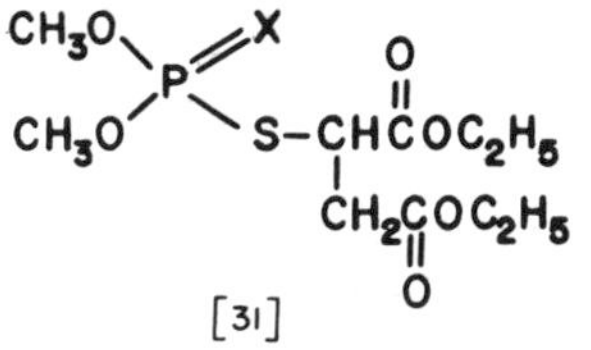

$$CH_3O \diagdown P \diagup X$$
$$CH_3O \diagup \diagdown S-CH\overset{O}{\overset{\|}{C}}OC_2H_5$$
$$CH_2\overset{\|}{\underset{O}{C}}OC_2H_5$$

[31]

a : X = S

b : X = O

Metabolism does not always result in a loss of biological activity and indeed the in vivo effectiveness of certain materials depends on metabolic activation. Thus the phosphorothionate insecticides, such as parathion [32a], are inactive as inhibitors of cholinesterase. The high toxicity of these compounds depends entirely on their in vivo oxidative conversion (de-sulfuration) to the corresponding phosphate analogs (paraoxon [32b] from parathion), which are potent inhibitors of cholinesterase.

$$C_2H_5O \diagdown P \diagup X$$
$$C_2H_5O \diagup \diagdown O \text{---} \langle \rangle \text{---} NO_2$$

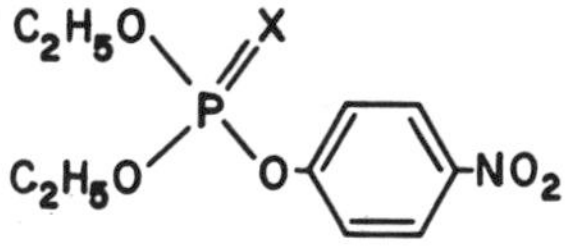

[32]

a : X = S

b : X = O

VI.   QUANTITATIVE ASPECTS OF STRUCTURE-
ACTIVITY INVESTIGATIONS

A.   The Search for New Materials

Up to this point consideration has been given to several physico-chemical, structural, and electronic properties, which determine the ability of chemicals to interact with various biological systems. But the biological activity of any substance, ultimately depends on a favorable combination of several of these properties. Except in isolated cases the consideration of any one factor alone will not provide a means of accurately predicting the potential activity of related materials or of designing new and more effective compounds, the practical goals of all structure-activity investigations.

The ultimate degree of sophistication will be attained when new compounds can be custom-designed for specific purposes through a thorough understanding of the structure, function, and mechanism of specific biochemical receptors. At the present time, however, we are far removed from this stage and the discovery and subsequent development of new materials rests almost entirely on empirical methods, typified by the huge screening programs employed by chemical industry.

In most cases the initial discovery that a chemical possesses some type of biological activity originates from either a chance observation or results directly from the empirical screening of tens of thousands of new materials. Having once obtained a lead in this way, the chemist attempts to improve the observed biological activity by effecting systematic structural variations in different parts of the molecule. The considerable degree of success which has been achieved by these methods speaks highly for the intuitive powers of the modern organic chemist. On the other hand, the ever increasing cost of drugs and agricultural chemicals is a reflection of the tedious and consequently expensive nature of the screening procedure and the continued need for more sophisticated chemicals must in turn be met by the development of more efficient techniques for their discovery.

Although in many cases the precise mechanism of action of biologically active substances remains unknown, there is often a good deal of information at hand regarding the changes in observed activity which result from specific variations in molecular structure. The need for more highly refined techniques in pharmaceutical development combined with easier access to modern computers has in recent years led to increasing interest in the quantitative aspects of correlations between chemical structure and biological activity. Two major approaches are presently being investigated, the empirical mathematical method developed by Free and Wilson (111) and the semiempirical approach advocated by Hansch (112, 113).

B. Empirical Methods

The Free and Wilson (111) approach defines the total biological activity of a chemical as a function of the sum of the individual activity contributions of each of its substituent groups. The method is, therefore, based on the assumption that in a series of analogs at least some of the substituent groups contribute to the total activity in an essentially additive manner. The contribution of any substituent towards the total activity is expressed in terms of a de novo constant derived mathematically from data on biological activity.

The method is perhaps best described by reference to the simple model discussed by Free and Wilson (<u>111</u>). Thus in a series of analgesic compounds of general structure [33], R_1 is either H or CH_3 and R_2 is $N(CH_3)_2$ or $N(C_2H_5)_2$.

$$-C_6H_5$$
$$NHCOCH-R_2$$
$$R_1$$

[33]

The observed LD_{50} values for each of the four analogs are shown in Table 6, from which the average contribution of each substituent can be calculated. Thus the contribution of H at R_1 is $1.705 - 1.475 = +0.23$ whereas the contribution of CH_3 at the R_1 position is $1.245 - 1.475 = -0.23$. In a similar way the average contribution of the $N(CH_3)_2$ and $N(C_2H_5)_2$ substituents at R_2 are $+0.41$ and -0.41 respectively.

The total activity of any of the four analogs can be expressed by Eq. (11)

$$LD_{50} = \mu + aH + aCH_3 + bN(CH_3)_2 + b(C_2H_5)_2 \qquad (11)$$

where μ is the overall average, aH and aCH_3 the contribution of H and CH_3 at position R_1, and $bN(CH_3)_2$ and $bN(C_2H_5)_2$ the contribution of $N(CH_3)_2$ and $N(C_2H_5)_2$ at position R_2. By substituting the actual LD_{50} values into this general equation, a set of simultaneous equations can be written from which the contribution of each substituent to the total activity can be calculated. The solution is actually simplified as a result of the mathematical symmetry built into the solution, so that $aH = aCH_3$ and $bN(CH_3)_2 = -bN(C_2H_5)_2$. This reduces the five unknowns in the general equation to only three which can be readily determined.

The method has useful potential when the activity of a large series of analogs is being investigated as it can provide guidelines to suggest the activity of untested members of the series. In this way it may supplement the intuition of the chemist by predicting untried combinations of substituents to be of high activity. Purcell (<u>114</u>), using the Free and Wilson method, found good activity correlations with a series of cholinesterase inhibitors. More recently Purcell and Clayton (<u>115</u>) have successfully applied similar

TABLE 6

Toxicity (LD_{50}) of Substituted Analgesics [33][a]

R_2	R_1		Mean
	H	CH_3	
$N(CH_3)_2$	2.13	1.64	1.885
$N(C_2H_5)_2$	1.28	0.85	1.065
Mean	1.705	1.245	1.475 (overall average)

[a]Data from Free and Wilson (111).

regression analyses to investigations of the antitumor activity of sixty-nine acetylenic carbamates. With the latter, good correlations were obtained between observed and calculated activity and the results predicted that several untested materials should have high antitumor activity. It will be of interest to see if these predictions prove true if or when the compounds in question are evaluated.

As pointed out by Hansch (112), one objection to this approach is that each of the empirical constants employed combines in one figure the effects of such important factors as lipid solubility, as well as steric and electronic effects. As a result, the constants derived from one series of analogs cannot be applied to others having a different type of biological activity. Furthermore, the Free and Wilson method does not take into account variations in biological activity resulting from either stereochemical configuration or from such physical properties as the degree of ionization (pK_a).

A series of empirical substituent constants have also been determined by Boček et al. (116) and Kopecký et al. (117) in attempts to express quantitatively the toxicity to mice of meta- and para-disubstituted benzenes. These investigators derived Eq. (12) where HH represents the unsubstituted compound (benzene) and XY a disubstituted derivative. Constants obtained

$$\log\frac{\left[LD_{50}\right]_{HH}}{\left[LD_{50}\right]_{XY}} = bX + bY + eXeY \qquad (12)$$

with the meta- and para-disubstituted derivatives had to be later modified
to obtain correlations with ortho-disubstituted benzenes (118).

Following the early work of Ferguson (Sec. IV), in which the efficiency
determining step is related to some physical property such as partition coefficient, Zahradnik and Chvapil (119) obtained a series of constants for
alkyl substituents which were independent of the type of biological response
studied and the class of compounds employed. Using compounds of the type
R-X where R is an alkyl chain and X a functional group such as OH or COO^-,
these workers suggested that quantitative correlation of structure and activity could be obtained from Eq. (13). In this equation τ_i is the biological

$$\log \frac{\tau_i}{\tau_{Et}} = \alpha \, \beta \tag{13}$$

activity associated with the ith member of the series and τ_{Et} is that of the
ethyl derivative which is used as a reference point. α and β are constants.
The constant β defines the activity of the substituent group R. This is independent of the functional group X and the biological object which are both
incorporated into the constant α. The actual β constants were determined
by putting the constant $\alpha = 1$ in a selected system involving the LD_{50} values
of a series of alcohols to white mice. Under these conditions $\tau_i/\tau_{Et} = \beta$.
The β constants thus obtained can be plotted against $\triangle \log \tau_i$ in other biological systems and when this is done a linear correlation is often obtained.

However, the approach taken by Zahradnik will only apply to those systems where biological activity can be defined with reference to a single
physical parameter such as partition coefficient. In this sense it is an extension of the Ferguson concept and cannot be applied to systems where activity depends on more specific steric and electronic effects.

A more general form of Eq. (13) has been subsequently derived in terms
of the free energy change of the investigated equilibrium (120). This is
shown in Eq. (14) where E_i and E_{ref} are the free energies of the ith and

$$E_i - E_{ref} = \phi \psi_i \tag{14}$$

reference members of the series and ϕ and ψ_i are constants. ψ_i is essentially similar to Zahradnik's β constant in Eq. (13), and $\phi\,(\alpha)$ character-
izes the system. If E_i and E_{ref} represent the rate or equilibrium constants for a chemical reaction, it can be seen that the equation is similar
to the Hammett (97, 121) or Taft (101, 122) equations in which case ψ_i will
be σ or σ^* respectively.

C. Semiempirical Methods

In attempts to obtain quantitative correlations between chemical structure and biological activity, the use of purely empirical substituent constants has very limited utility. Such constants are usually valid only for certain series of compounds acting on specific biological systems. If instead, constants which are directly related to fundamental physicochemical reference parameters can be employed, these would be expected to have more general application in the quantitative expression of biological activity. This latter approach is presently being investigated by Hansch and colleagues (112, 113, 123). Therefore, the basic problem becomes the development of equations which take into account each of the several major factors upon which biological activity depends.

Hansch and Fujita (123) begin from a fundamental equation that logically expresses the fact that the effectiveness of any biologically active substance depends on its initial concentration at some point outside the cell or organism, its rate of penetration to the receptor site, and subsequently its ability to interact with the receptor. The rate of biological response can therefore be expressed by Eq. (15) where A is the probability of a compound

$$\frac{d(\text{response})}{dt} = ACk_X \tag{15}$$

reaching the site of action, C is the external concentration and k_X an equilibrium constant describing its interaction with the receptor. Because biological data are usually described in terms of a constant response (LD_{50}, I_{50}, etc.), Eq. (15) can be more simply expressed by Eq. (16).

$$ACk_X = \text{constant} \tag{16}$$

The probability (A) of a material reaching its site of action is, as we have discussed, largely a function of its lipid-water partition coefficient. Fujita et al. (124), instead of using the latter directly, developed a comparative substituent constant π. This is a free-energy related constant for a given substituent and is therefore similar to Hammett's σ constant. π is defined by Eq. (17) where P_X is the octanol-water partition coefficient

$$\pi = \log \frac{P_X}{P_H} = \log P_X - \log P_H \tag{17}$$

of a substituted compound X and P_H is the coefficient of the unsubstituted compound. Thus the π value associated with an aromatic methyl group can

be readily obtained by subtracting the logarithm of the octanol–water partition coefficient of benzene from that of toluene. Log P_X itself is often employed instead of π.

It is assumed that maximum penetration and translocation through the biophase will occur when P_X for the biophase is 1 and there will consequently be an optimum value of $\pi(\pi_0)$ for certain substituents. At π values below or above the optimum therefore, biological response will be decreased, a fact which is suggested from the cut–off point phenomenon found with physical poisons (Sec. IV). The variation in biological activity with respect to π (or log P) was assumed to follow a normal gaussian distribution. As a result of this assumption A can be expressed by Eq. (18) where a and b are constants. Substituting Eq. (18) into Eq. (16) and taking the logarithm

$$A = f(\pi) = ae^{-(\pi-\pi_0)^2/b} \qquad (18)$$

yields Eq. (19). In Eq. (19) k_X is either an equilibrium or rate constant

$$\log \frac{1}{C} = -k\pi^2 + k_1\pi\,\pi_0 - k_2\pi_0^2 + \log k_X + k_3 \qquad (19)$$

describing the chemical interaction with the target receptor. Hansch and Fujita (123) propose that this can be expressed as in the Hammett equation, Eq. (20), where k_X is the rate or equilibrium constant for a compound with

$$\log \frac{k_X}{k_H} = \rho\sigma \quad \text{or} \quad \log k_X = \log k_H + \rho\sigma \qquad (20)$$

substituent X and k_H is the constant for the parent compound. As discussed earlier (Sec. V.E), σ is a constant defining the electronic character of substituent X relative to H and ρ defines the type of reaction under consideration.

Combining Eqs. (19) and (20) yields Eq. (21), where $k_4 = k_3 + \log k_H$

$$\log \frac{1}{C} = -k\pi^2 + k_1\pi\,\pi_0 - k_2\pi_0^2 + \rho\sigma + k_4 \qquad (21)$$

This constitutes the general equation developed by Hansch and Fujita (123), but, depending on the compounds under consideration, it can be considerably simplified into five forms, [Eqs. (22)–(26), Table 7]. In all cases the π_0 term is constant with any given series of compounds.

TABLE 7

Different Forms of the Hansch Equation[a]

Conditions	Relevant equation	
$\pi_0 >> \pi;\ \sigma \to 0$	$\log \dfrac{1}{C} = a\pi + b$	(22)
π_0 similar to $\pi;\ \sigma \to 0$	$\log \dfrac{1}{C} = -a\pi^2 + b\pi + c$	(23)
π insignificant; σ significant	$\log \dfrac{1}{C} = \rho\sigma + c$	(24)
$\pi_0 >> \pi;\ \sigma$ significant	$\log \dfrac{1}{C} = a\pi + \rho\sigma + c$	(25)
π_0 similar to $\pi;\ \sigma$ significant	$\log \dfrac{1}{C} = -a\pi^2 + b\pi + \rho\sigma + c$	(26)

[a]Data from Hansch and Fujita (123).   Constants a, b, and c are different in each equation.

In Eq. (22), its simplest form, the expression is essentially that describing the initial Meyer-Hemmi (37) concept correlating narcotic action with partition coefficients, and indeed is often written as Eq. (27), in which

$$\log \frac{1}{C} = \log P + \text{constant} \qquad (27)$$

form it is identical to this early hypothesis. Equation (22), relating biological activity to π or log P, will therefore effectively define the biological activity of structurally nonspecific toxicants or narcotics where specific and electronic effects can be neglected. High correlation coefficients have been obtained for a number of different systems including the toxicity of benzoic acids to mosquito larvae (123), the toxicity of phenols to a variety of Mycobacterium pyogenes (123), the inhibition of Avena cell elongation by phenoxyacetic acids (125), and the activity of penicillins (126).

The simplest form of the general equation has also been successfully applied to correlating the hydrophobic binding of certain compounds with

proteins. Excellent correlations were obtained with respect to the binding
of materials to ribonuclease and serum albumin (127) and also for the binding of barbiturates by rabbit brain (128).

The expression derived by McGowan, [Eq. (7), Sec. IV], is similar to
Eq. (22) except that parachor is used instead of π or log P. McGowan has
recently shown a relationship between π and parachor (129). Another constant which can be substituted for π or log P is the $\triangle R_M$ value introduced by
Green and Marcinkiewicz (130) which defines the effect of a substituent on
the chromatographic R_M value of the unsubstituted compound. In view of
the fact that chromatographic separation is a partitioning phenomenon, it
is not surprising that a linear correlation is found to exist between π and
$\triangle R_M$ (131), but as will be discussed shortly the use of R_M values may have
certain advantages in some structure-activity correlations. Good correlation has also been reported (132) between the π constants and the β constants
developed by Zahradnik and Chvapil (119). Some relationship would however
be expected in view of the fact that the equations of both Zahradnik and
Hansch are based on free-energy considerations.

As indicated in the initial assumption of Hansch and Fujita (123), the
correlation of π or log P and biological activity is found to deviate from
linearity over large ranges of P (113). This can be corrected using a π^2
term as in Eq. (23).

In considering those types of biological activity arising from interference with enzymatic reactions, electronic interactions are almost invariably involved. In these cases, consideration of π or log P alone is not
usually sufficient to successfully correlate structure and activity, so that
one of the equations including σ, Eqs. (24)-(26), must be employed. The
use of σ alone, Eq. (24), although extremely useful in determining the
chemical reactivity of materials in pure organic reactions, is seldom sufficient in considerations of reactions catalyzed by enzymes. This results
from the fact that even in simple in vitro reactions, where penetration
through cell membranes is avoided, the ability of most compounds to interact with the enzyme protein depends to a large extent on their hydrophobic binding properties and therefore will depend to some extent on π.

One of the few examples where satisfactory correlations have been obtained using σ alone is found in the in vitro anticholinesterase activity of
a number of para-substituted phenyl phosphates (68) (Sec. V. E). Under
these in vitro assay conditions the importance of π must be minimal and
remain relatively constant, as little improvement in the correlation is derived from using Eq. (25) rather than Eq. (24). In considering the correlation between structure and in vivo toxicity to house flies of these same

compounds the inclusion of the π term in the equation only slightly improved the observed correlation (123), a somewhat surprising fact in view of the several membranes which have to be successfully crossed before the compounds can reach their site of action.

Usually however, Eq. (25), which includes combinations of both π and σ terms, provides a much more satisfactory correlation between structure and biological activity (112, 113). In considering structure-activity relationships using whole animals or organisms, Eq. (26) often provides a better correlation (113). This includs a π^2 term to accommodate the parabolic relationship between log 1/C and π over large ranges of P.

Several additional refinements in the general equation have recently been discussed (113) and include linear combinations of terms to quantitatively define such properties as degree of ionization, and dipole moment, as well as attempts to satisfactorily express steric relationships between substrate and its site of action.

Such steric factors are often of extreme importance but are very difficult to define in quantitative terms. Intramolecular steric interactions can sometimes be accounted for by the use of Taft's Es parameter (122). The employment of R_M values may well provide a new method of quantitatively expressing steric factors, since stereoisomers can be separated chromatographically as can optical isomers if an asymmetric adsorbent is employed.

The Hansch approach to the correlation of chemical structure and activity has a great deal of flexibility and potential. It allows the derivation of equations that not only qualitatively express the several major factors determining the observed activity, but also make possible the quantitative evaluation of each of these factors in fundamental physicochemical terms.

Of particular utility in the design of new drugs is the term π_0 or log P_0. This is a constant for any series of compounds and as initially proposed defines the optimum partition coefficient required for maximum penetration and translocation to the site of action. Log P_0 can be determined for any series of materials and in the subsequent search for new and more active compounds of that type attempts should be made to incorporate substituents which will provide a log P value approaching that of log P_0. Furthermore, it should be possible to include in the molecule those substituents likely to provide the correct electronic distribution on which its chemical reactivity depends.

There is little doubt that in many cases we already have the potential to design chemicals with biological activity. It is equally clear that the next decade will see this potential dramatically increase as our basic knowledge of the interactions between chemicals and biological receptors becomes more sophisticated.

REFERENCES

1. W. A. Sexton, Chemical Constitution and Biological Activity, 3rd ed., Van Nostrand, Princeton, N.J., 1963, p. 517.

2. T. C. Daniels and E. C. Jorgensen, in Textbook of Organic Medicinal and Pharmaceutical Chemistry (C. O. Wilson, O. Gisvold, and R. F. Doerge, eds.), 5th ed., Lippincott, Philadelphia, 1966, Chap. 2.

3. N. J. Doorenbos, in Medicinal Chemistry (A. Burger, ed.), 2nd ed., Interscience, New York, 1960, Chap. 7.

4. A. Burger, in Medicinal Chemistry (A. Burger, ed.), 2nd ed., Interscience, New York, 1960, Chap. 6.

5. A. Burger, Proc. Intern. Pharmacol. Meeting, 1st, 1961 (B. Uvnäs, ed.), Vol. 7, Pergamon, Oxford, 1963, pp. 35–64.

6. H. Davson and J. F. Danielli, The Permeability of Natural Membranes, 2nd ed., Cambridge University, Cambridge, 1952.

7. L. S. Schanker, Ann. Rev. Pharmacol., 1, 29 (1961).

8. L. S. Schanker, Pharmacol. Rev., 14, 501 (1962).

9. B. B. Brodie, in Absorption and Distribution of Drugs (T. B. Binns, ed.), Livingstone, Edinburgh, 1964, pp. 16–48.

10. A. Goldstein, L. Aronow, and S. M. Kalman, Principles of Drug Action, Harper and Row, New York, 1968.

11. R. D. O'Brien, Insecticides: Action and Metabolism, Academic, New York, 1967, p. 332.

12. S. Mayer, R. P. Maickel, and B. B. Brodie, J. Pharmacol. Exptl. Therap., 127, 205 (1959).

13. B. B. Brodie, H. Kurz, and L. S. Schanker, J. Pharmacol. Exptl. Therap., 130, 20 (1960).

14.   L. S. Schanker, P. A. Shore, B. B. Brodie, and C. A. M. Hogben, J. Pharmacol. Exptl. Therap., 120, 528 (1957).

15.   M. E. Eldefrawi and R. D. O'Brien, J. Insect Physiol., 12, 1133 (1966).

16.   M. E. Eldefrawi and R. D. O'Brien, J. Exptl. Biol., 46, 1 (1967).

17.   M. E. Eldefrawi and R. D. O'Brien, J. Insect Physiol., 13, 691 (1967).

18.   A. Toppozada and R. D. O'Brien, J. Insect Physiol., 13, 941 (1967).

19.   M. E. Eldefrawi, A. Toppozada, M. M. Salpeter, and R. D. O'Brien, J. Exptl. Biol., 48, 325 (1968).

20.   R. D. O'Brien, Fed. Proc., 26, 1056 (1967).

21.   R. D. O'Brien and B. D. Hilton, J. Agr. Food Chem., 12, 53 (1964).

22.   R. D. O'Brien, J. Econ. Entomol., 52, 812 (1959).

23.   M. J. Kolbezen, R. L. Metcalf, and T. R. Fukuto, J. Agr. Food Chem., 2, 864 (1954).

24.   R. L. Metcalf and T. R. Fukuto, J. Agr. Food Chem., 13, 220 (1965).

25.   I. Yamamoto, Adv. Pest Control Res., 6, 231 (1965).

26.   R. D. O'Brien and C. E. Dannelley, J. Agr. Food Chem., 13, 245 (1965).

27.   W. P. Olson and R. D. O'Brien, J. Insect Physiol., 9, 777 (1963).

28.   A. A. Buerger and R. D. O'Brien, J. Cellular Comp. Physiol., 66, 227 (1965).

29.   W. J. Hayes, Arch. Indust. Health, 18, 398 (1958).

30.   W. S. Hoffman, W. I. Fishbein, and M. B. Andelman, Arch. Envir. Health, 9, 387 (1964).

31.   C. G. Hunter and J. Robinson, Arch. Envir. Health, 15, 614 (1967).

32. C. G. Hunter and J. Robinson, Food Cosmet. Toxicol., 6, 253 (1968).

33. J. M. Thorp, in Absorption and Distribution of Drugs (T. B. Binns, ed.), Livingstone, Edinburgh, 1964, pp. 64–76.

34. M. C. Meyer and D. E. Guttman, J. Pharmacol. Sci., 57, 895 (1968).

35. E. Overton, Vierteljahrsschr. naturforsch Ges. Zurich, 44, 88 (1899).

36. H. Meyer, Arch. Exptl. Pathol. Pharmakol., 42, 109 (1899).

37. K. H. Meyer and H. Hemmi, Biochem. Z., 277, 39 (1935).

38. R. Collander, Acta Chem. Scand., 5, 774 (1954).

39. A. Fühner, Arch. Exptl. Pathol. Pharmakol., 52, 69 (1904).

40. J. Ferguson, Proc. Royal Soc. (London), B127, 387 (1939).

41. J. Ferguson and H. Pirie, Ann. Appl. Biol., 35, 532 (1948).

42. J. C. McGowan, J. Appl. Chem., 1, S 120 (1951).

43. J. C. McGowan, J. Appl. Chem., 2, 323 (1952).

44. J. C. McGowan, J. Appl. Chem., 2, 651 (1952).

45. J. C. McGowan, J. Appl. Chem., 4, 41 (1954).

46. J. C. McGowan, J. Appl. Chem., 16, 99 (1966).

47. L. J. Mullins, Chem. Rev., 54, 289 (1954).

48. D. E. Koshland, Cold Spring Harbor Symp. Quant. Biol., 28, 473 (1963).

49. I. B. Wilson and F. Bergmann, J. Biol. Chem., 185, 479 (1950).

50. I. B. Wilson, J. Biol. Chem., 197, 215 (1952).

51. F. Bergmann and A. Shimoni, Biochem. Biophys. Acta, 9, 473 (1952).

52.   R. D. O'Brien, Toxic Phosphorus Esters, Academic, New York, 1960, p. 434.

53.   R. L. Metcalf and T. R. Fukuto, J. Agr. Food Chem., 15, 1022 (1967).

54.   R. D. O'Brien, in Drug Design (E. J. Ariëns, ed.), Vol. 2, Chap. 3, Academic, New York, 1971.

55.   R. B. Barlow, Introduction to Chemical Pharmacology, Methuen, London, 1955.

56.   D. F. Heath, Organophosphorus Poisons, Pergamon, New York, 1961, p. 403.

57.   B. R. Baker, Design of Active-Site Directed Irreversible Enzyme Inhibitors, Wiley, New York, 1967, Chap. 2.

58.   F. Bergmann, Disc. Faraday Soc., 20, 126 (1955).

59.   P. Bracha and R. D. O'Brien, Biochem., 7, 1545 (1968).

60.   P. Bracha and R. D. O'Brien, Biochem., 7, 1555 (1968).

61.   A. P. Brestkin, N. N. Godovikov, E. I. Godyna, M. I. Kabachnik, M. Ya. Mikhelson, E. V. Rosengart, and V. A. Yakovlev, Dokl. Akad. Nauk. S.S.S.R., 158, 880 (1964).

62.   W. P. Purcell, J. Med. Chem., 9, 294 (1966).

63.   T. R. Fukuto and R. L. Metcalf, J. Agr. Food Chem., 4, 930 (1956).

64.   T. R. Fukuto and R. L. Metcalf, J. Am. Chem. Soc., 81, 372 (1959).

65.   T. R. Fukuto, R. L. Metcalf, and M. Y. Winton, J. Econ. Entomol., 52, 1121 (1959).

66.   T. R. Fukuto, R. L. Metcalf, and M. Y. Winton, J. Econ. Entomol., 54, 955 (1961).

67.   T. R. Fukuto, Ann. Rev. Entomol., 6, 313 (1961).

68.   C. Hansch and E. W. Deutsch, Biochem. Biophys. Acta, 126, 117 (1966).

69. E. L. Becker, T. R. Fukuto, B. Boone, D. C. Canham, and
     E. Boger, Biochem., 2, 72 (1963).

70. C. F. Wilkinson, R. L. Metcalf, and T. R. Fukuto, J. Agr. Food
     Chem., 14, 73 (1966).

71. F. Matsumura and R. D. O'Brien, J. Agr. Food Chem., 14, 36
     (1966).

72. F. Matsumura and R. D. O'Brien, J. Agr. Food Chem., 14, 39
     (1966).

73. F. Matsumura and M. Hayashi, Science, 153, 757 (1966).

74. F. Matsumura and M. Hayashi, Mosquito News, 26, 190 (1966).

75. J. A. Moss and D. E. Hathaway, Biochem. J., 91, 384 (1964).

76. L. J. Mullins, Science, 122, 118 (1955).

77. S. B. Soloway, Adv. Pest Cont. Res., 6, 85 (1965).

78. R. L. Metcalf, Organic Insecticides, Wiley-Interscience, New York,
     1955, p. 392.

79. R. L. Metcalf and T. R. Fukuto, Bull. World Health Org., 38, 633
     (1968).

80. V. B. Schatz, in Medicinal Chemistry (A. Burger, ed.), 2nd ed.,
     Interscience, New York, 1960, Chap 8.

81. C. Kaiser, in Medicinal Chemistry (A. Burger, ed.), 2nd ed., Inter-
     science, New York, 1960, Chap. 9.

82. D. D. Woods, Brit. J. Exptl. Pathol., 21, 74 (1940).

83. R. B. Turner, in Principles of Insect Chemosterilization (G. C.
     LaBrecque and C. N. Smith, eds.), Appleton-Century-Crofts,
     New York, 1968, Chap. 5.

84. F. M. Huennekens, Biochem., 2, 151 (1963).

85. C. Heidelberger, Exptl. Cell Res. Suppl., 9, 462 (1963).

86. L. Crombie and R. M. Elliot, in Fortschritte der Chemie Organischer Naturstoffe (L. Zechmeister, ed.), Springer-Verlag, Berlin, 1961, pp. 121-164.

87. R. Barlow and J. Hamilton, Brit. J. Pharmacol., 25, 206 (1965).

88. H. S. Aaron, H. O. Michel, B. Witten, and J. J. Miller, J. Am. Chem. Soc., 80, 456 (1958).

89. A. J. J. Ooms and H. L. Boter, Biochem. Pharmacol., 14, 1839 (1965).

90. A. Hassan and W. C. Dauterman, Biochem. Pharmacol., 17, 1431 (1968).

91. R. L. Wain, N.Y. Acad. Sci. Ann., 144, 223 (1967).

92. A. Morello, E. Y. Spencer, and A. Vardanis, Biochem. Pharmacol., 16, 1703 (1967).

93. T. R. Fukuto, E. O. Hornig, R. L. Metcalf, and M. Y. Winton, J. Org. Chem., 26, 4620 (1961).

94. T. R. Fukuto, Adv. Pest Cont. Res., 1, 147 (1957).

95. R. D. O'Brien, Ann. Rev. Entomol., 11, 369 (1966).

96. W. N. Aldridge and A. N. Davison, Biochem. J., 52, 663 (1952).

97. L. P. Hammett, Physical Organic Chemistry, McGraw-Hill, New York, 1940.

98. E. Benjamini, R. L. Metcalf, and T. R. Fukuto, J. Econ. Entomol., 52, 94 (1959).

99. R. L. Metcalf and T. R. Fukuto, J. Econ. Entomol., 55, 340 (1962).

100. R. L. Metcalf, T. R. Fukuto, and M. Frederickson, J. Agr. Food Chem., 12, 231 (1965).

101. R. W. Taft, Jr., J. Am. Chem. Soc., 75, 4231 (1953).

102. T. R. Fukuto, R. L. Metcalf, M. Y. Winton, and R. B. March, J. Econ. Entomol., 56, 808 (1963).

103. R. F. Homer, G. C. Mees, and T. E. Tomlinson, J. Sci. Food Agr., 11, 309 (1960).

104. G. C. Mees, Ann. Appl. Biol., 48, 601 (1960).

105. G. Zweig and M. Avron, Biochem. Biophys. Res. Commun., 19, 397 (1965).

106. A. Calderbank, Biochem. J., 101, 2P (1966).

107. C. Hansch, J. Med Chem., 11, 920 (1968).

108. C. F. Wilkinson, J. Agr. Food Chem., 15, 139 (1967).

109. R. D. O'Brien, Adv. Pest. Cont. Res., 4, 75 (1961).

110. H. R. Krueger and R. D. O'Brien, J. Econ. Entomol., 52, 1063 (1959).

111. S. M. Free and J. W. Wilson, J. Med. Chem., 7, 395 (1964).

112. C. Hansch, Ann. Repts. Med. Chem. (C. K. Cain, ed.), Academic, New York, 1967, Chap. 34.

113. C. Hansch, Ann. Repts. Med. Chem. (C. K. Cain, ed.), Academic, New York, 1968, Chap. 35.

114. W. P. Purcell, Biochem. Biophys. Acta, 105, 201 (1965).

115. W. P. Purcell and J. M. Clayton, J. Med. Chem., 11, 199 (1968).

116. K. Boček, J. Kopecký, M. Krivucová, and D. Vlachová, Experientia, 20, 667 (1964).

117. J. Kopecký, K. Boček, and D. Vlachová, Nature, 207, 981 (1965).

118. K. Boček, J. Kopecký, and M. Krivucová, Experientia, 23, 1038 (1967).

119. R. Zahradnik and M. Chvapil, Experientia, 16, 511 (1960).

120. R. Zahradnik, Experientia, 18, 534 (1962).

121. H. Jaffé, Chem. Rev., 53, 191 (1953).

122. M. Newman, Steric Effects in Organic Chemistry, Wiley, New York 1956.

123. C. Hansch and T. Fujita, J. Am. Chem. Soc., 86, 1616 (1964).

124. T. Fujita, J. Iwasa, and C. Hansch, J. Am. Chem. Soc., 86, 5175 (1964).

125. R. M. Muir, T. Fujita, and C. Hansch, Plant Physiol., 42, 1519 (1967).

126. C. Hansch and E. W. Deutsch, J. Med. Chem., 8, 705 (1965).

127. C. Hansch, Il Farmaco, 23, 293 (1968).

128. C. Hansch and S. M. Anderson, J. Med. Chem., 10, 745 (1967).

129. J. C. McGowan, Nature, 200, 1317 (1963).

130. J. Green and S. Marcinkiewicz, J. Chromatog., 10, 389 (1963).

131. J. Iwasa, T. Fujita, and C. Hansch, J. Med. Chem., 8, 150 (1965).

132. J. Kopecký and K. Boček, Experientia, 23, 125 (1967).

Chapter 2

THE EMULSIFIER

Paul Becher

ICI America Inc.
Wilmington, Delaware

I. INTRODUCTION

Emulsifiers, or emulsifying agents, belong to that group of chemical compounds known as surface-active agents. They are therefore characterized by a structure in which the molecule is more-or-less clearly divided into distinct moieties. One moiety is hydrophilic or water-soluble, the other hydrophobic. Usually, and in all cases which will be of interest to us, this hydrophobic portion will, in fact, be lipophilic, i.e., exhibit solubility in organic liquids. In Fig. 1 we show the structures of a number of characteristic surface-active compounds and indicate the hydrophilic and lipophilic portions of the molecules.

Because of the hydrophilic-lipophilic (or polar) character of these molecules, they exhibit certain properties in solution which are not characteristic of other types of materials. In particular, they adsorb at interfaces (gas/liquid, liquid/liquid, liquid/solid) to form oriented monolayers. In addition, in the solution itself, they may aggregate to form micelles. The usefulness of these materials stems largely from these two effects.

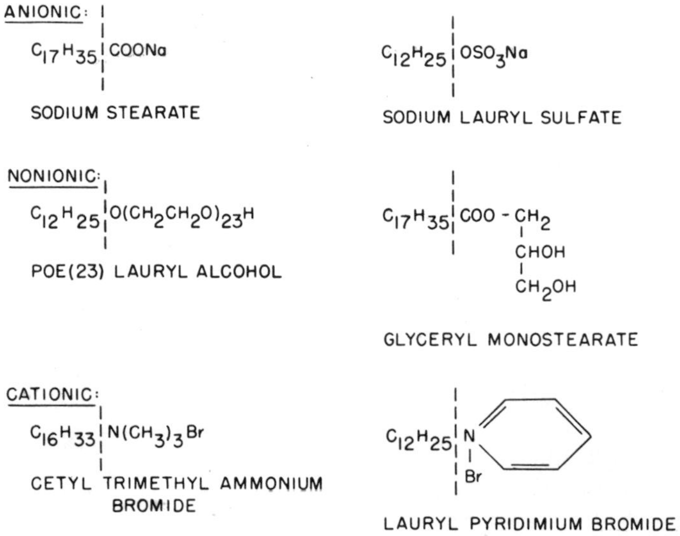

ANIONIC:

$C_{17}H_{35} \mid COONa$

SODIUM STEARATE

$C_{12}H_{25} \mid OSO_3Na$

SODIUM LAURYL SULFATE

NONIONIC:

$C_{12}H_{25} \mid O(CH_2CH_2O)_{23}H$

POE(23) LAURYL ALCOHOL

$C_{17}H_{35} \mid COO - CH_2$
$\qquad\qquad\qquad | \quad CHOH$
$\qquad\qquad\qquad\quad\ CH_2OH$

GLYCERYL MONOSTEARATE

CATIONIC:

$C_{16}H_{33} \mid N(CH_3)_3\, Br$

CETYL TRIMETHYL AMMONIUM
BROMIDE

$C_{12}H_{25} \mid N$ [pyridinium ring]
$\qquad\qquad\quad | \quad Br$

LAURYL PYRIDIMIUM BROMIDE

FIG. 1. Structural formulas of the various types of surface-active
agents. The portion of the molecule to the left of the dashed line is the
lipophilic moiety; that to the right the hydrophilic.

II. CLASSES OF EMULSIFIERS

Surface-active agents used as emulsifiers can be divided into five principal classes (1). These are: (1) anionic, (2) cationic, (3) non-ionic, (4)
ampholytic, and (5) water-insoluble.

Examples of the first four types are given in Fig. 1. The last type
may include such materials as finely-divided solids and vegetable gums,
e.g., tragacanth, gum arabic. Such water-insoluble materials may or
may not contain ionic groups.

The simple classification indicated above may be expanded in a more
detailed way, as follows (1) :

1. ANIONIC

a. Carboxylic Acids
    i. Carboxyl joined directly to hydrophobic group
    ii. Carboxyl joined through an intermediate linkage

b. Sulfuric Esters (Sulfate)
     i.   Sulfate group joined directly to hydrophobic group
    ii. Sulfate group joined through intermediate linkage

c. Alkane Sulfonic Acids
     i.   Sulfonic group joined directly to hydrophobic group
    ii. Sulfonic group joined through intermediate linkage

d. Alkyl Aromatic Sulfonic Acids
     i.   Hydrophobic group joined directly to sulfated aromatic nucleus
    ii. Hydrophobic group joined to sulfonated aromatic nucleus through
        intermediate linkage.

e. Miscellaneous Anionic Hydrophilic Groups
     i.   Phosphates and phosphoric acids
    ii. Persulfates, thiosulfates, etc.
   iii. Sulfonamides
    iv. Sulfamic acids, etc.

2. CATIONIC

a. Amine Salts (Primary, Secondary, and Tertiary)
     i.   Amino group joined directly to hydrophobic group
    ii. Amino group joined through intermediate linkage

b. Quaternary Ammonium Compounds
     i.   Nitrogen joined directly to hydrophobic group
    ii. Nitrogen joined through an intermediate group

c. Other Nitrogenous Bases
     i.   Nonquaternary bases (e.g., guanidine, thiuronium salts, etc.)
    ii. Quaternary bases

d. Non-nitrogenous Bases
     i.   Phosphonium compounds
    ii. Sulfonium compounds, etc.

3. NON-IONIC

a. Ether Linkage to Solubilizing Groups

b. Ester Linkage

c. Amide Linkage

d. Miscellaneous Linkages

e. Multiple Linkages

4. AMPHOLYTIC

a. Amino and Carboxy
    i. Nonquaternary
   ii. Quaternary

b. Amino and Sulfuric Ester
    i. Nonquaternary
   ii. Quaternary

c. Amino and Alkane Sulfonic Acid

d. Amino and Aromatic Sulfonic Sulfonic Acid

e. Miscellaneous Combinations of Basic and Acidic Groups

5. WATER-INSOLUBLE

a. Ionic Hydrophilic Group

b. Non-ionic Hydrophilic Group

Commercially, the most important emulsifiers are the anionic and
non-ionic agents, with the cationic agents a somewhat distant third. Ampholytic agents, although of considerable interest, have rather restricted
uses, and, to the writer's knowledge, have rarely or never been used in
herbicidal or pesticidal formulations.

III. SURFACE PROPERTIES OF EMULSIFIERS

As is well known, the surface of a liquid, or the interface between two
liquids is characterized by the existence of a surface (or interfacial) tension. Table 1 lists the surface tensions of a number of liquids of varying
chemical structure; Table 2 lists a number of interfacial tensions. The
surface tension may be thought of as arising from the unbalanced forces
acting on the liquid molecules in the neighborhood of the surface or interface, as represented schematically in Fig. 2.

The addition of a solute to a liquid may affect the surface tension in
varying ways. In the first instance, there may be a small increase in the

TABLE 1 (2)

Surface Tensions of Pure Substances at 20°C
(dyn/cm)[a]

Mercury	485.0	Chloroform	27.13
Water	72.80	Carbon tetrachloride	26.66
Acetylene tetrabromide	49.67	Ethyl caproate	25.81
Nitrobenzene	43.38	Methyl propyl ketone	24.15
Nitromethane	36.82	Di-isoamyl	22.24
Bromobenzene	36.26	n-Octane	21.77
Chloroacetone	35.27	n-Hexane	18.43
Oleic acid	32.50	Ethyl ether	17.10
Carbon disulfide	31.38	Caster oil[b]	39.0
Benzene	28.86	Olive oil[b]	35.8
Caprylic acid	28.82	Cottonseed oil[b]	35.4
n-Octyl alcohol	27.53	Liquid petrolatum[b]	33.1

[a]International Critical Tables, except as indicated.

[b]Halpern, A., J. Phys. Chem., 53, 896 (1949).

surface tension. This will be evidenced with nonpolar organic molecules, e.g., glycerine. A second effect, found usually with simple inorganic electrolytes, is a small depression of the surface tension. Finally, a dramatic effect, resulting in a substantial depression of surface or inter-facial tension is found when so-called surface-active compounds are the solute. Figs. 3 and 4 show the range of surface and interfacial tension lowering for the usual types of surface-active agents.

The lowering of the surface tension arises from the strong surface ad-sorption of the polar surface-active molecules, i.e., the concentration of surface-active molecules in the surface layer is much higher than in the bulk of the solution. This occurs because of the ability of the polar mol-ecules to orient at the surface or interface, with the lipophilic portion in the non-aqueous phase (air or oil). As a consequence of this, the original water-surface, which is a surface of high energy (as evidenced by its high

TABLE 2 (3)

Interfacial Tensions of Various Liquids Against Water at
20°C. (dyn/cm)[a]

Mercury	375.0	Chloroform	32.80
n–Hexane	55.10	Nitrobenzene	25.66
n–Octane	50.81	Ethyl caproate	19.80
Carbon disulfide	48.36	Oleic acid	15.59
Di–isoamyl	46.80	Ethyl ether	10.70
Carbon tetrachloride	45.0	Nitromethane	9.66
Bromobenzene	39.82	n–Octyl alcohol	8.52
		Caprylic acid	8.22
Acetylene tetrachloride	38.82	Chloracetone	7.11
Toluene	36.1	Methyl propyl ketone	6.28
Benzene	35.0	Olive oil[b]	22.9

[a]International Critical Tables, except as indicated.

[b]Sutheim, G. M., Introduction of Emulsions, Chemical Publishing Co.,
Inc., Brooklyn, 1946, p. 25.

surface tension, cf. Table 1), is reduced to the much lower energy of the
hydrocarbon (18–20 dyn/cm).

The extent of the adsorption of the surface–active agents at the interface is given by the well–known adsorption isotherm of Gibbs (4) :

$$\Gamma = \frac{-c}{RT} \frac{d\gamma}{dc} \tag{1}$$

where γ is the surface tension, c is the bulk concentration of surface–active agent, R and T have their usual meanings, and Γ is the surface concentration, in moles per square centimeter.

Thus, the more steeply the surface tension decreases with increasing
concentration of surface–active agent, the greater the surface concentration.

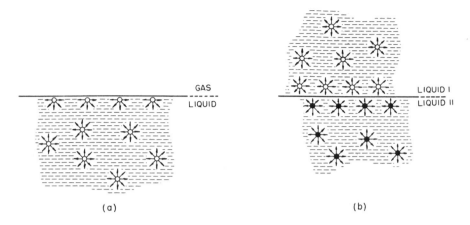

FIG. 2. (a) Forces acting on molecules at a gas-liquid interface. (b) Forces acting on molecules at a liquid-liquid interface.

The Gibbs equation governs the portion of the surface or interfacial tension curve in which the surface tension is decreasing, as shown in Figs. 3 and 4. However, examination of these curves will show that at some definite concentration (which is characteristic of the surface-active agent) the curve has a tendency to flatten out. At this point, the Gibbs equation no longer strictly applies, because the nature of the surface-active agent in the solution has changed as a result of **aggregation,** or micelle formation. This will be discussed in the following section.

In some cases, the surface tension-concentration curve shows a minimum in the region of micelle formation. This has been explained as due to the presence of trace impurities (6), although at one time it was taken as evidence of failure of the Gibbs equation.

IV. MICELLE FORMATION

Micelle formation is another example of the ability of the polar surface-active agents to orient themselves. At very low concentrations, the surface-active agent is in true molecular solution, like any simple compound. Because of their hydrophilic-lipophilic character, surface-active agents possess only minimal solubility in either water or organic liquids. However, when this solubility limit is reached, rather than precipitating (as would most molecular types) the molecules form aggregates, called micelles, in which the portion of the molecule which is compatible with the solvent is facing outside.

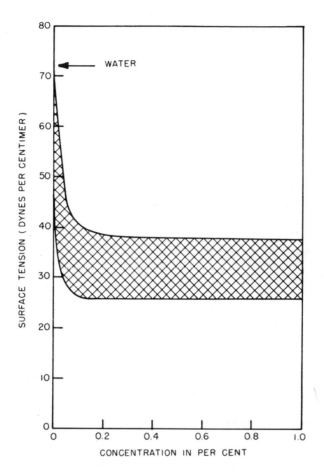

FIG. 3. Range of values of surface tensions found in solutions of most surface-active agents (5).

For example, with water as the solvent, the micelle is formed of molecules oriented so that the hydrophilic heads are pointed out into the aqueous phase, while the lipophilic tails are inside the micelle and hence removed from the aqueous environment. When a surface-active molecule is dissolved in an organic solvent, the micelles are oriented in the opposite sense, with the lipophilic tails out, and the hydrophilic heads in.

In Fig. 5, a number of possible shapes for micelles are given. The number of molecules in a micelle (as well as its shape) depends on the

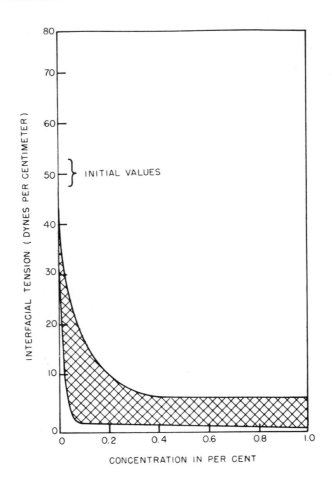

FIG. 4. Range of values of interfacial tension found in solutions of most surface-active agents with a typical oil phase (5).

particular surface-active agent, the solvent, the temperature, and the concentration of other ions. The number of molecules in a micelle, or aggregation number, can be determined by a number of methods, e.g., light scattering, but the determination of the shape of the micelle is still somewhat uncertain (11). The concentration at which micelle formation begins is called the critical micelle concentration, or c.m.c.

The way in which various properties of solutions of surface-active agents change in the neighborhood of the critical micelle concentration is shown schematically in Fig. 6.

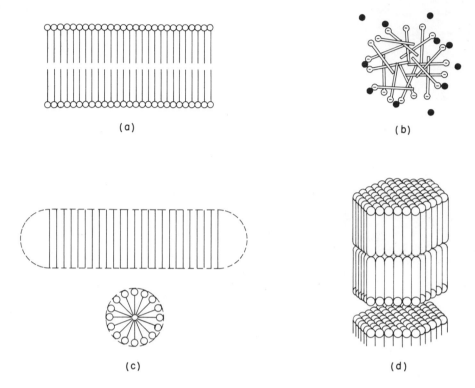

FIG. 5. Models of micellar structure. (a) McBain lamellar micelle (7).
(b) Hartley spherical micelle (8). (c) Debye "sausage" micelle (9). (d)
Philippoff version of McBain micelle (10).

Data for critical micelle concentrations of some selected surface-active agents of various structures are given in Table 3. Note that the c.m.c.
for non-ionics is significantly lower than for the other classes. Becher
(13) has recently reported an extensive tabulation of the micellar data on
non-ionic surface-active agents†.

V. HYDROPHILE-LIPOPHILE BALANCE (HLB)

As has been pointed out above, the polar, or hydrophilic-lipophilic nature of the molecules of surface-active agents is responsible for many of
the characteristic properties that these materials possess in solution.

†Cf. Note added in proof, p. 92.

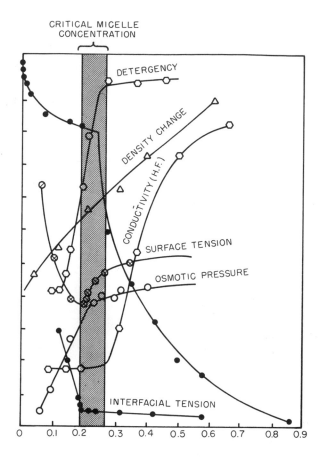

CRITICAL MICELLE
CONCENTRATION

FIG. 6. Changes in the properties of solutions of surface-active agents at the critical micelle concentration, from the data of Preston (12).

This is particularly true in connection with their use as emulsifiers. It would thus be useful to have some quantitative measure of this property.

Such a measure was proposed twenty years ago by Griffin (14), and called by him the HLB Method, where HLB stands for hydrophile-lipophile balance.

In this method, an HLB number is assigned to each surface-active agent, and is related by the scale to the suitable applications. Table 4

TABLE 3

Critical Micelle Concentration of Selected
Surface-Active Agents

Compound	C.m.c. (moles/liter)
Anionic:	
Sodium laurate	24.
Sodium myristate	6.
Sodium palmitate	1.5
Sodium lauryl sulfate	6.5
Sodium myristyl sulfate	**1.6**
Sodium cetyl sulfate	0.4
Cationic:	
Lauryl trimethylammonium chloride	16.
Lauryl trimethylammonium bromide	16.
Cetyl trimethylammonium chloride	1.3
Non-Ionic:	
POE (23) Lauryl alcohol	0.091
POE (9.5) Nonyl phenol	0.085
POE (30) Nonyl phenol	0.280

summarizes the ranges of application. The scale is devised so that the more hydrophilic (i.e., more water-soluble) emulsifiers have the higher HLB numbers.

The original method for determining HLB numbers involved a rather long experimental procedure (14), which, in fact, still has to be used in certain cases, e.g., for ionic materials. However, Griffin (15) has also provided us with a series of equations from which the HLB can be calculated from either analytical or composition data for non-ionic emulsifiers.

For most polyhydric alcohol fatty acid esters, approximate values can be obtained from the formula:

$$HLB = 20 \left(1 - \frac{S}{A}\right) \tag{2}$$

TABLE 4

HLB Ranges and Their Application (22)

Range	Application
3- 6	W/O emulsifier
7- 9	Wetting agent
8-18	O/W emulsifier
13-15	Detergent
15-18	Solubilizer

where S is saponification number of the ester and A is the acid number of the acid. Thus, for a glyceryl monostearate with S = 161 and A = 198, Eq. (2) gives an HLB value of 3.8.

For many fatty acid esters it is difficult to get good saponfication numbers. For these the following equation may be used:

$$HLB = \frac{E + P}{5} \qquad (3)$$

where E is the weight percentage of oxyethylene content and P is weight percentage of the polyhydric alcohol content.

When only ethylene oxide is used to form the hydrophilic portion, and for fatty alcohol ethylene oxide condensation products, Eq. (3) becomes:

$$HLB = E/5 \qquad (5)$$

where E has the same meaning as above.

A rough approximation to the HLB value can be obtained by observing the water solubility of the compound. Table 5 summarizes this approximate method. The assigned HLB numbers for a large number of commercial non-ionic emulsifiers are summarized in Table 6 in order of increasing HLB.

The situation regarding the theoretical interpretation of the HLB-scale has been reviewed by Becher (16).

TABLE 5 (17)

Approximation of HLB by Water Solubility

Behavior when added to water	HLB Range
No dispersibility in water	1–4
Poor dispersion	3–6
Milky dispersion after vigorous agitation	6–8
Stable milky dispersion (upper end almost translucent)	8–10
From translucent to clear	10–13
Clear solution	13+

Since, as is well known, mixtures of emulsifiers are frequently superior to single agents in stabilizing emulsions, it is helpful to be able to calculate the HLB of mixtures. Although Ohba (19) and Becher and Birkmeier (20) have shown that it is not precisely the case, the usual approximation is made that the HLBs are arithmetically additive. That is, in fact, not a very bad approximation, since the deviation from linearity is usually less than the uncertainty in the assigned HLB-numbers, i.e., ±0.5–1.0 HLB units.

Thus, if we wish to blend SPAN 60 (HLB = 4.7) with TWEEN 60 (HLB = 14.9) to achieve a resultant HLB of 10.0, what is involved in the solution of the simultaneous equation:

$$N_1 + N_2 = 100 \tag{5a}$$

$$(4.7N_1 + 14.9\ N_2) / 100 = 10.0 \tag{5b}$$

where N_1 and N_2 are the weight percents of SPAN 60 and TWEEN 60, respectively, in the mixture. It is readily seen that the percentages are 48.04 and 51.96, respectively.

Becher (21) has written a computer program, in CALL/BASIC language, which permits the rapid computation of the weight fractions. This is useful when a pair of emulsifiers is to be used to cover a range of HLB numbers in order, for example, to determine a required HLB number. A typical printout of this program is shown in Fig. 7.

TABLE 6

Calculated and Determined HLB Values of Surfactants (22)

Name	Manu-facturer[a]	Chemical designation	Type[b]	HLB[c]
SPAN 85	1	Sorbitan trioleate	N	1.8
ARLACEL 85	1	Sorbitan trioleate	N	1.8
SPAN 65	1	Sorbitan tristearate	N	2.1
ARLACEL 65	1	Sorbitan tristearate	N	2.1
Atlas G-1050	1	Polyoxyethylene sorbitol hexa-stearate	N	2.6
ATMUL 200	1	Lactylated mono- and digly-cerides of fat-forming fatty acids	N	2.6
Emcol EO-50	2	Ethylene glycol fatty acid ester	N	2.7
Emcol ES-50	2	Ethylene glycol fatty acid ester	N	2.7
ATMOS 300	1	Mono- and diglycerides of fat-forming fatty acids	N	2.8
ATMUL 84.	1	Mono- and diglycerides from the glycerolysis of edible fats	N	2.8
ATMOS 150	1	Mono- and diglycerides from the glycerolysis of edible fats	N	3.2
Emcol PO-50	2	Propylene glycol fatty acid ester	N	3.4
"Pure"	6	Propylene glycol monostearate	N	3.4
Emcol PS-50	2	Propylene glycol fatty acid ester	N	3.4
ATMUL 500	1	Mono- and diglycerides from the glycerolysis of edible fats	N	3.5
Emcol EL-50	2	Ethylene glycol fatty acid ester	N	3.6
Emcol PP-50	2	Propylene glycol fatty acid ester	N	3.7
ARLACEL C	1	Sorbitan sesquioleate	N	3.7
ARLACEL 83	1	Sorbitan sesquioleate		
Atlas G-2859	1	Polyoxyethylene sorbitol-4, 5-oleate	N	3.7
ATMUL 67	1	Glycerol monostearate	N	3.8
ATMUL 84	1	Glycerol monostearate	N	3.8
Tegin 515	5	Glycerol monostearate	N	3.8
Aldo 33	4	Glycerol monostearate	N	3.8
"Pure"	6	Glycerol monostearate	N	3.8
Emcol PM-50	2	Propylene glycol fatty acid ester	N	4.1
SPAN 80	1	Sorbitan monooleate	N	4.3
ARLACEL 80	1	Sorbitan monooleate	N	4.3
Atpet 200	1	Sorbitan partial fatty esters	N	4.3

TABLE 6 (continued)

Name	Manu-facturer[a]	Chemical designation	Type[b]	HLB[c]
Atlas G–3570	1	High-molecular-weight fatty amine blend	C	4.5
Emcol PL–50	2	Propylene glycol fatty acid ester	N	4.5
SPAN 60	1	Sorbitan monostearate	N	4.7
ARLACEL 60	1	Sorbitan monostearate	N	4.7
Emcol DS–50	2	Diethylene glycol fatty acid ester	N	4.7
BRIJ 72	1	Polyoxyethylene (2 mole) stearyl ether	N	4.9
BRIJ 92	1	Polyoxyethylene (2 mole) oleyl ether	N	4.9
Atlas G–1702	1	Polyoxyethylene sorbitol beeswax derivative	N	5
Emcol DP–50	2	Diethylene glycol fatty acid ester	N	5.1
TWEEN–MOS 100	1	Mono- and diglycerides from the glycerolysis of edible fats and TWEEN 80	N	5.2
BRIJ 52	1	Polyoxyethylene (2 mold) cetyl ether	N	5.3
Emcol DM–50	2	Diethylene glycol fatty acid ester	N	5.6
TWEEN–MOS 280 VS	1	Mono- and diglycerides from the glycerolysis of edible fats and TWEEN 65	N	5.9
Emcol DL–50	2	Diethylene glycol fatty acid ester	N	6.1
Glaurin	4	Diethylene glycol monolaurate (soap-free)	N	6.5
SPAN 40	1	Sorbitan monopalmitate	N	6.7
ARLACEL 40	1	Sorbitan monopalmitate	N	6.7
Atcor HC	1	High-molecular-weight amine blend	C	7.5
Atlas G–2684	1	Sorbitan monooleate polyoxyethylene ester mixed fatty and resin acids blend	N	7.8
Atlas G–2800	1	Polyoxypropylene mannitol di-oleate	N	8
Atlas G–1425	1	Polyoxyethylene sorbitol lanolin derivative	N	8
SPAN 20	1	Sorbitan monolaurate	N	8.6
ARLACEL 20	1	Sorbitan monolaurate	N	8.6

TABLE 6 (continued)

Name	Manu-facturer[a]	Chemical designation	Type[b]	HLB[c]
Atlas G–1234	1	Polyoxyethylene sorbitol esters of mixed fatty and resin acids	N	8.6
Emulphor VN–430	3	Polyoxyethylene fatty acid	N	9
Atlox 1087	1	Polyoxyethylene sorbitol oleate	N	9.2
TWEEN 61	1	Polyoxyethylene sorbitan monostearate	N	9.6
Atlas G–3284	1	Polyoxyethylene sorbitol tallow esters	N	9.6
Atlox 1256	1	Polyoxyethylene sorbitol tall oil	N	9.7
BRIJ 30	1	Polyoxyethylene lauryl ether	N	9.7
TWEEN 81	1	Polyoxyethylene sorbitan mono-oleate	N	10.0
Atlas G–1086	1	Polyoxyethylene sorbitol hexa-oleate	N	10.2
TWEEN 65	1	Polyoxyethylene sorbitan tri-stearate	N	10.5
TWEEN 85	1	Polyoxyethylene sorbitan tri-oleate	N	11.0
ARLACEL 165	1	Glycerol monostearate (acid stable, self-emulsifying)	N	11.0
Aldo 28	4	Glycerol monostearate (self-emulsifying)	A	11
Tegin	5	Glycerol monostearate (self-emulsifying)	A	11
Atlas G–1790	1	Polyoxyethylene lanolin deriva-tive	N	11
MYRJ 45	1	Polyoxyethylene monostearate	N	11.1
Atlas G–1096	1	Polyoxyethylene sorbitol hexa-oleate	N	11.4
P.E.G. 400 monooleate	6	Polyoxyethylene monooleate	N	11.4
P.E.G. 400 monooleate	7	Polyoxyethylene monooleate	N	11.4
RENEX 36	1	Polyoxyethylene (6 mole) tri-decyl ether	N	11.4
Atlas G–1045	1	Polyoxyethylene sorbitol laurate	N	11.5
S–541	4	Polyoxyethylene monostearate	N	11.6
P.E.G. 400 monostearate	6	Polyoxyethylene monostearate	N	11.6

TABLE 6 (continued)

Name	Manufacturer[a]	Chemical designation	Type[b]	HLB[c]
P.E.G. 400 monostearate	7	Polyoxyethylene monostearate	N	11.6
Atlas G-3300	1	Alkyl aryl sulfonate triethanolamine oleate	A	12
BRIJ	1	Polyoxyethylene (10 mole) stearyl ether	N	12.4
BRIJ 96	1	Polyoxyethylene (10 mole) oleyl ether		12.4
Atlas G-2090	1	Polyoxyethylene sorbitol oleate-polyoxyethylene amine blend	C	12.5
Atlas G-2127	1	Polyoxyethylene monolaurate	N	12.8
Igepal CA-630	3	Polyoxyethylene alkyl phenol	N	12.8
BRIJ 56	1	Polyoxyethylene (10 mole) cetyl ether	N	12.9
RENEX 690	1	Polyoxyethylene alkyl aryl ether	N	13
S-307	4	Polyoxyethylene monolaurate	N	13.1
P.E.G. 400 monolaurate	6	Polyoxyethylene monolaurate	N	13.1
Emulphor EL-719	3	Polyoxyethylene vegetable oil	N	13.3
TWEEN 21	1	Polyoxyethylene sorbitan monolaurate	N	13.3
RENEX 20	1	Polyoxyethylene esters of mixed fatty and resin acids	N	13.8
Atlas G-1441	1	Polyoxyethylene sorbitol lanolin derivative	N	14.
RENEX 30	1	Polyoxyethylene (12 mole) tridecyl ether	N	14.5
Atlox 8916P	1	Polyoxyethylene sorbitan esters of mixed fatty and resin acids	N	14.6
Atlas G-7586J	1	Polyoxyethylene sorbitan monolaurate	N	14.9
TWEEN 60	1	Polyoxyethylene sorbitan monostearate	N	14.9
TWEEN 80	1	Polyoxyethylene sorbitan monooleate	N	15
MYRJ 49	1	Polyoxyethylene monostearate	N	15.0
BRIJ 78	1	Polyoxyethylene (20 mole) stearyl ether	N	15.3

TABLE 6 (continued)

Name	Manu-facturer[a]	Chemical designation	Type[b]	HLB[c]
BRIJ 98	1	Polyoxyethylene (20 mole) oleyl ether	N	15.3
RENEX 31	1	Polyoxyethylene (15 mole) tri-decyl ether	N	15.4
Emulphor ON-870	3	Polyoxyethylene fatty alcohol	N	15.4
Atlox 8916T	1	Polyoxyethylene sorbitan esters of mixed fatty and resin acids	N	15.4
Atlas G-3780A	1	Polyoxyethylene alkyl amine	C	15.5
Atlas G-2079	1	Polyoxyethylene glycol mono-palmitate	N	15.5
TWEEN 40	1	Polyoxyethylene sorbitan mono-palmitate	N	15.6
BRIJ 58	1	Polyoxyethylene (20 mole) cetyl ether	N	15.7
Atlas G-2162	1	Polyoxyethylene oxypropylene stearate	N	16.0
Atlas G-1471	1	Polyoxyethylene sorbitol lano-lin derivative	N	16
MYRJ 51	1	Polyoxyethylene mono-stearate	N	16.0
Atlas G-7596P	1	Polyoxyethylene sorbitan monolaurate	N	16.3
TWEEN 20	1	Polyoxyethylene sorbitan monolaurate	N	16.7
BRIJ 35	1	Polyoxyethylene lauryl ether	N	16.9
MYRJ 52	1	Polyoxyethylene mono-stearate	N	16.9
Atlas G-1795	1	Polyoxyethylene lanolan de-rivative	N	17.0
MYRJ 53	1	Polyoxyethylene mono-stearate	N	17.9
MYRJ 53	1	Polyoxyethylene monostearate	N	17.9
		sodium oleate	A	18
Atlas G-3634A	1	Quaternary ammonium de-rivative	C	18.5
		potassium oleate	A	20

TABLE 6 (continued)

Name	Manu-facturer[a]	Chemical designation	Type[b]	HLB[c]
Atlas G-263	1	N-cetyl N-ethyl morpholinium ethosulfate	C	25-30
	1	Pure sodium lauryl sulfate	A	approx 40

[a]1 = Atlas Chemical Division, ICI America Inc., 2 = Emulsol Corporation Division of Witco Chem. Co., 3 = General Aniline and Film Corporation, 4 = Glyco Products Company, Inc., 5 = Goldschmidt Chemical Corporation, 6 = Kessler, division Armour Chemicals, 7 = W. C. Hardesty Company, Inc.

[b]A = anionic, C = cationic, N = non-ionic.

[c]HLB values, either calculated or determined, believed to be correct to ±1.

A. Required HLB

As has been indicated, the HLB measures the polarity of the molecule of surface-active agent. Griffin (13, 14) pointed out that this polarity had to match that of the oil being emulsified. In other words, each oil had a particular required HLB, which would be low for water-in-oil emulsion, and high if any oil-in-water emulsion was desired. The required HLB numbers for a selection of oils is given in Table 7.

Required HLB numbers are also additive, so that the required HLB for a mixture of oils can be calculated in exactly the same way as that of a mixture of emulsifiers (and, probably, to the same degree of certainty).

When one has an oil of unknown required HLB, the determination of the required HLB may be somewhat laborious. The usual technique is to make up a series of emulsions of varying HLB which bracket what one may suppose to be the required HLB. The thus-prepared emulsions are stored for some period of time (frequently overnight is enough) and examined for stability, which may be measured, for example, by the amount of creaming or oil separation. If the stability is then plotted against HLB, a bell-shaped

```
                    HLB CALCULATION

WHAT ARE THE HLBS OF YOUR EMULSIFIERS?
LOWEST HLB FIRST, PLEASE---
?  4.7,14.9

WHAT RANGE OF HLB DO YOU WANT COVERED?
?  5.0,12.0

WHAT HLB STEPS DO YOU WANT TO USE?
?  0.5

FOR EMULSIFIERS OF HLB 4.7   AND 14.9  -  THE COMPOSITIONS ARE:

        HLB            PERCENT 4.7           PERCENT 14.9
        5.00             97.06                  2.94
        5.50             92.16                  7.84
        6.00             87.25                 12.75
        6.50             82.35                 17.65
        7.00             77.45                 22.55
        7.50             72.55                 27.45
        8.00             67.65                 32.35
        8.50             62.75                 37.25
        9.00             57.84                 42.16
        9.50             52.94                 47.06
       10.00             48.04                 51.96
       10.50             43.14                 56.86
       11.00             38.24                 61.76
       11.50             33.33                 66.67
       12.00             28.43                 71.57

TIME      0 MINS.     0 SECS.
```

FIG. 7.   Printout of a computer program for the calculation of the composition of emulsifier blends.

curve will be obtained. The maximum in the curve corresponds to the required HLB.

It is not necessarily the case, however, that the emulsion made in this series will be the most stable emulsion possible. If the same series of emulsions is prepared using other pairs of emulsifiers, a family of bell-shaped curves will be found, shown in Fig. 8. Some of the systems will be less stable than the original, some more stable, but the maximum will be essentially at the same HLB. This is the required HLB for the oil under test.

Many attempts have been made to estimate the required HLB by more fundamental approaches. These have involved measurements of such diverse properties as interfacial tensions and dielectric constants (24). An

TABLE 7

Required HLB Values for Various Oils (23)

Oil phase	W/O emulsion	O/W emulsion
Acetophenone	–	14
Acid, dimer	–	14
Acid, lauric	–	16
Acid, linoleic	–	16
Acid, oleic	–	17
Acid, ricinoleic	–	16
Acid, stearic	–	17
Alcohol, cetyl	–	15
Alcohol, decyl	–	14
Alcohol, lauryl	–	14
Alcohol, tridecyl	–	14
Benzene	–	15
Carbon tetrachloride	–	16
Castor oil	–	14
Chlorinated paraffin	–	8
Kerosene	–	14
Lanolin, anhydrous	8	12
Oil:		
Mineral, aromatic	4	12
Mineral, paraffinic	4	10
Mineral spirits	–	14
Petrolatum	4	7–8
Pine oil	–	16
Wax:		
Beeswax	5	9
Candelilla	–	14–15
Carnauba	–	12
Microcrystalline	–	10
Paraffin	4	10

interesting approach has been used by Hayashi (25), who attempted to show a correlation between the Hildebrand solubility parameter (26), and the required HLB.

A more sophisticated treatment of this approach has recently been made by Beerbower and Nixon (27), who have considered the solubility parameters

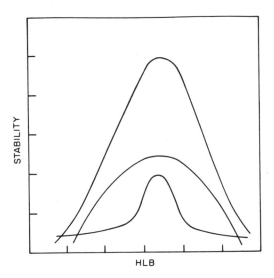

FIG. 8. The stability of emulsions as a function of HLB for several emulsifiers pairs.

of the hydrophilic and lipophilic portions of the molecule separately and have derived an equation which, in principle, permits the calculation of the required HLB from a knowledge of the molecular constitution of the oil phase. It will be interesting to see how this method stands up under the test of use.

B. Surface-Active Agents as Adjuvants

In addition to their use as emulsifying agents, it has been found that surface-active agents show adjuvant effects when used in conjunction with herbicides (28). Recently, Becher and Becher (29) have advanced a theoretical explanation of this phenomenon and demonstrated its dependence on the HLB of the surface-active agent used.

Their analysis was based on the following three considerations regarding the application of sprays: (1) the spray droplet must be delivered to the leaf; (2) once there, it must remain; and (3) not only must it remain, it must penetrate the leaf system. By considering, from this point of view, such things as reflection of the drop from the leaf surface, run-off from the leaf under the effect of gravity, spreading on the leaf surface, and penetration, it was concluded that optimum performance would occur when the quantity

$$\pi = \gamma_L \cos \theta \qquad\qquad (6)$$

was a maximum. In this expression, γ_L is the surface tension of the spray, and θ is the contact angle which the liquid makes with the leaf surface. The quantity π is thus the surface pressure at the liquid/solid interface.

Clearly, the addition of a surface-active agent will have an effect on both the surface tension and contact angle. Becher and Becher (29) measured the effect of variation of HLB on π for one per cent solutions of non-ionic surface-active agents on two model substrates, TEFLON and paraffin wax, and on two plant leaf surfaces, those of corn and soybean plants. Their experimental results are given in Table 8, and are shown graphically in Fig. 9. As can be seen, these curves all show a maximum in the surface pressure, varying slightly with the nature of the surface.

TABLE 8 (29)

Surface Tension, Contact Angle, and Film Pressure for
Aqueous Solutions as a Function of HLB

HLB	8	10	12	14	16	18
γ_L	32.8	33.5	40.3	40.4	40.0	43.7

Substrate:

TEFLON						
	61	60	64	67	72	74
π	15.9	16.8	17.7	15.8	12.4	12.1
Paraffin						
	54	53	63	71	68	67
π	19.3	20.2	18.3	13.2	15.0	17.1
Corn						
	57	58	66	63	57	65
π	17.9	17.8	16.4	18.3	21.8	18.5
Soybean						
	70	59	56	62	69	70
π	11.2	17.3	22.5	18.9	14.3	14.9

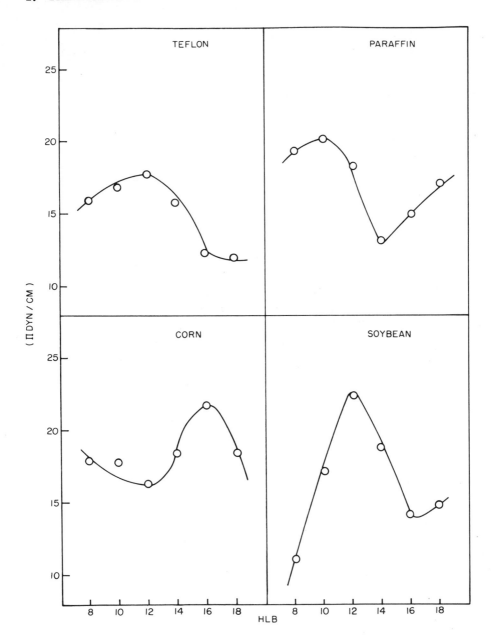

FIG. 9. Variation in $\pi = \gamma_L \cos \theta$ as a function of HLB for various solid substrates (29).

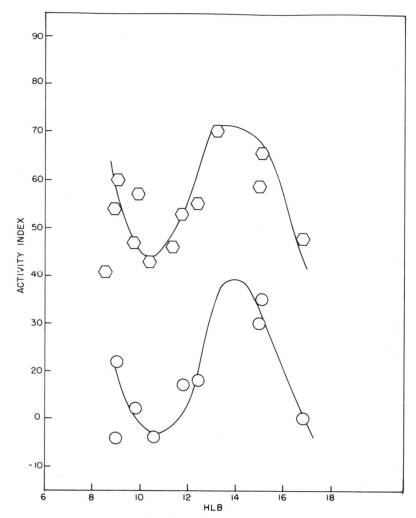

FIG. 10. Variation in kill by herbicide as a function of HLB for corn and soybean plants (29). Compare with corresponding curves of Fig. 9.

These curves should be compared with those of Fig. 10, which have been recalculated from the data of Jensen (30), who presented his results in a slightly different way. These curves, which show the herbicidal activity of 2,4-D on corn and soybean plants, are evidently very similar to the corresponding curves of Fig. 9.

REFERENCES

1. P. Becher, Emulsions: Theory and Practice, 2nd ed., Reinhold, New York, 1965, pp. 209-212.

2. P. Becher, Ref. 1, p. 7.

3. P. Becher, Ref. 1, p. 11.

4. J. W. Gibbs, Trans. Conn. Acad. Sci., 3, 391 (1876); Collected Works, 1, Yale, New Haven, 1928, pp. 229-231 (reprinted Dover, New York, 1961).

5. E. K. Fisher and D. M. Gans, Ann. N.Y. Acad. Sci., 49, 371 (1946).

6. G. D. Miles and L. Shedlovsky, J. Phys. Chem., 48, 57 (1944).

7. J. W. McBain, Colloid Science, Heath, Boston, 1950, pp. 255-259.

8. G. S. Hartley, Aqueous Solutions of Paraffin Chain Salts, Herman, Paris, 1936, p. 45.

9. P. Debye and E. W. Anacker, J. Phys. Colloid Chem., 55, 644 (1951).

10. W. Philippoff, J. Colloid Sci., 5, 169 (1950).

11. P. Becher, Ref. 1, pp. 41-46; P. Becher and H. Arai, J. Colloid Interface Sci., 27, 634 (1968).

12. W. C. Preston, J. Phys. Colloid Chem., 52, 84 (1948).

13. P. Becher, in Nonionic Surfactants (M. J. Schick, ed.), Dekker, New York, 1967, pp. 478-515.

14. W. C. Griffin, J. Soc. Cosmetic Chemists, 1, 311 (1949).

15. W. C. Griffin, J. Soc. Cosmetic Chemists, 5, 249 (1954).

16. P. Becher, Ref. 1, pp. 234-247; Am. Perfumer Cosmetics, 76 (9), 33, (1961); in Nonionic Surfactants (M. J. Schick, ed.), Dekker, New York, 1967, pp. 604-626.

17.   P. Becher, Ref. 1, p. 234.

18.   W. C. Griffin, in Kirk-Othmer Encyclopedia of Chemical Technology, 2nd ed., 8, pp. 128-130.

19.   N. Ohba, Bull. Chem. Soc., Japan, 35, 1016 (1962).

20.   P. Becher and R. L. Birkmeier, J. Am. Oil Chemists' Soc., 41, 169 (1964).

21.   P. Becher, unpublished results.

22.   P. Becher, Ref. 1, p. 233.

23.   P. Becher, Ref. 1, p. 249.

24.   P. Becher, Ref. 1, pp. 234-247.

25.   S. Hayashi, Yukagaku, 16, 554 (1967).

26.   J. H. Hildebrand and R. L. Scott, Regular Solutions, Prentice-Hall, Englewood Cliffs, 1962.

27.   A. Beerbower and J. Nixon, paper presented at Spring, 1969, ACS Meeting, Minneapolis.

28.   L. L. Jensen, W. A. Gentner, and W. A. Shaw, Weeds, 9, 381 (1961).

29.   P. Becher and D. Becher, in Pesticidal Formulations Research: Physical and Colloidal Chemical Aspects (J. W. Van Valkenburg, ed.), Advances in Chemistry Series No. 86, American Chemical Society, Washington, 1969, pp. 15-23.

30.   L. L. Jensen, J. Agr. Food Chem., 12, 223 (1964).

NOTE ADDED IN PROOF

A valuable compilation of critical micelle concentrations by P. Mukerjee and K. J. Mysels, Critical Micelle Concentrations of Aqueous Surfactant Systems, has recently appeared as NSRDS-NBS Bulletin 36, National Bureau of Standards.

Chapter 3

THE STABILITY OF EMULSIONS

Wade Van Valkenburg

The 3M Company
St. Paul, Minnesota

I. INTRODUCTION

An emulsion may be defined as a dispersion of one immiscible liquid
in a second, continuous phase. The system lacks stability if the dispersed phase is no longer homogeneously dispersed in the continuous
phase.

As the discontinuous phase separates in an emulsion it will first
"cream." A cream is that region of the emulsion which contains a higher
proportion of the dispersed phase than the average amount in the system.
It is still an emulsion since the integrity of the dispersed phase is still
intact.

If oil droplets form as the dispersed phase begins to coalesce, the oil
droplets get bigger. As soon as the oil droplets coalesce sufficiently that
they are visible as an intact layer, the emulsion has "oiled out."

To disperse one bulk phase in another phase, the dispersed phase must be subdivided into small droplets. This takes work which may be expressed as:

$$W = \gamma_{12} \, \Delta S \tag{1}$$

where W is work in dyne/cm, γ_{12} is the interfacial tension between the two phases in dynes/cm, and ΔS is the increase in surface area in cm^2.

When the interfacial tension is small (0. 01 dyne/cm or less), the work of subdivision is very small and one phase will spontaneously disperse in the second. This occurs when paraffin oil is emulsified in water containing 0. 001 M sodium oleate and 0. 001 M sodium chloride as the emulsifying agent. Flash or spontaneous agricultural emulsions are frequently obtained when a portion of an anionic-non-ionic blend of emulsifiers is composed of the calcium salt of dodecyl benzene sulfonic acid (1).

Flash emulsions are aesthetically pleasing both to the formulator and to the customer and at present are very popular. Flash emulsification, however, may not insure emulsion stability since there are many factors besides a low interfacial tension which foster a stable dispersion. Robinson (2) lists the factors which influence stability as shown in Table 1.

Although these variables all influence the stability of emulsions, the most important factors, from the formulator's standpoint, are the emulsifiers and the role they play in forming the stabilizing interfacial film. The emulsifiers and the interfacial film will constitute the main subject of this chapter.

II. PRACTICAL STABILITY CONSIDERATIONS

Factors relating to emulsion stability of concern to the consumer include: (1) that the formulation should meet any controlling government stability specifications, (2) that the composition should mix easily with water and remain homogeneous in application equipment, (3) that if the emulsion separates while standing in a spray tank it should easily reconstitute and (4) that the toxicant should deposit well on the target sprayed.

A. W. H. O. Stability Specifications

The World Health Organization has placed performance specifications on many of the common pesticide concentrates. The W. H. O. specification for a DDT emulsifiable concentrate states, "any creaming of the

TABLE 1

Factors Influencing Emulsion Stability

Internal phase

    (a) Volume concentration; interparticle interaction
    (b) Viscosity
    (c) Particle size and size distribution
    (d) Interfacial tension
    (e) Chemical constituents

Continuous phase

    (a) Viscosity
    (b) Polarity and chemical constituents

Emulsifying agent

    (a) Chemical constituents and concentration
    (b) Solubility in continuous phase, pH of aqueous phases
    (c) Physical properties of film around the particles, thickness of film; partial deformations; influence on the attractive forces between particles
    (e) Electroviscous effect, electrolyte concentration in aqueous media

Conditions

    (a) Temperature
    (b) Light
    (c) Pressure

---

emulsion at the top, or separation of sediment at the bottom, of a 100-ml cylinder shall not exceed 2 ml when the concentrate is tested as described in Annex 12 in Specifications for Pesticides" (3). Annex 12 is reproduced here as follows:

Annex 12

EMULSION STABILITY TEST

1. Special reagent

    Standard hard water. Water of the following composition, designed to provide a hardness of 342 parts per million, calculated as calcium carbonate

        Calcium chloride, anhydrous..........0.304 g
        Magnesium chloride, hexahydrate .....0.139 g
        Distilled water, to make ....1 liter

2. Procedure

Two methods are described; the choice will depend upon the manufacturer's instructions for preparing the diluted emulsion for use in the field. Method a applies to concentrates which have to be added in water, and method b to concentrates to which water must be added.

a. Into a 250-ml beaker having an internal diameter of 6-6.5 cm and a 100-ml calibration mark, pour 75-80 ml of standard hard water, brought to a temperature of $30^{\circ}C \pm 1^{\circ}C$. By means of a Mohr-type pipet add 5 ml of the concentrate, while stirring with a glass rod, 4-6 mm in diameter, at about four revolutions per second. The concentrate should be added to the water at the rate of 25-30 ml per minute, with the point of the pipet 2 cm inside the beaker, the flow of the concentrate being directed toward the center, and not against the side, of the beaker. Make up to 100 ml with standard hard water, stirring continuously, and immediately pour into a clean, dry 100-ml graduated cylinder. Keep at $29-31^{\circ}C$ for 1 hour and examine for any creaming or separation.

b. Pour 5 ml of the concentrate into a 250-ml beaker having an internal diameter of 6-6.5 cm and a 100-ml calibration mark. Add sufficient standard hard water, brought to a temperature of $30^{\circ}C \pm 1^{\circ}C$, to make 100 ml. The water should be added at the rate of 15-20 ml per minute, using a dropping funnel and stirring by hand with a glass rod, 4-6 mm in diameter, at about four revolutions per second. Transfer the emulsion immediately to a clean, dry 100-ml graduated cylinder. Keep at $29-31^{\circ}C$ for 1 hour and examine for any creaming or separation.

In the case of formulations used at concentrations substantially lower than 5%, further emulsion stability tests must be carried out at the concentrations used in the field. Depending upon the manufacturer's instructions for dilution, prepare the test emulsion by adding either the appropriate quantity of the concentrate to standard hard water at $30^{\circ}C \pm 1^{\circ}C$, or the standard hard water to the appropriate quantity of the concentrate, to make 100 ml of emulsion. Keep at $29-31^{\circ}C$ for 1/2 hour and examine for any creaming or separation. (Reproduced from Ref. 3.)

B. Market Considerations

If a formulation is designed for the lawn and garden market, an emulsion should be quite a stable one. Most home applicators have minimal agitation and, therefore, to maintain homogeneity and prevention of overdosing, a spontaneously formed stable emulsion is required. A stable emulsion, in this case, would show no visible separation or creaming upon standing 15 minutes or longer.

Emulsions applied from commercial equipment with paddle or jet agitation may be less stable. In these instances it is frequently desirable to have lower concentrations of emulsifiers to reduce the possibility of foam generation. Visible creaming or separations from these emulsions may occur in as little as two minutes in the absence of agitation. With the less stable emulsions particular care should be taken to formulate so that a separated phase reemulsifies with minimal agitation.

A fairly unstable emulsion has definite advantages if one is spraying foliage to "run off;" i.e., spraying to the point of complete wetting with the spray beginning to fall off the foliage to the ground. In this case definitely greater deposits of toxicants from an oil phase are left on foliage for the "quick breaking" emulsions. In cases of fractional coverage of surfaces (low gallonage sprays) the stability of the emulsions is less important to deposit and coverage than wetting characteristics of the spray.

C. Laboratory Evaluations

Laboratory testing of emulsion stability may be measured by determining the rate of cream separation. This may be measured using a cone-shaped centrifuge tube or a calibrated cylindrical tube. An alternate method is to measure changes in optical density in various portions of an emulsion. The optical density of the cream layer increases while the optical density of the balance of the emulsion decreases as the mixture loses its homogeneity. The stability of oil in water emulsions should be checked using several types of water which may include soft water (under 100 ppm Ca and Mg hardness), 342 ppm hardness (WHO specs), 1000 ppm hardness and alkaline water containing sodium carbonate. Performance in water from 40 to 90°F should also be determined prior to giving a final approval to a formulation. In addition, since chemical reactions may occur between components of a composition, repeated stability tests should be undertaken after the formulation has been stored cold (0°F or above), hot (125°F) for at least 90 days. It is wise also to check stability of formulations after they have been stored in the container of choice for extended periods of time.

III. THE OIL-WATER INTERFACE

The emulsifiable concentrate may contain the pesticide, an organic solvent or diluent, and an emulsifier. In a gallon of formulation there may be 100 to 200 grams of surface-active agent. If the gallon of concentrate is all emulsified into 10μ oil droplets, there will be 7.2×10^{12} drops, and 2×10^7 cm^2 of new surface. To stabilize this amount of emulsion, a total of 6.7×10^{21} molecules of surfactant, each 30Å cross sectional area must form a monomolecular film at the interface. For a surfactant of molecular weight 600, this is equivalent to six grams of material and three to six percent of the original emulsifier (4).

A dynamic equilibrium exists in this system as follows:

Micelles $\rightleftarrows$ Monomer $\rightleftarrows$ Monomer $\rightleftarrows$ Monomer $\rightleftarrows$ Micelles
(O)   (O) Film (IF)   (W)   (W)

(O) refers to micelles and surfactant monomers in the oil phase, (IF) the interface; and (W) refers to the water phase. The stability of an emulsion is dependent upon a suitable quantity of surfactant congregating at the interface. The concentration of emulsifiers at the interface in excess of the concentration of emulsifier in the bulk phase is called the "surface excess" (Γ) and this property is dependent upon concentration, temperature, and the change in surface tension (γ) with concentration:

$$\Gamma = -\frac{C}{RT}\frac{\delta\gamma}{\delta C} \qquad\qquad (2)$$

Factors affecting the micelle, monomer, and film equilibria as well as the structure of the stabilizing film have a profound effect on the stability of emulsions. These factors are discussed in greater detail in the following sections.

Micelles

Surface active agents are in monomeric form in organic and aqueous solvents when their concentration is very low (ca 10^{-5} to 10^{-2} moles/liter). The normal use concentration of emulsifiers (5-10% by weight) in organic concentrates is above a critical micelle concentration. The degree of aggregation has been found for alkali dinonylnapthalenesulfonates to be related to the solubility parameter of the solvent (6). The surfactant chain tends to extend itself to allow greater interaction between solute and solvent as the solubility parameter of the repeating unit in the surfactant approximates the solubility parameter of the solvent. In other words, "like dissolves like." The greater the discrepancy in solubility parameter of surfactant and solvent the more the surfactant wants to take itself out of its hostile environment. This it can do by forming micelles and interacting with itself. Taking in to account toxicology, toxicant solubility, and solubility parameter effects of a toxicant, the formulator does not always

have complete freedom in selection of the ingredients of a composition.
Perhaps the emulsifier system of his choice does not yield a suitable
stable emulsion. Moderate alterations in emulsifier performance can be
accomplished by minor additions (2-10%) of a polar or nonpolar solvent.
In the event the emulsifier is too soluble in the system a nonpolar solvent
will tend to force it toward agglomeration and increase the surface excess
at the interface. In the event micellization is too great in the organic system, small amount of a polar organic solvent will tend to decrease this aggregation to optimize solubility relations.

Interactions of Polar Molecules

An explanation for the aggregation and variation in the degree of aggregation of polar surfactants in organic solvents has been proposed by Fowkes
(7). The behavior and interaction of polar molecules in organic solutions
may be predicted by Pearson's hard and soft acid and base theory (8).
Lewis acids which are electron acceptors are "hard" when they have low
polarizability. "Soft" acids have high polarizability and prefer to react
with "soft bases" which are electron donors.

Fowkes proceeds to point out how the Pearson theory predicts micellization in organic solvents. Sulfonates are weak "hard" bases which do not
react strongly with cations. Thus, with the addition to the organic system
of small amounts of water (a hard base) water tends to react with the hard
cations of sodium, potassium, calcium, etc. Thus, the interaction between
the sulfonates and the hydrated cations is greater than interaction with un-
hydrated cations resulting in greater aggregation or larger micelles.

The carboxylate group in a surface-active molecule is a strong hard
base. It interacts strongly with hard acid cations resulting in a tendency
to form large micelles. The addition of water decreases this anion-ca-
tion interaction in the carboxylate case causing a decrease in the size of
the micelle.

Hydrogen bonding solvents such as chloroform, methanol, etc. act as
hard Lewis acids and can compete effectively with cations for the hard base
carboxylate group. This results in a sharp reduction in micelles and may
actually totally eliminate the formation of micelles.

Thus, if the interaction between the anionic portion of a surface-active
molecule and the cation is decreased, the size of micelles is decreased
and there is a tendency to increase the monomer concentrations in organic
solutions. Increased cation-anion interactions increase a tendency to aggregate and form micelles. By judiciously altering this solubility relationship of the surface-active molecule one can optimize conditions for an

appropriate "surface excess" at the oil-water interface and obtain maximum emulsion stability for a chosen system.

Temperature and agitation affect the degree of agglomeration in organic solvents (9). In o-xylene, sorbitan monostearate micelles increase in molecular weight as temperature rises. In addition the critical micelle concentration rises with increasing temperature. The enthalpy of micellization by dye solubilization at 25°C was found to be -1760 calories. With this very small heat of micellization it is not surprising that above the CMC agitation caused a temporary decrease in surface tension of the o-xylene solution. When agitation ceased, the temporary excess concentration of surfactant monomer ceased, the surfactant again agglomerated, and surface tension rose to a steady state value.

Solvation effects in nonaqueous solvents are much less than that which occurs in highly polar water. In a water system Kresheck (10) found by a temperature jump technique an activation energy of 1750 cal/mole for micellization of dodecylpyridinium iodide. These authors hypothesize a mechanism of deaggregation in which water is ordered around a monomer as it leaves the micelle. In organic systems the activation energy for micellization is much smaller. Hence in both organic and aqueous systems very little energy is involved in the micellization process.

In summary the amount of surfactant available for the formation of a stabilizing interfacial film is dependent upon the physical state of the surfactant in solution. Solubility parameters, temperature, and agitation affect the concentration of surfactant available as a "surface excess." Micellization in the organic phase need not be eliminated as in this case the surfactant may become too soluble to suitably migrate to the interface. There are optimum solubility relationships which have been found by experience but which have not been fully documented quantitatively.

IV. CLOUD POINTS AND PHASE INVERSION TEMPERATURES

Every emulsifier is composed of a combination of hydrophilic and lypophilic segments. The relative amounts of hydrophilic and lypophilic segments are characterized by the HLB as described in the previous chapter. Maximum stability of an emulsion may be obtained by choosing an emulsifier system of optimum HLB (11).

The HLB system is excellent as an indication of the solubility characteristics of the emulsifier and is very useful for preliminary screening of potential emulsifiers. However, the HLB is based on the pure surfactant and does not take into account molecular interactions which take place in the chosen system. An understanding of these interactions may be accomplished through a discussion of cloud points and phase inversion temperatures.

A. Non-Ionic Agents

Emulsifiers are usually blends of non-ionic and anionic surfactnats. The performance of the blend is influenced by the variable solubility characteristics of the non-ionic portion of the system. Non-ionics generally have "reverse" solubility relationships. In water, they are more soluble at lower temperatures than at high. They are more soluble at higher concentrations than at low ones due to formation of micelles.

A typical solubility diagram for pure $R_{12}O(CH_2CH_2O)_6H$ in water is shown in Fig. 1 as determined by Balmbra et al. (12).

Above 20% weight percent the tendency is for water to dissolve in the surfactant. Below that the surfactant dissolves in water. The cloud point is achieved when the solution passes as a function of temperature from a single isotropic liquid into two separate isotropic liquids.

The cloud point of a non-ionic surfactant is affected by the presence of organic liquid as shown in Table 2 (13).

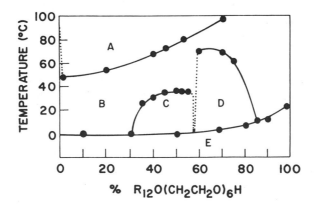

FIG. 1. Phase diagram of pure $R_{12}O(CH_2CH_2O)_6H$ vs water. Estimated boundaries, dashed lines. A, two isotropic liquids; B, one isotropic liquid; C, middle phase; D, neat phase, E, solid plus ice.

TABLE 2

Cloud Points in Aqueous Solution of Commercial Polyoxyethylene (9.2)
Nonyl Phenyl Ether Saturated with Various Oils

Oil	Cloud point, $^{\circ}$C
None	56
Liquid paraffin	80.4
Hexadecane	80
Decane	79
Heptane	71.5
Cyclohexane	54
Perchloroethylene	31
Ethylbenzene	30.5
Benzene	Below 0

Note that aliphatic oils, being poor solvents, increase the solubility of
the surfactant in water by stabilizing the micellar form. In the absence of
an oil at 60°C two isotropic liquids exist. However, when the nonyl phenol
is saturated with decane, only one isotropic liquid exists at 60°C. In the
presence of aromatic solvents, the cloud point is decreased, indicating a
lowering of solubility and a decrease in size and stability of the micelle.

Shinoda (14) has reported on "The Correlation Between Phase Inversion Temperature in Emulsion and Cloud Point in Solution of Non-Ionic
Emulsifier." Recall that above the cloud point of a surfactant solution
two phases exist. This may be regarded as containing water as a discontinuous phase and detergent as a continuous phase. Thus, there is
an analogy between a W/O emulsion and a W/D or water in detergent
emulsion.

Confirming the results indicated earlier Shinoda measured the cloud
points of a nonyl phenol containing 9.6 moles of ethylene oxide with and
without the addition of organic solvents. The results may be found in
Fig. 2.

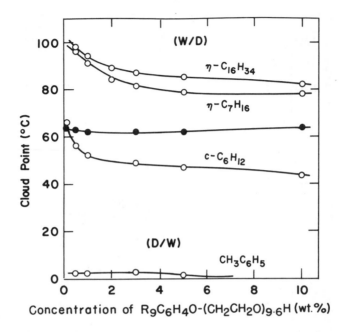

FIG. 2. The cloud points of polyoxyethylene (9.6) nonyl phenyl ether solution saturated with different hydrocarbons. Filled circles express the cloud points in the absence of oil.

Results were more easily obtained in the presence of the organic solvent. The presence of aliphatic solvents raises the cloud point and aromatic solvents lower the cloud point. That is, the better the solvent for the surfactant the lower the cloud point. Above each line a two phase system indicates a W/D emulsion. Below the line a D/W emulsion exists where the solvent is incorporated in the micelle. In this system the surfactant solution is saturated with the organic solvent.

At a 50-50 ratio of organic solvent to water Shinoda measured the phase inversion temperature (O/W $\longrightarrow$ W/O) as a function of weight percent in water of the nonyl phenol 9.6 mole ethylene oxide adduct. The results are shown in Fig. 3.

Note the similarity of effects of solvents on the phase inversion temperature when compared to cloud point determinations. Aliphatic hydrocarbons raise the phase inversion temperature while aromatic solvents lower it.

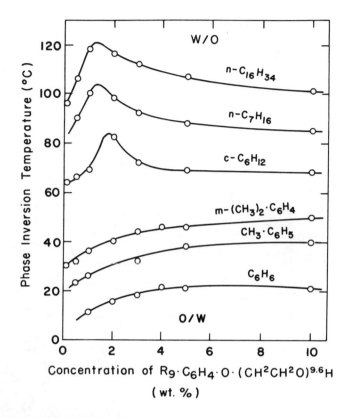

FIG. 3. The effect of different hydrocarbons on the phase inversion temperature of emulsions (volume ratio = 1) vs the concentration of polyoxyethylene (9.6) nonyl phenyl ether (wt. % in water).

The correlation shown by Shinoda between cloud points and phase inversion temperatures is shown in Fig. 4.

Correlations are far better in the presence of organic solvents than when surfactants alone are dispersed in water. Since these deviations with the surfactants alone occur mostly with the very water soluble surfactants, this may indicate varying degrees of hydration of the surface-active agents.

PIT Adjustment

Optimum stability of an emulsion can be obtained when the use temperature of the emulsion is 20 to 65 degrees centigrade below the phase

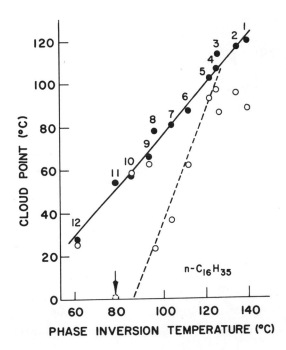

FIG. 4. The correlation between the phase inversion temperatures in emulsion and the cloud points of emulsifiers in the absence of, O, or saturated with hexadecane, ●; 1, Tween 40; 2, Tween 20; 3, Tween 60; 4, $R_{12}O(CH_2CH_2O)_{10}H$; 5, i-$R_9C_6H_4O(CH_2CH_2O)_{14}H$; 6, i-$R_9C_6H_4O$ $(CH_2CH_2O)_{9.6}H$; 7, i-$R_{12}C_6H_4O(CH_2CH_2O)_{9.0}H$; 8, i-$R_9C_6H_4O$ $(CH_2CH_2O)_{7.4}H$; 9, Pluronic L-44; 10, Pluronic L-64, 11, i-$R_9C_6H_4O$ $(CH_2CH_2O)_{6.2}H$; 12, Pluronic L-62.

inversion temperature (PIT) (15). Hence ways and means of adjusting the PIT are important to the formulator.

By increasing the hydrophilic chain length of a non-ionic surfactant the PIT is raised (14). This is exemplified in Fig. 5.

Intermediate PIT's, of course, may be obtained through the blending of surfactants.

As indicated previously, the PIT will be raised by poor aliphatic solvents and lowered by aromatic solvents which are good solvents for the non-ionic surfactants. Arai and Shinoda (15) demonstrate this in their work shown in Fig. 6.

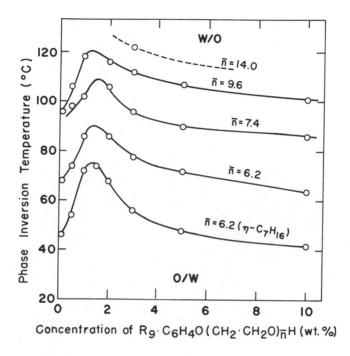

FIG. 5. The effect of the hydrophilic chain length of the emulsifier on the phase inversion temperature in hexadecane–water emulsions (volume ratio = 1).

B. Anionic Agents

Phase reversal (O/W $\longrightarrow$ W/O) of emulsions stabilized by anionic surface–active agents may be accomplished through temperature changes, salt addition, phase volume changes and the blending of anionic agents.

Anionic surfactants behave quite differently than the non–ionic agents discussed in the previous section. Whereas non–ionic materials have "reverse solubility"; (i.e., decreased solubility with an increase in temperature), anionics, like other polyelectrolytes, increase in solubility with an increase in temperature. At low temperatures the surfactants are in a hydrated monomeric form. As temperature increases solubility gradually increases until a characteristic temperature known as the Krafft point (17) is reached. At and above this temperature solubility increases rapidly due to the formation of micelles (18).

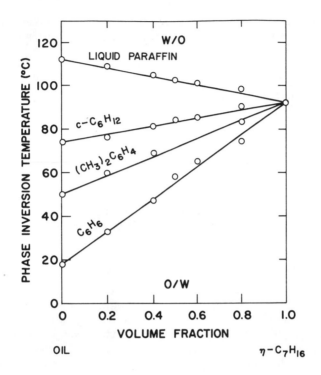

FIG. 6. The effect of the mixture of n-heptane with various oils on the PITs of emulsions stabilized with polyoxyethylene (9.6) nonylphenylethers (3 wt. % for water).

Thus, it is not surprising that phase reversal might be obtained with an anionic stabilizing agent by cooling instead of heating. Wellman (19) found that sodium stearate-stabilized O/W emulsions (benzene/water) underwent phase reversal on cooling. Phase reversal was also accomplished by the addition of oleic acid to the sodium oleate which causes the interfacial film to become more hydrophobic.

Bancroft's rule (20) concerning solubility relationships of emulsifiers is quite valid for anionic surfactants. He states that surfactants soluble in water but not in oil produce O/W emulsions and oil soluble surfactants produce W/O emulsions.

Water soluble sodium oleate placed in an olive oil-water emulsion stabilizes an O/W type emulsion. Calcium oleate, being more oil soluble,

stabilizes an emulsion of the W/O type. Clowes (21) inverted an olive oil-
water emulsion stabilized by sodium oleate through the addition of calcium
chloride.

Kremenev (22) studied the phase reversal of benzene emulsions stabilized by sodium oleate. Inversion was accomplished through the addition of sodium chloride. When the concentration of NaCl increased over
0.08 moles/liter the O/W emulsion became unstable. At 0.42 moles/liter
a stable W/O emulsion was obtained. Then Kremenev decreased the salt
concentration through supplementary addition of soap solution until at a
NaCl concentration of 0.14 moles/liter a stable O/W emulsion was once
again obtained.

Kremenev points out that the role of the added sodium chloride is dehydration of the sodium oleate at the oil/water interface.

C. Discussion

Fischer (23) states that a stable O/W emulsion is fostered by a "colloid
hydrate" at the oil-water interface. Taking this in to account Hartman (24)
states that O/W emulsions of intermediate stability will break when diluted
with water because the stabilizing soap is converted from a hydrated form
to a true solution. Heating the solution brings the same result. The addition of calcium or barium to an alkali metal soap is said to decrease the degree of hydration at the interface. By the same token, oleic acid added to
a sodium oleate soap decreases the hydration of the surfactant film at the
interface.

Nikitina (25) investigated the stability of concentrated xylene emulsions
stabilized by sodium butylnapthalenesulfonate. Measuring the zeta potential of the dispersed droplets she found that the zeta potential remained relatively constant while stability varied greatly with changes in concentration
of the emulsifier. The strength and thickness of the interfacial film as
measured by "surface yield values" correlated well with emulsion stability.
Stability is fostered by formation and thickening of the interfacial film.
Conditions which decreased the degree of hydration of the interfacial film
decreased the stability of the O/W emulsion. Nikitina thus concluded that
zeta potential is not the overriding variable affecting stability but that
thick stable interfacial films promote emulsion stability.

The stability of O/W emulsions stabilized by non-ionic agents has been
related to cloud points and phase inversion temperatures. The emulsion is
least stable right at the phase inversion temperature but has optimum stability when evaluated at temperatures 20 to 65 degrees below the PIT.

The cloud point, and hence PIT, is related to the dehydration energy (26) of the non-ionic surfactant and of the interfacial film. It thus appears that when the temperature of use of an emulsion is more than 20 degrees below the PIT the interfacial film is sufficiently hydrated to enhance stability.

Oil in water emulsions stabilized by anionic agents are also rendered more stable when the interfacial film is suitably hydrated. Dehydration of this interfacial film by cooling or salt addition makes the emulsion less stable.

Although emphasis in this chapter has been on O/W emulsions, similar considerations hold for the stability of invert or W/O emulsions. Saito and Shinoda (27) have shown that invert emulsions stabilized by non-ionics are most stable when they are evaluated at temperatures 20 to 40 degrees above the phase inversion temperature. In this instance, as in systems where anionic agents are used, a thick stable interfacial film occurs on the organic side of the interface, due to solvent interaction with the lypophilic portion of the emulsifier. Little structure exists on the aqueous side of the interface, due to a lack of hydration of the polar portions of the molecule.

V. CONCLUSIONS

The pesticide formulator has a keen interest in emulsions stabilized by blends of anionic and non-ionic surface-active agents. The anionic agents of interest include calcium, amine, alkanol amine and alkali salts of alkyl aryl sulfonates. The calcium salt of dodecyl benzene sulfonic acid has increasing use because it fosters good spontaneity and stability of O/W emulsions. This salt in combination with a very hydrophilic non-ionic agent undoubtedly yields an interfacial film of an optimum degree of hydration.

In any developing science, scientists must first work with, know, and understand simplified pure systems. Hence much work has been done on pure organic solvents, and surfactants of a single type. To this author's knowledge no reports have appeared in the literature taking knowledge learned from the simplified systems and extending and applying it to the complicated multicomponent systems of interest to the pesticide formulator. This, then, becomes the burden of the physical chemist in the agricultural formulation laboratory. Some suggestions and recommended procedures follow.

It appears that the stability of an O/W emulsion is very closely related to the degree of hydration of the interfacial film. It thus becomes important to us to establish tests that measure this degree of hydration, so that we may be able to compare one emulsifier system with another.

Determination of the phase inversion temperature may work for the majority of systems. Measurement may be by a visual technique or by the differential thermal analysis method recently reported (28). Pesticide formulations should perform satisfactorily up to 40°C. This means a minimum PIT of 60°C.

In the event a toxicant is unstable at 60°C and above adversely affecting the PIT measurement it may be desirable to measure the temperature at which the emulsion has a reproducible instability. By stirring an emulsion mildly one could determine the temperature at which a certain percentage of the oil phase separates out as an oily layer.

One may also take advantage of the effect of salts on dehydrating the interfacial film. The apparent change of HLB as affected by salt concentrations in the aqueous phase has recently been reported by Shinoda (29).

Dashevskaya (30) has indicated that dehydration of polyethyleneoxide non-ionics by salts follows the Hofmeister lyotropic series: $SO_4 => PO_4 \equiv > Cl^-; Na^+ > K^+ > Mg^{++}$. Thus, one might set up a series of waters containing 0.1 to 1.0 M Na_2SO_4 and compare the stabilities of the O/W emulsions. That emulsion least affected by Na_2SO_4 would have the greatest hydration layer at the interface and would be the most stable.

REFERENCES

1. H. L. Sanders, E. A. Knaggs, and M. L. Nussbaum, Reissue U.S. Pat. 24, 184 dated July 24, 1956.

2. J. R. Robinson, Am. Perfumer and Cosmetics, 83(8), 27 (1968).

3. Specifications for Pesticides, World Health Organization, Geneva, 1961.

4. W. Van Valkenburg, Solvent Properties of Surfactant Solutions (Kōzō Shinoda, ed.), Dekker, New York, 1967, p. 271.

5. J. W. McBain, Colloid Science, Heath, Boston, 1950.

6. R. C. Little and C. R. Singleterry, J. Phys. Chem., 68, 3453 (1964).

7. F. M. Fowkes, Solvent Properties of Surfactant Solutions (Kōzō Shinoda, ed.), Dekker, New York, 1967, p. 65.

8. R. G. Pearson, J. Am. Chem. Soc., 85, 3533 (1963).

9. C. W. Brown, D. Cooper, and J. C. S. Moore, J. Colloid Interface Sci., 32, 584 (1970).

10. G. C. Kresheck, E. Hamori, G. Davenport and H. A. Scheraga, J. Am. Chem. Soc., 88, 246 (1966).

11. R. W. Behrens, Weeds, 12(4), 255 (1964).

12. R. R. Balmbra, J. S. Clunie, J. M. Corkill, and J. F. Goodman, Trans. Faraday Soc., 58, 1661 (1962).

13. K. Shinoda, Solvent Properties of Surfactant Solutions, Dekker, New York, 1967, p. 29.

14. K. Shinoda and H. Arai, J. Phys. Chem., 68, 3485 (1964).

15. K. Shinoda and H. Arai, J. Colloid and Interface Sci., 30, 258 (1969).

16. H. Arai and K. Shinoda, J. Colloid and Interface Sci., 25, 396 (1967).

17. F. Krafft and H. Wiglow, Ber., 28, 2566 (1895).

18. A. W. Ralston and C. W. Hoerr, J. Amer. Chem. Soc., 68, 851 (1946).

19. V. E. Wellman and H. V. Tartar, J. Phys. Chem., 34, 379 (1930).

20. W. D. Bancroft, Applied Colloid Chemistry, McGraw Hill, New York, 1926.

21. G. H. A. Clowes, J. Phys. Chem., 20, 407 (1916).

22. L. Y. Kremenev and N. I. Kuibina, Colloid J. (U.S.S.R.), 17, 31 (1955), U. S. Army Translation AD 682, 586.

23. M. H. Fischer, Oil and Soap, 13, 31 (1936).

24. R. J. Hartman, Colloid Chemistry, Houghton Mifflin, Boston, 1947.

25. S. A. Nikitina, O. S. Mochalova, and A. B. Taubman, Colloid J. (U.S.S.R.), 30, 100 (1968).

26. W. C. Griffin, Official Digest (Fed. Paint Varnish Production Clubs), June 1956.

27. H. Saito and K. Shinoda, J. Colloid and Interface Sci., 32, 647 (1970).

28. S. Matsumoto and P. Sherman, J. Colloid and Interface Sci., 33, 294 (1970).

29. K. Shinoda and H. Takeda, J. Colloid and Interface Sci., 32, 642 (1970).

30. B. I. Dashevskaya, M. K. Gluzman, and G. M. Fridman, Colloid J., 30, 359 (1968).

Chapter 4

AGRICULTURAL FORMULATIONS WITH LIQUID FERTILIZERS

Paul Lindner

Witco Chemical Company
Chicago, Illinois

I. LIQUID FERTILIZERS—COMPOSITION AND USE

The intensive farming of land removes the nutrients from the soil which then have to be replenished by means of fertilizers. Historically, fertilizers that have been developed to replenish nutrients include superphosphates and ammonia sulfates from coke ovens, calcium cyanamide, and other synthetic fertilizers. With growing sophistication, the trend is toward more concentrated fertilizers and easier methods of application.

One line of development was directed toward liquid fertilizers. The convenience and uniformity of application, avoidance of dust and feasibility

113

of late application were the prime reasons for increased use of liquids. But the main use and development has taken place in the last 20 years, although some liquid fertilizers were in use earlier.

The main categories of liquid fertilizers are the following:

1. Anhydrous ammonia.
2. Aqueous ammonia.
3. Ammonium nitrate solutions.
4. Ammonium nitrate-urea combinations.
5. Ammonium nitrate-anhydrous ammonia-urea combinations.
6. Mixed liquid fertilizers containing nitrogen-phosphorus-potassium nutrients.

A. Nitrogen Solutions

Use of anhydrous ammonia (82-0-0) is developing very rapidly because of its economy and convenience of application. It is flammable and explosive when its concentration in air is between 16 and 25%; it is toxic and irritating, and requires pressure equipment and special handling. However, farmers and dealers soon learn how to handle it and thereby obtain the significant savings that goes with its use. Stored under pressure of 250 psi it has a density of 5.14 lb/gal at 60°F. Since it is so volatile, it must be applied with a pressure injector into the soil at the depth of 6 to 8 in. where it is held in a colloidal complex with the soil (1, 2).

Even though the application equipment for anhydrous ammonia is very costly, it is cheaper to apply when the acerage is high. Soil retention is very good and 95% was found to be within 2 in. of point of application in silt loam and black clay soils. In sandy soils, however, there is evidence of movement as far as 8 in. and with some losses (3). Fall application is recommended (1), and it has been shown to have the additional effect of killing fungi and nematodes in the zone of application (3).

Transportation costs for aqueous ammonia are more expensive since it contains only 20 - 24% nitrogen (24.4% - 29.4% NH_3) against 82% nitrogen in anhydrous ammonia. It has to be pumped into the soil, but it does have the advantage that other materials can be added to it. Surface application, because of high vapor pressure of ammonia, may cause some losses.

Ammonium nitrate solutions are marketed as 19-0-0 and 21-0-0, equivalent to 57 and 63% ammonium nitrate solutions, respectively (saturation temperatures are 30 and 48°F). Since Allied Chemical was one of the first to introduce names to the different types of liquid fertilizers (4), it is customary in the industry to use their nomenclature. Ammonium nitrate solutions are often called "Feran" type solutions; urea-ammonium nitrate solutions are the "Uran" type solutions; ammonium nitrate-ammonia solutions are the "Nitrana" type solutions; urea-ammonia-ammonium nitrate combinations are the "Urana" type solutions. NPFI (5), another nomenclature system for nitrogenous materials only, allows companies to use their own tradenames: the first figure gives the total nitrogen content and the three figures in parentheses give the percentages, rounded off, of ammonia-ammonium nitrate-urea. Nitrana U, which contains 37.1% nitrogen in a blend containing 15.8% NH_3, 58.5% NH_4NO_3, and 7.7% urea, may be designated as Solution 371 (16-58-8). This system is not too popular because of interference with the $N-P_2O_5-K_2O$ system. Although the unit production cost of nitrogen in ammonium nitrate is about 90% higher than ammonia-nitrogen, the low vapor pressure and ease of using conventional spraying equipment is, under some conditions, attractive. On the other hand, ammonium ions and urea are bound in complexes with the colloidal soil, while nitrate ions move quite freely, raising the possibility of loss (6, 8). The nitrification bacteria, such as Pseudomonas, reduce the nitrate ions to nitrogen (7) with a loss of the nutrient, while ammonia ions and urea must first be oxidized to nitrite ion by Nitrosomas and to nitrate ion by Nitrobacter.

By far the most popular nitrogen solutions are the urea-ammonium nitrate solutions. Although urea nitrogen is slightly more expensive (by 15%) than that of ammonium nitrate, the higher nitrogen content of dry urea (46% nitrogen against 33%) and better water solubility overcome that difference. In fact, the most popular blend, Uran 32 solution, contains nearly as much nitrogen as ammonium nitrate; this equalizes transportation costs. In addition, solution evaporation, crystallization, and other finishing processes are avoidable in production; therefore the cost per unit of nitrogen at the producing plant is 25% less than for solid ammonium nitrate. The most popular blends of this type are the 32-0-0 solution with a salting-out temperature of 29°F, 28-0-0 with a salting-out temperature of 1°F, and Uran 34 with a salting-out temperature of 39°F. Lately TVA (9), for the purpose of adjustment of nitrogen suspensions, suggested the use of a 37-0-0 suspension containing attapulgite clay for stabilization of the suspension against recrystallization. All of these solutions have the optimum low-temperature stable ratio of ammonium nitrate-to-urea of 1.3/1.

Ammonium nitrate-ammonia solution or Nitrana types are combinations that are built up to contain more nitrogen nutrient and have mostly low saturation temperatures, but the vapor pressure is inherently high. They have to be stored in pressure equipment and often applied by soil

injection. But they are regularly referred to as low-pressure solutions.
The vapor pressures at the cut-off temperature of storage at 104°C are
7-15 psi at the 37-0-0 to 41-0-0 grade and up to 100 psi at the 44-0-0 to
57-0-0 grade. The most popular are the 41% nitrogen blends. Combinations of ammonium nitrate-ammonia-urea of the Urana type are also used
and show similar behavior (10).

Other solutions of this type also contain ammonium carbonate and ammonium sulfate (Uramons), ammonium carbonate and formaldehyde, or
ammonium carbamate. Carbon dioxide diminishes the vapor pressure;
urea, ammonium carbamate, and especially formaldehyde slow down the
release of the nitrogen to the bacteria and plant; and sulfate may improve
sulfur deficiency.

B. N-P-K Liquid Fertilizers

Mixed nutrient fertilizers are graded by the amounts of primary nutrients; the contents are expressed in per cent of nitrogen, phosphorus
pentoxide (P_2O_5 or P_4O_{10}), and potassium oxide (K_2O). The N-P-K value
is given in full percents. They are either shipped as such for short distances or prepared in local formulating or mixing plants. N-P-K solutions
are prepared generally with an ammonium phosphate or polyphosphate solution with urea, ammonium nitrate, or Uran 32 type solutions as the additional nitrogen source and potassium chloride (62% K_2O) as the potash
source.

There are two sources of phosphoric acid. The wet process uses
phosphate rock and sulfuric acid which gives an impure, mostly green
or black phosphoric or polyphosphoric acid. Furnace acid is derived
from reduced rock. In the latter, elemental phosphorus (P_4) is distilled
off and is oxidized to P_4O_{10} and hydrolyzed to phosphoric or polyphosphoric acid. Given the current scarcity of sulfur, if electrical power
were cheaper, i.e., 2 mills per kW, the furnace process could become
much more attractive. It will also be more economical if shipped as elemental phosphorus. This may be the only way to make phosphoric acid
economically from low grade rock. There also are recommended processes which are of a mixed character.

For a clear, liquid fertilizer, the phosphoric acid is ammoniated to
the optimum solubility ratio, 8-24-0 (11) for orthophosphoric acid and
11-37-0 (12) or 10-35-0 (13-16) for polyphosphoric acid, and then shipped
to cold formulating plants. Hot formulating plants may do their own neutralization. Wet process superphosphoric acid contains 72-76% P_2O_5,
45-55% of it as polyphosphates. Furnace process superphosphoric acid

contains 76-80% P_2O_5, 51-85% of it as polyphosphoric acids. The polyphosphoric part is in the form of pyrophosphoric, tripolyphosphoric, and
tetrapolyphosphoric acids.

Adjustments for an increase of nitrogen in the N : P ratio will result
in better solubility characteristics if urea is used rather than ammonium
nitrate. That is, if we use as a base, furnace grade orthophosphoric acid
(54% P_2O_5) neutralized to 8-24-0 ammonium phosphate (molar ratio N : P
1 : 1. 55), potassium chloride (0-0-62), urea (46-0-0), we can make a 9-9-9
grade fertilizer, while if we use ammonium nitrate (33-0-0), the top 1 :1 :1
grade is 7-7-7 (17). Formulators with cold mixing equipment therefore
prefer a 32-0-0 liquid solution such as Uran 32 as a starting liquid for the
adjustments.

As to grades used, Achorn, Anderson, Jr., and Hargett from TVA (18)
reported the ten most frequently mentioned clear liquids in their survey as
follows: (1) 6-18-6, (2) 7-21-7, (3) 4-10-10, (4) 5-10-10, (5) 9-9-9, (6)
12-6-6, (7) 15-5-5, (8) 8-8-8, (9) 10-34-0 and (10) 8-24-0.

Different manufacturers and different areas use a legion of different
grades of liquid N-P-K fertilizers. The requirements of the farmer will
depend on the soil analysis, fall fertilization (1, 2), pop-up use (30, 31),
lay-by-use, starter fertilizer plowdown, broadcast, follow-up fertilization, the type of crop, climatic condition, irrigation, and other factors.

C. Suspension and Slurry Fertilizers

Another group of N-P-K liquid fertilizers are the suspension and slurry
fertilizers (19). These new fertilizer types were developed in 1962 by TVA
engineers. Addition of swelling clays such as attapulgite, Attagel 150, and
Minugel 200 used in amounts of 1-3% keeps the crystals in suspension.
Without addition of clays the crystals settle rapidly in slurries and constant mixing is required. The basic ammonium phosphates used are
10-35-0 or 11-37-0, and in summer a grade of 12-40-0 (20, 21) can be
used. In the preparation of the suspensions, the necessary amounts of
water and Uran 32 type solutions are mixed with the clay for 5 min, the
ammonium superphosphate solution is added under stirring, and finally
the crystalline potassium chloride (22-29).

The ten suspension fertilizer grades used most frequently according to
the survey by Achorn et al. (18) in 1967 were: (1) 7-21-21, (2) 14-7-7,
(3) 6-18-18, (4) 6-20-20, (5) 18-6-6, (6) 3-10-30, (7) 3-9-27, (8) 10-30-10,
(9) 7-21-15, and (10) 14-14-14.

Another group of liquid fertilizers are the nitrophosphates; they are not used much because of the corrosiveness of their strongly acid formulations (32).

D. Secondary Nutrients and Micronutrients

Besides the fertilizers which introduce the primary nutrients necessary to grow crops, there are secondary nutrients which are necessary in some soils. These are calcium, magnesium, sulfur, chlorine, and sodium. From these, calcium and magnesium are introduced separately, i.e., in form of gypsum, limestone, and dolomite, etc. Sodium and chlorine are very seldom deficient in the soil. Sulfur could be introduced as gypsum, ammonium sulfate, or elemental sulfur. Sulfur is also easily (20, 32, 45) introduced in anhydrous ammonia or in aqueous ammonia in the form of elemental 20 mesh sulfur or as sodium polysulfide with aqueous ammonia.

Another group of elements necessary for the normal growth of crops are the micronutrients, depletion of which in the soil may cause grave pathological changes in the crops leading to lower yields. Sparr et al. (33, 34), Aldrich (35), Berger (36, 37), and Pratt et al. (43) show the effect and the geography of the deficiencies in the U.S. The main micronutrients are boron, zinc, iron, manganese, molybdenum, copper, and cobalt. These materials can be applied in the form of salts, oxides, chelates, by themselves or with fertilizers (38–42). When the polyphosphates were introduced, their sequestering function made the addition of some micronutrients to mixed fertilizers a distinct possibility (32). Duis (44) shows how $ZnSO_4$ can be added to different formulations. Polon and Fyffe (22) show suspension formulations made with MgO, borax, sulfur, sulfomag and sulfur, zinc oxide, zinc sulfate, copper sulfate, manganese sulfate, and iron sulfate. Scott, Wilbanks, and Burns (20) suggested addition of secondary nutrients such as sulfur and magnesium and of micronutrients such as boron, zinc, manganese, iron, molybdenum, and copper to suspension fertilizers and also one multimicronutrient combination. Farm Chemicals (46) in 1967 in an interview reported that incorporation of micronutrients into liquid fertilizer is difficult when the phosphate is orthophosphate. When a 11-37-0 solution is used in which at least 50% of the P_2O_5 is of the polyphosphate type, the solubility of micronutrients is enhanced. Two per cent zinc oxide, 1.5% copper sulfate, 1% iron sulfate, 0.9% borax, and 0.2% manganese oxide can be dissolved in 11-37-0 liquid fertilizer. It can also dissolve a mixture containing 1% each of boron, copper, and zinc plus 0.2% each of manganese and molybdenum for a total of 3.4% micronutrients. Allied reported that citrus growers need a liquid mix with 4-5% magnesium and this could be supplied with suspensions.

National Sulphur Company is producing Thiovite, ammonium thiosulfate, which can be added to liquid fertilizers. The same report gives many other ways to introduce the micronutrients with different fertilizers.

The growing liquid fertilizer market in the U.S. and the estimate for the next few years are shown in Table 1 which is a combination of information from many sources (9, 47a–47r). These are mostly based on U.S.D.A. reports and an extrapolation of the expected growth. This phenomenal growth of liquid fertilizers has attracted interest from other countries (48, 49, 51, 52).

II. PESTICIDES

Combinations of toxicants and fertilizers have resulted in savings in time and labor in the application process. The combinations with solid fertilizers had many drawbacks. The formulations had to be custom made since the requirements for toxicants and fertilizer depend on different problems: nutrients versus infestations. The separate additions with solid fertilizer are not possible in one application since the farmer cannot blend solids and it is too time consuming for the formulator.

With liquid fertilizers the toxicants can be added to the spray tank in the form of wettable powders or liquid formulations. Dewey (50) drew attention to the dangers and difficulties of using combined formulations which are put together by the formulator. Other papers show the decomposition of insecticides when combined for a long time with fertilizers. This can be avoided when the liquid fertilizer and the required toxicant are mixed just prior to use.

III. PHYSICAL CHEMISTRY OF FERTILIZER SOLUTIONS

In order to investigate the ways of applying the fertilizer and the insoluble toxicant or the combining of two phases to achieve a uniformity of distribution, we must first investigate the individual phases and the interactions between phases.

For economical reasons the liquid fertilizers are highly concentrated solutions or suspensions of plant nutrients, but their characteristics will put them in a few different classes. The high salt concentrations distinguish their behavior from the more common water dispersions of toxicants in the form of emulsions and wettable powder suspensions. The solvated ions and ion pairs behave differently in concentrated solutions

TABLE 1

U.S. Consumption of Fertilizers (in million tons)

Year	Primary nutrient	N	P$_2$O$_5$	K$_2$O	Total fertil.	Total liquid	%	NH$_3$ Anh.	%	NH$_3$ Aq.	%	Other N-sol	%	Fluid mixes	%	% Increase of nutrient
1959–1960	7.46	2.74	2.57	2.15	24.9	2.2	8.8	.71	2.9	.43	1.7	0.65	2.6	0.4	1.6	
1960–1961	7.85	3.03	2.65	2.17	25.6	2.7	10.5	.81	3.2	.43	1.7	0.95	3.7	0.5	2.0	5.0
1961–1962	8.45	3.37	2.8	2.27	26.6	3.3	12.4	.95	3.6	.50	1.9	1.2	4.5	0.6	2.4	7.0
1962–1963	9.5	3.93	3.1	2.5	28.9	4.1	14.1	1.22	4.2	.58	2.0	1.5	5.2	0.8	2.7	12.5
1963–1964	10.5	4.4	3.37	2.73	30.7	4.8	15.7	1.4	4.6	.78	2.5	1.7	5.5	0.9	3.0	10.5
1964–1965	11.0	4.6	3.5	2.8	31.8	5.4	17.0	1.55	4.9	.82	2.6	1.9	5.9	1.14	3.6	5.0
1965–1966	12.4	5.3	3.9	3.2	34.5			1.93	5.6	1.0	2.9	2.4	6.9			12.7
1966–1967	14.0	6.0	4.3	3.6	37.1											12.9
1967–1968	14.8	6.6	4.4	3.85	38.3	8.1	21.1	2.98	7.8	.86	2.3	2.6	6.8	1.65	4.3	5.7
1969–1970	16.1	7.5	4.6	4.0	39.6	10.0	25.3	3.1	7.8	.9	2.3	2.9	7.3	2.65	6.7	8.8
1970–1971	16.7	7.8	4.8	4.1	41.0	11.5	28.0	3.5	8.5	1.1	2.7	3.3	8.0	3.45	8.4	3.7
Projected																
1974–1975	19.9	9.5	5.6	4.8	45.4	14.6	32.2	4.5	9.9	1.45	3.2	4.3	9.5	4.3	9.5	19.2

and in diluted solutions. Instead of speaking of concentration of an ingredient i (c_i), we have to talk about activity $\gamma_i c_i = a_i$, when γ_i is the activity coefficient.

A. Water Structure and Liquid Fertilizers

We also have to take into account the structure of water, its character as a solvent and the changes in its structure caused by dissolved salts, added surfactants, and clays. Because of the high dielectric constant of water, the salts added to it are mostly dissociated into cations and anions. These strongly charged entities will be solvated with quite strong electrical forces by the water molecules, and Bonner and Woosley (53) calculated solvation numbers of several salts for ion-solvent solvation. Harned and Owen (54) give different values, since besides ion solvation there is also ion pair solvation, and in cases of weaker salts, the solvation of undissociated salts. The solubility of the solvated entities may well be a function of the solvation and the dielectric constant (55).

The structure of water (56-61) is basically described as a mixture of hydrogen bonded water molecules and water polymers. Whereas in ice every molecule of water is immobilized by being tetrahedrally hydrogen bonded into the lattice, liquid water has a number of its bonds broken down which gives it mobility but it is still partially polymerized. There are different models of the water structure which explain the behavior of water.

The vacant lattice point model of Forslind (62, 63) suggests a partial breaking of the bonds with some lattice points vacant and some molecules taking interstitial positions with an increase in density.

The flickering cluster model of Frank (57) postulates partial formation of clusters of water molecules hydrogen bonded with tetrahedral bonds which are forming and collapsing in groups. These short-lived, flickering clusters have a high volume-to-surface ratio with a greater stability of the hydrogen bond in the interior of the cluster. This model was analyzed statistically in detail by Nemethy and Sheraga and the thermodynamic data fits with assumed structure. At 20°C, 30% of the molecules are monomers, 23% are single, 4% are double, and 20% are triple bonded at the cluster boundary, and 23% are quadruple bonded inside the clusters.

The water hydrate model by Pauling and Frank and Quist (64-66) suggests that water has a loose cage-like polyhedron structure with big cavities in which are located more loose unbonded water molecules and the cages can also be occupied by other materials to form clathrates and gas hydrites. This model was analyzed in a statistical thermodynamic

treatment by Frank and Quist (66) who predicted the free rotation of inter-
sticial molecules.

The distorted-bond model assumed a bending or distortion instead of a
breaking of hydrogen bonds in melting ice. This model gave exactness in
calculation of the dielectric constant for a range of temperatures from 0 to
83°C.

Frank and Evans (57) suggested that nonpolar molecules dissolved in
water seek or form and stabilize regions of ice-like structures called the
iceberg structures. The dissolution of nonpolar molecules in water increases the hydrogen bonding of water molecules surrounding them. The
hydrogen bond energy is about 2.67 kcal/mole (53).

Ions formed by introducing salts, acids, or bases into water interfere
with the regular water structure by breaking down some tetrahedral lattice and bonding more strongly to water. Ions with a radius not greater
than 1.3 Å will fit into an octahedral water shell, changing the outer structure little but immobilizing the nearest water molecules by polarization.

The effect of an ion on water structure is proportional to its polarizing
power or to its charge divided by its radius, with a range of 2 to 5 Å of
immobilized water. But the effect is not limited to the hydration shell.
By the polarization of the electric fields, ions interfere with the formation of hydrogen-bonded clusters and shorten the half-life of existing clusters in a much bigger shell (62, 63). This effect will depend on the balance
between the polarizing spherical influence of the electrical field of the ion
and the tetrahedrally oriented hydrogen bonding of water. In cases of ions
which have a tetrahedral structure, such as perchlorate ion or ammonium
ion, they may fit into the water structure without greatly disturbing the
structure. In particular, the hydrogen bonds of NH_4^+ and H_2O are of about
the same strength. In dilute solutions relatively small ions like Li^+, Na^+,
H_3O^+, Ca^{2+}, Ba^{2+}, Mg^{2+}, Al^{3+}, OH^-, and F^- have a net effect of structure breaking, increasing entropy and fluidity of water, and the affected
water molecules become more mobile. In concentrated solutions Li^+ and
Na^+ have a net effect of structure breaking also.

Clay also has an effect of increasing order and rigidity in the water
lattice and reducing thermal vibration, which is apparently effective at a
distance of at least 300 Å. The structure building is generally assumed
to increase the viscosity of water. Holtzer and Emerson (67) point out
that changes in water structure may not be the only cause of deviations
from Einstein's viscosity equation

$$[\eta]/2.5 \, v_h = 1$$

where $[\eta]$ is the intrinsic viscosity and v_h is the specific volume of the spheres of the solute particles. Deviations may be the result of a non-spherically symmetric particle and a modified volume factor (v_h). The picture is even more complicated at higher concentrations of the solute which Moulik (68) calls the "beyond Einstein region" where extrapolation to zero concentration does not give the intrinsic viscosity.

When the salt solution reaches saturation, all monomer H_2O is presumed to be removed. The general model of the system will be a competition for the solubilizing, bound water molecules by the ions, ion pairs, ion triplets, nondissociated salts, hydrogen bonding molecules (such as urea), clay particles, and the hydrophilic parts of surfactants. Naturally the competition will be won by those entities which release more free energy by their bonding with the water.

B. Salting-out and -in Behavior of Salts and Urea

The high bonding energy between ions and water, the energy of dissociation of a salt, and the high dielectric constant of water combine to make solvated ions the most stable species. And this leads to the problem of the availability and the activity of water.

Adams and Gibson (69) calculated the vapor pressure of different solutions of ammonium nitrate in water and came up with an empirical equation fitting these values at $25^{\circ}C$

$$\log \frac{P}{P_0 X_1} = 0.00093 + 0.31317 X_2^{3/2} - 0.1580 X_2^{5/2}$$

where P/P_0 is the relative vapor pressure, X_1 is the mole fraction of water, and X_2 is the mole fraction of ammonium nitrate assumed to be wholly dissociated so the average molecular weight is taken as 40. Since the vapor pressures are small, we can then give the difference in chemical potential of water at a given salt concentration μ_1 and the chemical potential of pure water at the same temperature μ_w by the equation

$$\mu_1 - \mu_w = \frac{RT}{M_1} \ln \frac{P}{P_0} = \frac{2.303 RT}{M_1} (10g\, X_1 + 0.00093 + 0.3137 X_2^{3/2} - 0.1580 X_2^{5/2})$$

Since in salt solutions $P < P_0$, the chemical potential of water in the salt solution will always be smaller than in pure water and the fraction

$$\frac{\mu_1}{\mu_w} = 1 - \frac{RT}{M_1\mu_w} \ln \frac{P_o}{P}$$

will be the measure of the driving force of water for chemical reactions in comparison with pure water or the availability of water. Bone (70) in a talk given in February 1968 to the Division of Oil Chemists of the ACS (unpublished data) connected this value to different factors in foods and its effect on food spoilage, dough viscosity, and solvation.

According to the Gibbs equation (61), for a binary system at constant temperature and pressure, $x_1 d\mu_1 = -x_2 d\mu_2$, where x_2 is the mole fraction of the salt and μ_2 its chemical potential. By assuming μ_2 at saturation point as zero, we can calculate μ_2 for different concentrations. In the same way, having the activities of the salts, water activity can be calculated.

Ise and Okubo (71, 72) calculated mean activity of polyelectrolytes by the isopiestic method. This method introduced by Bousfield (73) and improved upon by Sinclair (74), Robinson and Sinclair (75), and Scatchard et al. (76) makes possible the exact (±2%) calculation of the osmotic coefficient of the solution ϕ. The mean activity coefficient γ^* can be calculated (54) from the rearranged Gibbs-Duhem equation if one value was obtained from the emf measurement

$$\ln \frac{\gamma_1^*}{\gamma_2^*} = (\phi_1 - \phi_2) + 2 \int_{m_1}^{m_2} \frac{1 - \phi}{\sqrt{m}} \, d\sqrt{m}$$

where m is the molarity. And this was used to calculate the activity coefficients of polyphosphates with different gegenions.

Rush and Johnson (77) investigated isopiestically triple component systems such as $LiClO_4$-$NaClO_4$-H_2O. They measured the osmotic coefficients and determined activity coefficients. A deviation function δ could be used to demonstrate the validity of the Brønsted rule for obtaining the activity coefficients of electrolytes in mixed solutions.

Bower and Robinson (78) investigated the ternary system of urea-sodium chloride-water isopiestically and found that the effect of urea on the activity coefficient of sodium chloride is generally that of "salting in," although at lower concentrations of urea there is a very small salting out effect. This effect of urea is very important. It would indicate that addition of urea increases the water activity when combined with ionic salts. The same effect is observed in other combinations of urea.

High salt concentrations increase (79) the surface tension of the solution with polyvalent cations and anions causing a greater increase than monovalent ones; the increase is linear with the concentration.

Another effect of increased salt concentration is the change of the surface potential of water; this was tested by Jarvis and Scheiman (79) using the radioactive electrode technique. The surface potential increases with salt concentration and depends on the ion species involved. Some combinations cause the potential to become positive, some negative. This appears to be related to the water structure in the vicinity of the ions. Since the forces near the interface may be nonuniformly distributed, they will change the surface potential of water. Water molecules will preferentially associate with anions near the surface, and one would anticipate a negative change with increase of salt concentration. The positive change caused by some salts will then possibly be caused by cations preferentially approaching the surface and changing the orientation of the water molecules on the surface. These anions which give the positive potential change are known to be the water structure-making anions as opposed to the structure-breaking ions which give the negative potential (79). Surface potential will determine whether a surfactant will be salted out of solution. Some mixed salt solutions may have opposite effect and cancel each other. A decrease in water activity occurs simultaneously with an increase in surface tension, except that in small salt concentrations Jones and Ray (80) reported an apparent minimum surface tension.

An optimum correlation of the concentration characteristic for the specific surfactant, salt concentration, and the oil phase foster the formation of black films for that system, yielding a maximum stability (81).

C. Surface Behavior and Surfactants

Anhydrous ammonia has a surface tension, σ, of about 20 dyn cm^{-1} at 20°C according to Lange's (82) Handbook of Chemistry, and the interpolation gives for 29.4% aqueous ammonia, the value of $\sigma_{18^o} = 69.5$ dyn cm^{-1}, while water at the same temperature has a surface tension $\sigma_{18^o} = 73.05$ dyn cm^{-1}.

No investigation has been done on emulsification or suspension of materials in anhydrous ammonia other than sulfur (20, 32, 45).

The behavior of surfactants as wetters and emulsifiers will greatly depend on the water activity, surface tension, and alkalinity or acidity of the liquid fertilizers.

There are two ways to incorporate materials which are insoluble, or in which solubility is lower than required, into the liquid phase, such as the liquid fertilizers: (1), a second liquid phase, oily, when the necessary addition materials are oil soluble, is incorporated by means of emulsification, and (2) a solid suspensions, stabilized by surfactants and protective colloids such as clays.

Emulsion Formation and Stability

The theory of emulsion formation and stability (83) is based on the fundamental work using Gibbs' thermodynamic absorption law:

$$-d\sigma = RT \sum_{i=2}^{n} \Gamma_i d (\ln\gamma_i C_i)$$

where σ is the surface tension, Γ_i is the surface excess of i species, γ_i is its activity, and c_1 is its concentration. In cases of concentrated salt solutions such as liquid fertilizers, the behavior of water as a solvent, its activity as well as the water structure, determines the formulation used in industry. There is no work done in treating the effects of water structure quantitatively. Nemethy and Sheraga (58-60) treated it semiempirically using an effect they called "hydrophobic bonding"; other people (83, 84) prefer to use the term "hydrophobic association."

The long-range forces in emulsification are the London-van der Waals forces and electrical forces over the electrical double layer, which may depend on surface potential. The short-range forces are the ionic, covalent, hydrogen bonding, dipole and quadrupole interaction, hydrophobic association, and Bohr repulsion forces.

There is a controversy about the depth of the effect of hydration of the emulsion droplets in the water phase. Derjaguin (85) claims that the influence of hydrogen bonding of water to hydrophilic surfaces extends for a considerable distance from the interface. Many others (83) are of the opinion that it is limited to only a few molecules of water in the form of the Stern layer. Alcohols and carboxylic acids are weak hydrogen bonding materials and their effect on changing the water structure in their vicinity should not be as large as with some other surfactants. Strong ionic groups are highly hydrophilic and form strong hydrogen bonds, but the most highly charged water in the Stern layer is not too thick (a few angstroms).

Ether oxygen will be expected to have a weaker hydrogen bond than those in the alcohol group, but if many ethylene oxide units are combined in one molecule the water soluble ethoxylated chain will extend into the water phase and by its geometrical position extend the Stern layer by about 5 Å when extended and 2 Å when coiled for each ethylene oxide unit. With 10 ethylene oxide groups this will amount to about 20 Å. This protective water layer, stabilized by the strong hydrophilic effect of anionic surfactants, is the reason why non-ionic-anionic combinations make better emulsifiers at a lower use level than the hydrophilic ionic surfactants or non-ionic polyethoxylated surfactants with a strongly extending Stern layer function taken separately.

The stabilizing effect on the interface by Marangoni–Gibbs elasticity effect and Plateau viscosity effect (86) accounts for the surface tension gradient and the resulting elasticity of the surface can be given by the general Gibbs elasticity relation:

$$E = 4RT \sum{}' \frac{\Gamma_i^2}{c_i} \frac{1 + \dfrac{d\,\ln\gamma_i}{d\,\ln c_i}}{h + 2\dfrac{d\Gamma_i}{dc_i}}$$

where h is the thickness of the film.

Van der Tempel, Lucassen, and Lucassen-Reynders (87) investigated the application of the surface thermodynamics to the Gibbs elasticity and the effect of mixing emulsifiers on changes in elasticity. The same way as the activity of the i component in the bulk of solution is expressed by the relation

$$\mu_i^\alpha = \zeta_i^\alpha + RT \ln \gamma_i^\alpha C_i$$

where α denotes the bulk and ζ_i^α is a constant characteristic for i, such that at infinite dilution $\mu_i = \zeta_i^\alpha$, the surface will be expressed by a similar equation:

$$\mu_i^s = \zeta_i^s + RT \ln \left(\gamma_i^s \frac{\Gamma_i}{\Gamma_\infty} \right) - \frac{\sigma}{\Gamma_\infty}$$

where s denotes the element i in the surface; Γ_i the absorption of the i species on the surface, or the surface excess; and Γ_∞ the saturation absorption equals the sum of solvent and surfactant absorption in moles cm^{-2}, so that Γ_i/Γ_∞ is the mole fraction of i species on the surface. ζ_i^s-like in bulk is the value of μ_i^s at infinite dilution.

Van den Tempel (88) made a thorough investigation of the stability of oil-in-water emulsions. Upon investigation he found that the zeta potential falls with an increase in salt concentration. The effect is much stronger with replacement of monovalent cations by polyvalent cations. This may be attributed to specific absorption in the Stern layer which increases the concentration of the gegenions in that layer and diminishes the effectiveness of the layer in protecting the emulsion droplets from coalescence.

Sonntag, Netzel, and Klare (89) investigating the stability of emulsions found that increasing the salt concentration diminishes the equilibrium distance between emulsion droplets stabilized with non-ionic emulsifiers. A cyclohexane-water emulsion stabilized by a nonyl phenol adduct with 20 moles of ethylene oxide has a critical distance of droplet separation of 235 Å. Presence of 0.01 M KCl compresses the electrical double layer, allowing a closer approach of droplets and ultimate coalescence. All the effects are specific to the gegenion (90), but the increased concentration of any salts will affect the electrical double layer. Ionic surfactants in the form of carboxylic soaps are less stable in salt solutions than ethylene oxide non-ionic surfactants. Certain other surfactants such as dodecylbenzene sulfonates in very hard waters containing calcium or magnesium salts will exchange the alkali ion in the Stern layer for divalent ions. Although the sulfonate ion is strongly hydrogen bonding, the surfactant has a much stronger hydrophobic association or lipophilicity than the hydrophilic bonding and will not be a good emulsifier in strong salt solutions. It is a question of hydrophilic-lipophilic balance. Ethylene oxide adducts, the non-ionic surfactants, should have their hydrophilic part stronger than their lipophilic part. But in spite of lingering opinion (91) that non-ionics have considerable insensitivity to the presence of hard water while ionic surfactants have a limited value in hard water, the contrary is true in high salt concentrations. The multiple, but weak, hydrogen bondings of the non-ionic materials in the presence of the diminished water activity indicates their lack of usefullness in fertilizer solutions. The backbone of the materials useful for emulsification in high salt concentrations are strongly hydrophilic ionic emulsifiers.

D. Grouping of Liquid Fertilizers and Surfactants in Liquid Formulations

With the water activity in mind, we can group the liquid fertilizers into classes:

1. Anhydrous ammonia: with low surface tension is not as of now used as the base for emulsification.

2. Aqueous ammonia: high water activity, low surface tension, moderate changes in water structure, and regular type emulsifiers can handle the emulsification.

3. Ammonium nitrate–urea–water system (Uran type): quite high water activity. Urea moderates the system by increasing the water activity. Most probably forms a complex. The emulsification can be carried out by modified non–ionic–anionic blends.

4. High vapor pressure nitrogen solutions containing ammonia, ammonium nitrate, water, with or without urea (Nitrana and Urana types): as in the previous system, ammonia diminishes surface tension and increases water activity. The emulsification can be carried out with modified non-ionic–anionic blends.

5. Ammonium nitrate solutions (Feran type): low water activity and high surface tension. Emulsification requires highly hydrophilic ionic emulsifiers as the backbone of the emulsifier.

6. N–P–K liquid fertilizers: low water activity and high surface tension requires strong ionic emulsifiers.

With differing ingredients even at the same degree of saturation, the emulsification problems will vary since each material may affect the structure of water in different ways. And naturally the requirements for higher hydrophilic emulsifier systems will diminish with dilution of the fertilizers.

E. Liquid Fertilizers and Wettable Powders

Wettable powders are formulated from materials which have low solubilities, making a liquid formulation uneconomical. Generally, the flocculation is much faster than the creaming rates of an emulsion, and circulating pumps are desirable. Most wettable powders contain fillers (92) of high surface areas (93) and certain fillers swell in water, increasing the viscosity, and in this way slowing the flocculation and separation in the water phase. The wetting agents used are anionic in character and only seldom non–ionics.

In general, three factors are necessary for the utility of wettable powders: (1) wettability—to wet the powder rapidly with the aqueous phase (94, 95) without clumping out to form aggregates; (2) dispersability—to suspend the powder in the aqueous phase with minimum flocculation; and (3) redispersibility of the flocculated powders (prevention of clumping of the separated flocculant).

In wetting of the powders, we have to take into account the fact that they behave like capillaries, and the speed and uniformity of wetting will depend on the kinetics of the wetting processes in the capillary system.

By combining the Laplace and Poiseuille equations (96, 97) and the Rideal-Washburn equation (61), we can arrive at the formula

$$Q = \frac{Ca^2 \rho gh + 2Ca\sigma_{a/w} \cos \theta}{8kl}$$

where Q is the quantity of liquid entering the wettable powder per unit area and unit time, C is the volume fraction of wet powder, a is the effective pore radius size, ρ the density of the liquid, g is the gravitational constant, h is the length of the liquid column, l is the length of the wet column, $\sigma_{a/w}$ is the air-liquid surface tension, θ is the contact angle, and k is the factor introduced to allow for the tortuous path in the capillary.

If we forget about the part played by gravity, by rearranging the equation and eliminating Q and C we get

$$\frac{l^2}{t} = \frac{a\sigma_{a/w} \cos \theta}{2k^2 \eta}$$

The wetting $\frac{l^2}{t}$ is then proportional to the packing factor a/k^2 and to $\sigma_{a/w} \cos \theta$.

Addition of wetting agents will diminish the surface tension $\sigma_{a/w}$ and the wetting angle θ, with the effect of increasing $\cos \theta$. So addition of surfactants will with one factor tend to diminish the speed of wetting and with the other will improve the wetting of the surface. To solve this dilemma we first have to optimize the equation. For hydrophile surfaces we use either only small amounts of surfactants to reach the optimum or none. For a hydrophobic powder even a small cut in $\sigma_{a/w}$ is by far overcome by the increase in $\cos \theta$.

Simultaneously, we are using, if possible, those surfactants which diminish the interfacial tension $\sigma_{s/w}$ more strongly than the surface tension $\sigma_{a/w}$ and so θ is diminished and $\sigma_{a/w}$ only moderately affected. These surfactants are mostly anionic in structure such as alkyl naphthalene sulfonate, dioctyl sulfosuccinate, fatty acid esters of isethionic acid, fatty acid amides of N-methyltaurine, and others. Because of the character of the surfactants, the wettable powders are to a degree more wetted by liquid fertilizers of quite high water activity.

Another group of materials used in wettable powders are the dispersants. In this capacity, lignin sulfonates and other polyelectrolytes are used which prevent flocculation and also prevent any flocculated materials from clumping. These materials are also to a degree compatible with higher water activity liquid fertilizers. In many cases the wetters will also carry out this function and no special dispersants are needed.

In the same way as we have to modify emulsifiers for emulsificable concentrates for use with liquid fertilizers, the wettable powders for this purpose should also be modified.

As an illustration, Klosterboer and Bardsley (98) show the dramatic effect of a decrease in suspendability of linuron with the increased concentration of ammonium nitrate in solution.

IV. RECOMMENDATIONS OF COMBINED USES OF TOXICANTS WITH LIQUID FERTILIZERS

It is definitely recognized that growth-hormone type herbicides have increased activity in the presence of nitrogen-carrying fertilizers (99, 100). This effect is less pronounced with other herbicides, but it stands to reason that a plant that is enhanced by fertilizers during the growing period will become more susceptible to herbicidal action even when the plants are not normally suspectible. So care must be taken when the fertilizer-toxicant combination is placed in the soil during the preplanting application (101, 102). Dewey (50) warned the formulators and fertilizer manufacturers to be sure that the combinations are applied uniformly (good and stable formulations), that the toxicants do not deteriorate in the formulation, and that excessive doses of toxicants are not used. But all these problems can be avoided if the toxicants are well formulated and added in the right amounts to the liquid fertilizers just prior to application.

The usage of combinations of toxicants and liquid fertilizer is growing because of the economical advantages and improved yields reached by this application (51, 103, 104). Naturally, nonpersistant toxicants cannot be applied with fall fertilizer application unless the effect, i. e., on nematodes, will be carried over to the next season. Grimes suggests using toxicants in combination with pop-up fertilizers. He warns of the necessity of some equipment agitation requirements and of the night carry-over problem, especially if the formulations are not optimized to overcome this problem. One should also be concerned with state laws in the U.S. with regard to transportation on public roads of mixtures containing toxicants with fertilizers. Some states fail to recommend the use of such mixtures; this is in most cases the result of insufficient research into the effect of such combinations. In fact, the limited research which has been done shows that where such combinations have been used according to the recommendations, the biological effectiveness has been as good or better than when the toxicants were used separately. Worsham (105) stated that mixing pesticides with liquid fertilizers was practical and where not limited by regulations is a step in furthering their efficiency by proper use. Grimes (106), Farm Chemicals (107), Agricultural Chemicals (108), and TVA (109) recommended the combined use. Table 2 gives a list, although not a complete one, of toxicants which are used in combination with liquid fertilizers. However, it does not necessarily mean that commercial formulations are indiscriminately applicable in all their recommended uses.

TABLE 2

Pesticides Recommended with Fertilizers

Pesticide	Nitrogen solutions	Mixed N-P-K fertilizers Liquid	Dry	Pesticide	Nitrogen solutions	Mixed N-P-K fertilizers Liquid	Dry
Herbicides				Paraquat	+		
Atrazine	+			Planavin	+		
Simazine	+			Premerge	+		
Caparol	+			Telvar	+		
CIPC	+		+	Treflan	+	+	
Dacamine	+			Insecticides			
2,4-D	+			Aldrin	+	+	
Eptam	+	+	+	Chlordane	+	+	+
Knoxweed	+			Diazinon	+	+	+
Vernam	+			Disyston	+	+	+
Lorox	+	+		Dieldrin	+	+	
DSMA + Karmex	+			Endrin	+	+	
Alanap	+			Guthion	+	+	
Banvel D	+			Heptachlor	+	+	
Dacthal	+			Malathion	+	+	
Cotoran	+			Nemagon	+	+	
Fenac	+			Parathion	+	+	
Herban	+			Trithion			+
Ramrod	+		+	2,4-D + Parathion			+

In order to allow the uniform distribution of oleogineous formulations with the liquid fertilizers, different materials may be added either into the oil phase or into the water phase to diminish the interfacial tension; this permits the formation of a more or less stable emulsion. To facilitate such an emulsion with materials added to the oleogineous phase, there are many patents (110-118) which describe such use.

Table 3 is based on McCutcheon's Detergents and Emulsifiers (119). It shows materials that diminish interfacial tension in liquid fertilizers and that may be useful for pesticide-fertilizer combinations.

TABLE 3

Trade names	Manufacturer	Class and formula	Use and remarks
Aerosol 22	American Cyan-amid Industrial Chem. Div.	Tetra sodium N(1, 2 dicarboxyethyl) N-octadecyl sul-fosuccinamate	30% solution. Wetter for salt solutions (120)
Aerosols AY and MA	American Cyan-amid Industrial Chem. Div.	Sodium dialkyl sulfosuccinate	Wetting agents for salt solutions (121)
Alkapents BD 100, DPH 60, DNP 100, TD 60, TD 100	Wayland Chem. Div.	Complex organic phosphates in acid form	Emulsifiers for pesti-cides-liquid fertilizer combinations
Dowfax 2A1, 3B1	Dow Chem. Co.	Disodium alkyl diphenyl ether di-sulfonate	Wetting agents for salt solutions
Emcol AD5-13	Witco Chem. Co.	Anionic-non-ionic blend	Emulsifier for pesti-cide-nitrogen solution combinations
Emcol PS and CS series	Witco Chem. Co.	Complex organic phosphated es-ters	Emulsifiers for pesti-cide-liquid fertilizer mixtures
Emcols H-A, H-B, H-C, H-JP1, H-JP2	Witco Chem. Co.	Complex sul-fonates	Emulsifiers for pesti-cide-liquid fertilizer combinations

TABLE 3 (continued)

Trade names	Manufacturer	Class and formula	Use and remarks
Gafac RS-610	General Aniline & Film Corporation	Complex organic phosphate esters	Emulsifiers for pesticide-liquid fertilizer combinations
Monawet SNO 35	Mona Ind., Inc.	Like Aerosol 22	35% solution. Wetter for salt solutions
Petro AGS, P	Petrochemicals Co., Inc.	Sodium alkyl naphthalene sulfonates	Surfactant constituents of wettable powders and fertilizers
Schercowet DHS	Scher Bros. Inc.	Sodium dihexyl sulfosuccinate	Wetting agents for salt solutions
T-Mulz 3HF, 4AF	Thom.-Hay. Chem. Co.	Anionic-nonionic blend	Emulsifiers of pesticide-liquid fertilizer combinations
Compex	Colloidal Prod. Co.	Blend	Compatibility agent for liquid fertilizer-pesticide mixtures added to the fertilizers (122)

REFERENCES

1. V. Sauchelli, Agr. Chem., 23-7, 46 (1968).

2. C. N. Sawyer, Chem. Eng. News, May 6, 1968.

3. A. V. Slack, The Chemistry and Technology of Fertilizer, ACS Monograph No. 148, pp. 513-537 (1960).

4. G. C. Matthiesen, "Nitrogen Solutions," Farm Chem., 129 (9)45, (10) 63, (11)26 (1966).

5. Anonymous, Agr. Chem. 17-2, 29 (1962).

6. E. O. Huffman and A. W. Taylor, J. Agr. Food Chem., 11, 182 (1963).

7.   W. H. Fuller, J. Agr. Food Chem., 11, 188 (1963).

8.   R. D. Munson and W. L. Nelson, J. Agr. Food Chem., 11, 193 (1963).

9.   W. C. Scott and J. A. Wilbanks, Chem. Eng. Prog., 63-10, 58 (1967).

10. W. R. Schantz, C. L. Hart, and H. H. Tucker, Farm Chem., 130, (4)53 (1967).

11. D. G. Rands, Agr. Chem., 21-5, 23 (1966).

12. Anonymous, Agr. Chem., 21-12, 14 (1966).

13. T. M. Kelso, J. J. Stumpe, and P. C. Williamson, Commer. Fert. Plant Food Ind., March, 1968, p. 1.

14. A. V. Slack, Farm Chem., Nov., 1962.

15. A. V. Slack, J. M. Potts, and H. B. Shaffer, J. Agr. Food Chem., 12, 154 (1964).

16. A. V. Slack, "Liquids '65", Farm Chem., 128, (2)17, (3)23, (4)25, (5)21 (1965).

17. L. B. Nelson, Agr. Chem., 17-5, 39 (1962).

18. F. P. Achorn, J. F. Anderson, and N. H. Hargett, Solutions, May-June, 1968, p. 8.

19. J. A. Wilbanks, Natl. Fert. Sol. Assn. Liq. Fert. Round-Up Proc., July, 1967, p. 6.

20. W. C. Scott, J. A. Wilbanks, and M. R. Burns, Solutions, March-April, 1967.

21. W. H. Kibbel, R. J. Fuchs, and A. R. Morgan, Solutions, Sept.-Oct., 1968, p. 34.

22. J. A. Polon and H. B. Fyffe, "Suspensions", Farm Chem., 131, (1)48, (2)48 (1968).

23. J. Silberberg, Commer. Fert. Plant Food Ind., August, 1966.

24. A. V. Slack and H. K. Walters, Agr. Chem., 17-11, 36 (1962).

25. J. A. Wilbanks, Agr. Chem., 21-9, 37 (1966).

26. Anonymous, Farm Chem., 131, (6)42 (1968).

27. Anonymous, Solutions, Jan-Feb. 1967.

28. W. C. Scott, J. A. Wilbanks, and M. R. Burns, Solutions, Nov-Dec., 1965.

29. T. P. Hignett and E. L. Newman, Farm Chem., 130, (9)30 (1967).

30. Anonymous, Agr. Chem., 22-4, 22 (1967).

31. K. D. Jacob, Farm Chem., 130, (3)46 (1967).

32. Anonymous, Agr. Chem., 19-11, 24 (1964).

33. M. C. Sparr, E. O. Schneider, and L. J. Sullivan, Solutions, Jan.-Feb., 1968.

34. M. C. Sparr, C. W. Jordan, and J. R. Turner, Solutions, March-Apr., 1968.

35. S. R. Aldrich, Solutions, Jan.-Feb., 1963.

36. Anonymous, Agr. Chem., 17-1, 16 (1962).

37. K. C. Berger, J. Agr. Food Chem., 10, 178 (1962).

38. G. L. Bridger, M. L. Salutsky, and R. W. Starostka, J. Agr. Food Chem., 10, 181 (1962).

39. L. Chesnin, J. Agr. Food Chem., 11, 118 (1963).

40. E. J. Haertl, J. Agr. Food Chem., 11, 108 (1963).

41. E. R. Holden, N. R. Pace, and J. I. Wear, J. Agr. Food Chem., 11, 188 (1962).

42. A. Wallace, J. Agr. Food Chem., 11, 103 (1963).

43. P. F. Pratt and F. L. Bair, Agr. Chem., 19-9, 39 (1964).

44.   J. H. Duis, Solutions, Nov.-Dec., 1968.

45.   J. J. Mortvedt, Agr. Chem., 20-6, 39 (1965).

46.   Anonymous, Farm Chem., 130, (2)68 (1967).

47.   (a) W. Scholl, G. W. Schmidt, and C. A. Wilker, Agr. Chem., 18-5,
      30 (1963); (b) Anonymous, Ibid., 19-4, 26 (1964); (c) W. Scholl,
      G. W. Schmidt, and H. P. Toland, Ibid., 19-4, 28 (1964); Farm Chem.,
      May, 1965, p. 16; (d) W. Scholl, G. W. Schmidt, C. A. Wilker, and
      H. P. Toland, Agr. Chem., 19-1, 16 (1964); (e) W. Scholl,
      G. W. Schmidt, and H. P. Toland, Ibid., 20-5, 26 (1965); (f)
      H. H. Shepard and J. N. Mahan, Ibid., 21-6, 35 (1966); (g) W. Scholl
      and H. P. Toland, Ibid., 21-4, 23 (1966); (h) Anonymous, Farm
      Chem., Jan., 1967, p. 56; (i) Agr. Chem., 22-8, 17 (1967); (j) 23-7,
      16 (1968); (k) C. D. Spencer and H. P. Toland, Ibid., 22-7, 25
      (1967); (l) Anonymous, Farm Chem., June, 1968, p. 36; (m) Jan.,
      1968, p. 29; (n) Feb., 1968, p. 24; (o) Dec., 1968, p. 32; (p) 135(1)
      16, (1972), (q) F. P. Achorn, Fert. Sol., 14(2) 26 (1970).

48.   L. A. Soubies, Solutions, July-Aug., 1968.

49.   L. A. Soubies and J. P. Baratier, Solutions, Nov.-Dec., 1966, p. 48.

50.   J. E. Dewey, Agr. Chem., 21-5, 27 (1966).

51.   J. Schonberg and A. M. Mendoza, Bol. Soc. Quim. Peru, 29, 159
      (1963).

52.   C. Stenseth, ForskingForsok Landburket, 14(5), 735 (1963).

53.   O. D. Bonner and G. B. Woolsey, J. Phys. Chem., 72, 899 (1968).

54.   H. S. Harned and B. B. Owen, The Physical Chemistry of Electro-
      lytic Solutions, 3rd ed., Reinhold, New York-London, 1958.

55.   A. S. Quist and W. L. Marshal, J. Phys. Chem., 72, 1536 (1968).

56.   J. L. Kavanau, Water and Solute-Water Interaction, Holden-Day, Inc.,
      San Francisco, 1964.

57.   H. S. Frank and M. W. Evans, J. Chem. Phys., 13, 507 (1945).

58.   G. Nemethy and H. A. Sheraga, J. Chem. Phys., 36, 3382 (1962).

59.   G. Nemethy and H. A. Sheraga, J. Chem. Phys., 36, 3401 (1962).

60.   G. Nemethy and H. A. Sheraga, J. Phys. Chem., 66, 1773 (1962).

61.   J. T. Davies and E. K. Rideal, Interfacial Phenomena, Academic,
      New York–London, 1961.

62.   E. Forslind, Acta Polytech., 115, 9 (1952).

63.   E. Forslind, Proceedings Second International Congress on Rheology,
      Butterworth, London, 1953.

64.   L. Pauling and R. E. Marsh, Proc. Natl. Acad. Sci. U.S., 38, 112
      (1953).

65.   L. Pauling, Science, 134, 15 (1961).

66.   H. S. Frank and A. S. Quist, J. Chem. Phys., 34, 604 (1961).

67.   A. Holtzer and M. F. Emerson, J. Phys. Chem., 73, 26 (1969).

68.   S. P. Moulik, J. Phys. Chem., 72, 4682 (1968).

69.   L. H. Adams and R. E. Gibson, J. Am. Chem. Soc., 54, 4520
      (1932).

70.   D. Bowe, unpublished paper, Chicago AOCS Meeting, Feb., 1968.

71.   N. Ise and T. Okubo, J. Phys. Chem., 71, 1287 (1967).

72.   N. Ise and T. Okubo, J. Phys. Chem., 72, 1370 (1968).

73.   W. R. Bousfield, Trans. Faraday Soc., 13, 401 (1918).

74.   D. A. Sinclair, J. Phys. Chem., 37, 495 (1933).

75.   R. A. Robinson and D. A. Sinclair, J. Am. Chem. Soc., 56, 1830
      (1934).

76.   G. Scatchard, W. J. Hamer and S. E. Wood, J. Am. Chem. Soc.,
      60, 3061 (1938).

77.   R. M. Rush and J. S. Johnson, J. Phys. Chem., 72, 767 (1968).

78.   V. E. Bower and R. A. Robinson, J. Phys. Chem., 67, 1524 (1963).

79.   N. L. Jarvis and M. A. Scheiman, J. Phys. Chem., 72, 74 (1968).

80. G. Jones and W. A. Ray, J. Am. Chem. Soc., 59, 187 (1937).

81. H. Sonntag and H. Klare, Tenside, 2, 33 (1965).

82. A. L. Lange, Handbook of Chemistry, 10th ed., McGraw-Hill,
     New York, 1961.

83. J. A. Kitchner and P. R. Musselwhite, "The Theory of Stability of
     Emulsions," in Emulsion Science (P. Sherman, ed.), Academic,
     New York-London, 1968, Chap. II.

84. G. S. Hartley, Aqueous Solutions of Paraffin Chain Salts, Herman
     et Cie, Paris, 1936.

85. B. V. Derjaguin, Symp. Soc. Exptl. Biology, 19th, Swansea,
     Butterworths, London, p. 55 (1965).

86. B. V. Derjaguin and A. S. Titijevskaya, Proc. Intern. Congr. Sur-
     factant Actions 2nd, London, 1, 211 (1957).

87. M. van der Tempel, J. Lucassen, and E. H. Lucassen-Reynders,
     J. Phys. Chem., 69, 1798 (1965).

88. M. van der Tempel, Comm. No. 225 of the Rubber-Stichting, Delft.,
     Rec. Trav. Chem. Pay-Bas, 72, 419, 433, 442 (1953).

89. H. Sonntag, J. Netzel, and H. Klare, Kolloid Zeitschrift, 211, 121
     (1966).

90. P. Mukerjec, K. J. Mysels, and P. Kapanan, J. Phys. Chem., 71,
     4166 (1967).

91. R. W. Behrens, Weeds, 12, 255 (1964).

92. D. E. Weidhaas and J. L. Brann, Jr., Handbook of Insecticide Dust
     Diluents and Carriers, Dorland, Caldwell, N.J., 1955.

93. U. Hofmann, Angew. Chem. (Intern. Ed.), 7, 681 (1968).

94. J. H. Schulman and J. Leja, Trans. Faraday Soc., 50, 598 (1968).

95. Y. Tamai, K. Makuuchi and M. Susuki, J. Phys. Chem., 71, 4176
     (1967).

96. J. J. Bikermann, Surface Chemistry for Industrial Research,
     Academic, New York, 1947.

97. "Wetting," Monograph No. 25, Society of Chemical Industry, London, 1967.

98. A. D. Klosterboer and C. E. Bardsley, Weed Sci., 16, 468 (1968).

99. G. C. Klingman, North Carolina Agr. Exptl. Sta. Res. Farming, 12, 3 (1954).

100. A. D. Klosterboer and C. E. Bardsley, Proc. Assoc. Southern Agr. Workers, 63, 91 (1966).

101. R. Aycock and J. N. Sasser, Plant Disease Reporter, 45(8), 620 (1961).

102. J. G. Kantzes, W. R. Jenkins and R. A. Davis, Plant Disease Reporter, 43, 1231 (1959).

103. N. I. Kir, Sb. Nauch. Rabot. Asp. Vses. Nach.-Issled. Inst. Khlop, 5, 202 (1964).

104. P. V. Protasov, G. I. Yarovenko, I. N. Kir, and I. I. Protopopova, Khim. v Sel'sk. Khoz., 1964(5), 26.

105. A. D. Worsham, Solutions, Nov.-Dec., 1968, p. 10.

106. W. H. Grimes, Agr. Chem., 22-8, 32 (1967).

107. Anonymous, Farm Chem., 130-6, 108, June, 1967.

108. Anonymous, Agr. Chem., 19-12, 46, Dec., 1964.

109. National Fert. Develop. Center TVA, Muscle Shoals, Ala. "Production of Suspension Fertilizers," May 1968 (TVA Library No. TR 88).

110. P. L. Lindner, U.S. Pat. 2,976,208 (1956).

111. P. L. Lindner, U.S. Pat. 2,976,209 (1956).

112. P. L. Lindner, U.S. Pat. 2,976,211 (1958).

113. P. L. Lindner, U.S. Pat. 3,080,280 (1961).

114. P. L. Lindner, U.S. Pat. 2,236,626 (1961).

115. P. L. Lindner, U.S. Pat, 3,236,627 (1961).

116. P. L. Lindner, U.S. Pat. 2,284,187 (1962).

117. P. L. Lindner, U.S. Pat. 3,408,174 (1962).

118. A. Stefcik and F. E. Woodward, U.S. Pat. 3,317,305 (1963).

119. McCutcheon's Detergent and Emulsifiers 1968 annual, J. W. McCutcheon, Inc., Morristown, N.J., 1968.

120. K. L. Lynch, U.S. Pat. 2,438,092 (1948).

121. A. O. Jaeger, U.S. Pat. 2,028,091 (1936).

122. S. E. Ainsworth, Natl. Fert. Sol. Assn. Fert. Round-Up Proc., July, 1967.

Chapter 5

FORMULATION OF PESTICIDAL DUSTS, WETTABLE
POWDERS AND GRANULES

James A. Polon

Group Leader - Industrial Products Research
Minerals & Chemicals Division,
Engelhard Minerals & Chemicals Corporation,
Menlo Park, Edison, New Jersey

I. INTRODUCTION

In its strictest sense, the term "to formulate" denotes the act of expressing in precise form. In the chemical field, formulating is the act of
finding the proper combination of ingredients of a product and expressing
their concentrations or proportions mathematically in a formula. In the
agricultural pesticide field, the meaning of the word has been extended to
the actual manufacture of pesticidal products. Thus, there are "formulators" who combine pesticides with carriers, diluents, and surface-active
agents, according to a predetermined formula, to produce a final consumer
product.

This chapter is concerned primarily with the art of laboratory formulation of pesticidal dusts, wettable powders, and granules, but this treatise
could not be considered complete without a discussion of the methods and
equipment required for manufacture of the final products. Laboratory formulation is still an art, but certain principles concerning desired product
properties, and formulating methods to attain these properties are slowly
being evolved. Attempts will be made in this chapter to review some of
these principles and to discuss other methods of formulation and manufacture, though not as yet commercially practiced, which are potentially of
great value in this industry.

The prime purpose of formulation is the dilution of the high concentration pesticide down to a level at which it will be toxic to the pest, but will
not cause damage to desirable plants, fish, birds, animals, and natural
habitat. The forms which the products take upon dilution include dust concentrates, field-strength dusts, granular products, wettable powder concentrates, water-soluble powders and liquids, emulsifiable concentrates,
flowable emulsions, and flowable suspension concentrates.

The choice of the form of the pesticidal product depends primarily on
the purposes for which the pesticide is used. Will it be an insecticide,
herbicide, fungicide, nematocide, acaricide, molluscicide, algaecide,
rodenticide, attractant, repellant, or plant-growth regulator? Will the
product be applied to the pest by fogging, spraying, or dusting? Will it
be applied by airplane, applied on or incorporated into the soil, or applied
to ponds or waterways?

Probably the second-most important factor governing the choice of
product form is the pest involved. Generally, many different forms of
pesticides can be used to treat a single type of infestation, e.g., both
liquid concentrates after dilution, and granules can be incorporated into
the soil for soil insects. But one form may be found to be superior to
another type because of better control of a particular pest.

Another factor in choosing the form of the product is the properties of the concentrated chemical. Insolubility of the pesticide in low-cost, non-phytotoxic solvents can exclude formulation as emulsifiable concentrates, or even perhaps granular products; low melting point of the pesticide concentrate may present problems in manufacture of wettable powders.

Local weather conditions can also dictate the type of material to be used. For instance, where high wind and drift conditions may endanger nearby crops, the heavier granular products would be superior to dusts, wettable powder concentrates, or emulsifiable concentrates. Certainly the choice of the pesticide form is dictated by the ability to use and availability of the necessary equipment to apply the product.

There are many other factors governing the choice of the form of the product, too extensive to mention here. However some mention must be made of the economics involved in one form over another. Obviously, if there are three products of the same form being used in the same area on the same pest, a fourth product of a different and more expensive form cannot be competitive.

II. CARRIERS

By definition, an inert ingredient is one without active properties, primarily chemical. It is generally agreed that the inert materials used in the manufacture of dry pesticidal products do not have any inherent pesticidal activity in the same sense as the organic poisons, but they are far from inert. It has been found that some carriers and diluents react chemically with the impregnated pesticide, decomposing it and rendering it pesticidally inactive. This carrier/pesticide incompatibility problem will be discussed later. It has also slowly been determined over the last twenty years that the carrier can indeed play an active physical role in the product. The carrier can impart certain desirable physical properties to the product, making the choice of carrier extremely important. In view of the importance of carrier properties in formulating, we are obliged to completely review all available types of carriers and their general properties.

The term "carrier" is generally used to denote the inert ingredient used to dilute the pesticide. However, some restrict its use to those inert materials with high absorptivities, whereas the term "diluent" has been used with materials having low or medium sorptive capacities.

A. Classification of Carriers

Inert materials can be divided into two general categories: inorganic minerals and botanicals. Probably the only attempt to completely list the sources and properties of inert materials was made by Watkins and Norton (1) in 1947 and revised by Weidhaas and Brann (2) in 1955. Since then, the number of carriers available in the market has increased and, more important, a granular inert market has developed in the last 15 years. Table 1 gives a classification of the types of inerts available today.

Inert materials can also be divided into two other classes: dusts and granules. The use of dust diluents started in the late thirties as simple diluents for inorganic pesticides. With the advent of the organic pesticides, more sorptive carriers were necessary and the use of various fullers earths were initiated. This was the real start of the pesticide carrier industry. More-refined clay carriers were soon tailor-made specifically for the pesticide industry. Today, a multitude of dust inerts is available from a variety of origins (both mineral and botanical), properties, and price ranges.

The use of granular pesticide products developed in the late forties because of a need to overcome some application difficulties. Dust products presented drift problems causing damage to adjacent crops. Control of corn borer required a carrier that would enter the whorls of the corn plant. Application of pesticides to the surface of the soil for control of insects and weeds required discrete particles that would penetrate dense ground foliage. The granular carrier market has grown rapidly, especially in the last five years, to a market now approaching 150,000 tons per year, whereas the use of dusts has declined. The most popular mineral granule is made from the clay mineral attapulgite, followed by montmorillonite and bentonite clays, granular diatomaceous earths, and finally vermiculite (which is used almost exclusively for the home and garden market). The most important botanical carriers are granular corn cobs, followed by granular walnut shells. Appendix A lists some of the more important carriers along with their suppliers, and similar information is given in Appendix B for diluents.

B. Properties of Carriers

The properties of inerts are important since they can impart or alter the characteristics and performance of the final pesticidal product. The choice of the inert should be governed by a combination of (1) the desired properties of the pesticidal product to be formulated, (2) the properties of the inerts, and (3) the economics involved in use of a particular inert.

TABLE 1

Classification of Inert Materials

I. **Minerals**

A. Elements
    1. Sulfur

B. Silicates
    1. Clays
        a. Palygorskite group
            (1) attapulgite
            (2) sepiolite
            (3) palygorskite
        b. Kaolinite group
            (1) anauxite
            (2) dickite
            (3) kaolinite
            (4) nacrite
        c. Montmorillonite group
            (1) beidellite
            (2) montmorillonite
            (3) nontronite
            (4) saponite
        d. Illite group
            (1) mica
            (2) vermiculite
    2. Pyrophyllites
    3. Talcs

C. Carbonates
    1. Calcite
    2. Dolomite

D. Sulfates
    1. Gypsum

E. Oxides
    1. Calcium
        a. Calcium lime
        b. Magnesium lime

    2. Silicon
        a. Diatomite
        b. Tripolite

F. Phosphates
    1. Apatite

G. Indeterminate
    1. Pumice

II. **Botanicals**

A. Citrus pulp

B. Corn cob

C. Ground grains

D. Rice hulls

E. Soybean

F. Tobacco

G. Walnut shell

H. Wood

III. **Synthetics**

A. Inorganic
    1. Precipitated hydrated calcium silicate
    2. Precipitated calcium carbonate
    3. Precipitated hydrated silicon dioxide

B. Organic

In this section, attempts are made to list the properties of inerts, in general, to discuss their significance, and to describe how to characterize them by various tests. Since some of the test methods and ranges of characteristics are different for powder inerts and granular inerts, the discussion of the properties of each type will be separated.

1. Properties of Dusts

a. Sorptive Capacity. Probably the most important characteristic of any inert is its sorptive capacity. The prime purpose of use of any carrier or diluent is to spread a pesticide concentrate over a large number of particles that can be applied uniformly in the field over large areas. Therefore, an inert must not only possess sufficient sorptive capacity to initially absorb the toxicant during production, but it must also have the ability to keep the product in a free-flowing condition throughout storage and during use.

Sorption of liquids by inerts takes place by two distinct actions: (1) adsorption which is the pickup of liquids on the surface (either external or internal) by physical or chemical attraction, and (2) absorption which is the soaking up of a liquid into the internal pores of the inert (similar to the pores of a sponge). It is indeed difficult to separate these two different modes of sorption, and it is generally agreed that adsorption takes place first, followed quickly by absorption, which accounts for the predominant amount of sorptivity. In the case of inerts, the sorptive capacity is related to the pore volume and pore size distribution of the inert. Even finely powdered inerts have an internal pore system. A given inert can absorb a specific volume of liquid, and this volume would be constant for all liquids if their properties such as viscosity, molecular size, and wetting characteristics were the same. Density is a prime factor when a comparison of the sorptivity of a given inert for various liquids is made on a weight basis.

Probably the most widely used test to determine the sorptive capacity of dust inerts is the ASTM Rubout Method (3). In this test linseed oil is added dropwise from a buret onto a sample of inert being worked constantly with a spatula until the powdered inert becomes plastic and can be curled with the tip of the spatula. Another similar test is the Gardner-Coleman Sorptivity Test for Pigments and Extenders (4, 5). In this test the oil is again added from a buret to the inert in an evaporating dish being mixed with a stirring rod until the inert becomes wet and can be rolled into a ball with the stirring rod. Both of these tests yield data which are in excess of the true sorptive capacity of the clay, since wet endpoints are obtained. Furthermore, the practical sorptivity capacity limit, when considering impregnation of toxicants, should be that point at which dry flowability of the

product starts decreasing. For high-sorptive carriers this point may be
as high as 90% of the true sorptivity, whereas with low-sorptive diluents
it may be as low as 50% of the true sorptivity.

There are three methods available in this industry that yield practical
pesticide sorptive capacities of powdered inerts. However, all three of
these methods use special instruments that are not readily available. The
first, devised and used by the Floridin Company (6), is described as an
instrument with a long-stemmed metal funnel rigidly supported over a steel
disk rotating at 900 rpm. The clearance between the spout of the funnel and
the disk can be adjusted to a predetermined rate of flow using a "standard"
dust. The second method uses an apparatus devised by Malina and The
Velsicol Chemical Co. (7); it consists of a metal feed funnel fitted with a
60-mesh screen and a receiver, both attached to a variable frequency vi-
brator which can be adjusted to give a standard rate of flow using a standard
dust. The third method operating on a similar principle is used by Engelhard Minerals and Chemicals Corporation (8). This apparatus has a motor-driven vibrating funnel whose spout height can be adjusted over a screen
which also vibrates. In all three methods the pesticide sorptive capacity of
a dust inert can easily be found by first preparing a series of samples containing increasing concentrations of a toxicant, and then determining the
rate of flow of each impregnated sample through the apparatus. The pesticide sorptive capacity is considered to be that point at which flowability
breaks off. Comparisons of the sorptive capacities of various carriers
and diluents can be made using a single oil or solvent.

   b. Flowability. The flowability of an impregnated pesticidal product
is dependent upon the inherent flow characteristics of the inert. While the
flow characteristics of impregnated products can be damaged if the sorptive capacity of the inert is exceeded, a properly formulated product should
have the same flowability as the inert. The flow properties of inerts have
not been systematically studied and catalogued, but this property is believed to be related not only to the fineness and particle size distribution
of the inert, but more fundamentally to the particle shape. It is well known
that the mineral inerts vary in their crystalline shape. For instance, (1)
attapulgite has needle-like crystals, (2) talc has both fibrous and platy crystals, (3) pyrophyllites are thick platelets, (4) kaolins are thin platelets,
and (5) diatomaceous earths have very irregularly shaped and sized particles.

   Malina (7) presented data on the flowability indices of each of these five
different inerts as follows: attapulgite 100, talc 112, pyrophyllite 89, kaolin 12, and diatomaceous earth 8. Following an adjustment for differences
in density, the flow index of each inert was calculated with respect to attapulgite which was assigned a flow index of 100. The talcs rapidly loose

flowability when impregnated. Flowability of the inert is important for
ease of handling and storage as a raw material, flow of materials through
blenders, minimizing clogging in the grinding mills, and packaging of the
final product. Product flowability is also important during application,
e.g., flow out of bags, field applicators, shaker canisters, etc.

   c. Particle Size Distribution. The particle size distribution of inerts
is not only important during production of dusts and wettable powder concentrates, but also to the performance of products in the field. While
finely divided particles are desirable, a narrow particle size distribution is also important. Wide distributions can result in products containing particles impregnated with varying concentrations of toxicant giving
nonuniform control after application. Wide distribution products are also
subject to separation due to differences in particle size and density. Most
of the dust carriers available in todays market are at least 85% finer than
a 325-mesh screen (U.S. Standard Sieve Classification).

   The methods used for determining particle size distribution by screening through sieves are described by T. D. Oulton (9). A Ro-Tap Testing
Sieve Shaker available from W. S. Tyler Co., Cleveland, Ohio, is widely
used to shake the nest of screens during the test. However, very erratic
results can be obtained in determining the weights of materials held on
200-, 230-, 270-, and 325-mesh screens, since these fine screens tend
to "blind" or plug. The problems involved are described by Whitby in an
ASTM Symposium (10). A new sieving apparatus called Allen-Bradley
Sonic Sifter made by Allen-Bradley Co., Milwaukee, Wisconsin operates
on an air pulsing principle which minimizes this problem. Determination
of mesh distribution by wet methods are more reliable, but care must be
taken not to fracture the particles during preparation of the dispersion
prior to the wet screening. Oulton (11) also describes the proper method
for wet sieve analysis.

   For the subsieve size range there are four basic methods: (1) water
sedimentation, (2) air elutriation, (3) permeametry, and (4) microscopic
count. The water sedimentation methods which are based on Stoke's Law
include the Andreason pipet method (12-14), Casagrande hydrometer method (15, 16), and a sedimentation decanting/separation method described by
Lesveaux et al. (17). The Andreason and Casagrande methods are used
widely for determination of particle size distributions of unimpregnated
inerts, especially the more refined materials; however, they are not
easily adaptable to impregnated products because of the wide variation
in density of the different particle sizes of a particular sample. The
Lesveaux method, while still subject to the same factors, more realisti-
cally simulates the sedimentation of a wettable powder under field conditions, and uses a microscopic examination to measure the separated particles.

There are many air elutriation methods based on the principles of Roller (18). The biggest source of error in employing one of these methods is the difficulty in determining the proper particle density, especially the densities of particles that should be conditioned with the same relative humidity air actually used in the test.

An instrument that is used in the pesticide industry for evaluation of wettable powder concentrates is the Fisher Sub-Sieve Sizer. This instrument is based on the same principle of permeametry as one earlier developed by Gooden and Smith (19). Measuring the flow rate of gas through a powder bed under a controlled pressure differential, the Fisher instrument yields the average particle size of a powder directly. This instrument can only be used on dry powders, thus its value for determining the average particle size of wettable powder concentrates is doubtful, since data on wettable powders should be obtained after the product is dispersed in water. This instrument does find use, however, as a production quality control unit during the manufacture of wettable powder concentrates.

The method yielding the most reliable results in determining the particle size distributions of inerts and dust and wettable powder products, is microscopic examination. Loveland (20) and Irani and Callis (21) describe the procedures involved in making a microscopic count. The method can be divided into three steps: (1) slide preparation, (2) particle observation, and (3) actual counting and sizing of individual particles. The preparation of a slide for an examination is very important. Inerts and dusts can be examined dry, or dispersed in almost any liquid that allows sufficient difference in refractive index between the particle and the vehicle. Wettable powder concentrates can also be examined dry as an indication of extent of milling. However, it is best to examine wettable powders dispersed in water as they normally would be in field use to determine the effectiveness of wetting and/or dispersing agents used in the formulation. The concentration required on a slide is that which gives enough particles to count in a field without being too crowded. The minimum number of particles to be measured should be that number which is determined not to change the results significantly by counting additional particles; usually more than 400 particles must be measured. Use of a calibrated stage micrometer to determine magnification is mandatory. In addition to the visual methods commonly used, images can also be photographed or projected on a screen for counting and measuring, these additional methods being described by Loveland (22).

d. Bulk Density. The degree of coverage of the pesticide over an area, wind drift, penetration of foliage, sinking in water, ease of handling through production and application equipment, cost of packaging and shipping, etc., are all factors that are affected by the bulk density of inert. For many commercial pesticides, the bulk densities provided by the attapulgites and montmorillonites within the 30-40 lb/ft^3 range are ideal. Products made from the lighter inerts are subject to wind scattering and drift, and poor placement of the impregnated particles at the site where the pest is to be controlled. Inerts such as the vermiculites are not used in the commercial pesticide market because of difficulties in wide-scale application, but do enjoy a home-owners' pesticide market where light-weight product bags have appeal. When inerts heavier than 40 lb/ft^3 are used, the products do not have good coverage, and difficulties in metering through application equipment are encountered.

The density of a carrier or diluent is expressed as true density, apparent or particle density, or bulk density, often indiscriminately. Oulton (23) describes these three densities as follows: (1) true density, the weight per unit volume of the solid part of the inert structure; (2) apparent or particle density, the weight per unit volume of the solid part of the inert and its pores; and (3) bulk density, the weight per unit volume of the solid part of the inert, its pores, and the void volume between individual particles.

The bulk density is of primary interest with respect to properties of inerts and pesticidal products. There are basically two types of bulk densities: (1) loose-packed bulk density is the weight per unit volume of an inert determined under reproducible conditions of free fall, and (2) tightly-packed bulk density (more commonly known as tamped volume weight) is the weight per unit volume of an inert when packed or tamped to a minimum volume. An instrument used for determination of the loose-packed density is the Scott Volumeter No. 13-355 sold by the Fisher Scientific Co., Pittsburgh, Pa. The method described by Gardner (24) consists of fluffing a sample of the material through a coarse mesh screen and allowing it to pass over several glass baffles into a 1-in.^{3} box and then weighing it. The tamped volume weight is determined by placing 50 g of material in a 100-ml graduated cylinder and recording the volume after some period of controlled tamping. A machine developed by Engelhard Minerals & Chemicals Corporation (25) to obtain reproducible and controlled tamping is described as a motor-driven, cam-operated platform which repeatedly lifts and drops the material, contained in a graduated cylinder, a fixed distance at a definite rate and for a measured period of time. The platform is dropped 1/16 in. at a rate of 250 times a minute for precisely five minutes.

2.   Properties of Granular Inerts

A "granular" inert is defined as a carrier or diluent that is limited to a size range of 4-mesh to 80-mesh (U.S. Standard Sieve Series) (26). (Inerts with finer particle sizes than 80-mesh are considered to be dusts.)

For any given granular material (e.g., a product labeled 16/30 mesh) at least 90% by weight of the product must fall within the designated mesh range, and the remaining 10% maximum may be distributed on either end of the designated mesh range. Because of the large difference in size between the dust and granular inerts, one might expect that they would have some different properties and might also require different methods of testing.

a. Sorptive Capacity. There are very few methods available to determine the sorptive capacities of granular inerts. One which was developed by Polon (27) is based on the same principles as the methods used by Floridin (6), Velsicol (7), and Engelhard (8) for the sorptive capacities of dust inerts. The method is simple in that it uses two 60° glass powder funnels of 500-ml capacity, one mounted on a ring stand 1 in. directly over the second funnel. The procedure is as follows:

(1) Measure out 400 ml of granules, tamped in a 500-ml graduated cylinder.

(2) Place a finger over the spout of the upper funnel and pour the granules in the upper funnel.

(3) Place a finger over the spout of the lower funnel and allow the granules to flow from the upper funnel into the lower funnel.

(4) Using a stop watch, time the flow of granules out of the lower funnel into a receiving beaker.

Granular inerts can be impregnated stepwise with an oil, solvent, or pesticide solution, measuring the flowability after each impregnation. The pesticide sorptive capacity is considered to be that point at which the flowability starts dropping off.

Another method developed by Polon (28) that determines the saturated sorptive capacity of granular inerts deserves mention because of its high degree of reproducibility. The method is described as follows:

(1) Weigh 12 g of granular inert into a tared Gooch crucible fitted inside with a disk of 80-mesh screen.

(2) Insert the crucible into an adapter in a vacuum flask connected to a vacuum source, but do not turn on the vacuum.

(3) Pour in an excess of mineral oil over the granules and keep the level in excess for 20 min.

(4) Turn on the vacuum and filter off the excess oil for 5 min.

(5) Remove crucible and place in special adapter in centrifuge cup, and centrifuge for 5 min at 2000 rpm.

(6) Remove crucible, wipe outside dry with paper towel, and weigh to determine oil retained.

The details of the procedure and equipment required for this test can be obtained from the author (28). This method has value when comparing the total saturated sorptive capacities of a number of inerts. However, a pesticide solution can be substituted for the mineral oil if care is taken against exposure of the laboratory personnel to the toxicant.

b. Flowability. The flowability of granular inerts and products is important for the same reasons described under the discussion of flowability of dusts. However, very little work has been conducted to develop testing methods for this property. The test method, described above by Polon (27) for sorptive capacity, can also be used to determine not only the flowability of inerts but also that of impregnated granular products.

c. Particle Size Distribution. The test for particle size distribution of granular products is the same as that described by Oulton (9), but different weights of material and times of shaking are used depending on the range of sizes of the granular or dust products. Sieves used in all of the screening tests should conform to ASTM Specification E-11 (29) and be designated according to the U.S. Standard Sieve Classification. The NACA Granular Pesticide Committee has recommended that all inerts and pesticidal products be described using the U.S. Sieve Series.

Granular carriers today come in a variety of mesh size ranges, such as 8/16, 8/20, 16/30, 18/35, 20/40, 25/50, 30/60, 16/60, etc. The first number indicates the screen number with the large holes through which most of the material will pass, and the second number indicates the screen number with the small holes on which most of the material will be retained. As stated previously, the NACA Granular Pesticide Committee has recommended that at least 90% of the material lie between the two designated screen sizes.

Often it occurs that two inerts, although labeled by the same screen sizes, can have substantially different distributions within the same screen limits. For example, two inerts may be labeled 30/60 mesh, but may have different distributions within the 30 to 60 mesh range. One may have a majority of granules in the 30 to 40 range, while the other may be preponderantly 50 to 60 mesh. Neither one of these materials follow the idealized bell-shaped distribution curves. The first product with the distribution to the coarser sizes is said to be negatively skewed, while the second one with the distribution to the finer sizes is said to be positively skewed. This nonideal distribution is important to the formulator since some breakdown in particle size occurs during production, and use of a positively-skewed inert can result in a granular product finer than the desired range. If this occurs the formulator should start with a coarser inert from the same supplier, or shift to a different supplier. In general, the coarser products available in the market, such as 8/16 and 16/30, are used for the production of herbicides for aquatic weed control; the medium size products such as 18/35 and 20/40 are used extensively for soil incorporated insecticides; and 25/50 and 30/60 meshes are used for herbicides and insecticides applied above the ground.

d. Bulk Density. As with the dusts, the bulk densities of granular products can be measured in two ways: (1) loose-packed bulk density (fall volume weight) and (2) tightly-packed bulk density (tamped volume weight). The apparatus described by Oulton (9) for fall volume weight determination consists of a 250-ml graduated cylinder cut off square at the 100-ml mark, a 1/2-in. inside-diameter glass tube 24 in. long, and a 1-pint funnel set up so the funnel feeds into the top of the glass tube with the bottom of the tube centered exactly 1 in. above the top of the cut-off cylinder. Approximately 125 ml of material to be tested is poured into the top of the funnel. The excess sample is carefully cut off from the top of the cylinder with a straight edge such as a spatula. The material in the cylinder is weighed and the fall volume weight calculated.

The tamped volume weight for granules is determined in a similar manner to the test described above for dusts, except the amount of material and tamping periods are different. This method and apparatus for tamping is also described by Engelhard (25).

It should be remembered that impregnation of inerts with pesticides increases the bulk density over that of the inert. This can readily be understood by examining what happens when granules are impregnated as indicated in Table 2. All of the liquids are absorbed into the internal pores of the granules. The weight of the product is increased without adding any volume, thus the bulk density of the product is higher than that of the inert. A similar action takes place when toxicants in liquid form are impregnated onto dust inerts. However, when solid toxicants and dust inerts are ground together, the solid toxicant not only adds weight, but also adds volume to the product.

e. Particles/Pound. The ability of a granular inert to spread the toxicant evenly and effectively over the area to be controlled is related to the

TABLE 2

Increase in Density When Formulating 20% Aldrin Granules

	State	Weight %	Weight (lb)	Volume added to product
60% Aldrin solution in heavy aromatic naphtha	Liquid	33.67	18.1	--
50% Urea solution in water	Liquid	1.33	0.4	--
Attapulgite granules, AA RVM type, 30/60 mesh	Solid	65.00	35.0	1.0
	Totals	100.00%	53.5 lb	1.0 ft^3

number of distinct impregnated particles per unit weight. The number of particles per pound is related directly to the bulk density and mesh distribution of the product. Thus, the bulk density and mesh distribution of unimpregnated inerts must be carefully considered when formulating granular pesticidal products. Table 3 shows data on the particles per pound contained in various grades of attapulgite carriers supplied by Engelhard (30). The method of particle count developed by Engelhard is described by Gwyn (31).

f. Water Disintegrability. The water disintegrability of granules is important to a degree to the release of certain toxicants from granules. The mechanisms of granular toxicant release include: (1) granular disintegration, (2) dissolving of toxicant in water, (3) displacement of toxicant by water, (4) volatilization of toxicant, and (5) combinations of two or more of the preceding. The rate of disintegrability of granular inerts can be controlled to a slight degree during manufacture of some of the mineral types, such as the clay minerals.

During manufacture, the mined material is dried or calcined depending on the product properties desired. These grades are usually labeled as "dried" and "calcined," or RVM and LVM, respectively. The latter designations are used by the attapulgite manufacturers and denote the amount of

TABLE 3

Particles/Pound/Square Foot for Attapulgite Granules

U.S. Mesh	Number of granules/lb		Number of particles/ft^3 [a]	
	RVM	LVM	RVM	LVM
8/16	145,000	166,000	3.3	3.8
16/30	1,150,000	1,300,000	26.4	29.9
18/35	2,708,000	3,244,000	62.2	74.5
25/50	7,722,000	8,672,000	177.5	199.3
30/60	12,300,000	13,600,000	282.7	312.5

[a]Based on application rate of 1 lb/acre.

volatile matter remaining in the clay, i.e., regular (or intermediate) volatile matter (dried), and low volatile matter (calcined). The RVM materials tend to disintegrate readily while the LVM materials resist disintegration in contact with water.

Attapulgite granules are manufactured by two basic methods: (1) raw clay is dried or calcined followed by grinding and screening to size, and (2) raw clay is pugged, extruded, dried or calcined, followed by grinding and screening to size. Extrusion increases surface area, porosity, sorptive capacity, and granular slaking in water. Non-extruded granular attapulgite products are designated as "A," and extruded products as "AA." Thus, there are four different grades available: AA RVM, A RVM, AA LVM, and A LVM, decreasing in water disintegrability in the order shown. The degree of water disintegrability differs greatly between the RVM and LVM grades, while there is only a slight difference between A and AA grades. Since AA RVM and A LVM represent extremes in water disintegrability, these two grades are the most commonly used.

The most suitable test for water disintegrability is one developed by the United States Department of Agriculture workers in testing and writing federal specifications for granular insecticides for fire ant control (32) and adapted by Engelhard (33) for the testing of granular inerts. This test consists of inverting a 12-g sample of granular inert in 100 ml of distilled water in a 100-ml, glass-stoppered, graduated cylinder, end over end at a rate of 30 inversions per min for a total of 3600 revolutions. The contents are then washed through a 60-mesh U.S. Series Sieve with water, then the residue on the screen is collected, dried, and finally weighed. The water disintegrability is reported as the weight % breakdown through the 60-mesh screen. In general, the AA RVM attapulgite granules exhibit 85-90% breakdown, whereas A LVM granules disintegrate only 10-15%. The difference between "A" (non-extruded) and "AA" (extruded) granules is usually less than 5% breakdown.

The above test can only be used as a guide for comparison of granular inerts or for various granular formulations of a particular toxicant. The disintegration of granular inerts is slightly altered and influenced by the type and concentration of impregnated toxicant, the quantity and type of impregnation solvent, and other additives such as deactivators and surfactants. The relationship between data obtained by the above disintegration tests and actual breakdown results occurring under field conditions is not known. There has been very little investigation to date of the influence of such factors as rainfall and soil moisture content on the granular disintegration and its effect on efficiency of control by the toxicant.

g. Dry Hardness. The property of the inert to resist attrition during all of the production, shipment, and handling steps by the inert manufacturer, formulator, and even the final user is called dry hardness. Inert production costs restrict the manufacture of granules with narrow size distributions, and products usually contain a small amount of fines that are difficult to remove. During shipment to the formulator and placement in his storage facilities, a soft carrier could break down to form additional fines. Furthermore, and perhaps more important, attrition can take place during impregnation, especially if poor equipment or poor methods are used. Fortunately, during most impregnations, loose fines are picked up and plastered to the larger granules by the toxicant spray. However, it is still important that the product be screened to scalp out both fines and larger agglomerates, if a high quality product is desired. Excessive handling during shipment to the customer can also cause attrition; therefore, the formulator should prudently choose and use a fines limiting screen to compensate for this possible source of attrition. Equipment has been well designed for the application of granular pesticides, and attrition from this source is minimal.

Usually the hardness of the mineral carriers is somewhat related to the water disintegrability described above; that is, for the attapulgites, "A"

grades are harder than "AA," and LVM grades are harder than RVM. However, with wise formulating practices, both RVM and LVM grades can be
used without encountering excessive attrition.

The number of tests for checking the hardness of granular inerts is
large, but probably the test that has had the widest distribution and use is
one described by Engelhard (34). The test consists of prescreening a test
sample to remove fines, subjecting a weighed amount of the sample to the
action of ten 5/8-in. diameter steel balls in the pan from a nest of 8-in.
screens being gyrated on a Ro-Tap for 30 min, rescreening the fractured
sample using the same size screen used in the prescreening test, and finally weighing the residue remaining. The data is calculated as "% hardness". For LVM attapulgite products approximately 65% hardness is obtained, while approximately 35% hardness is obtained for RVM grades. In
this test, granular products are subjected to severe attrition to obtain
measurable breakdown; it is not likely that attrition encountered in the
field would be this drastic.

h. Dustiness. Another property of granular inerts whose significance
has not been thoroughly defined is dustiness. Fines outside the limiting
screen, e.g., material finer than 60 mesh in a 30/60 grade, include the
dust. Dust can be defined as the cloud that remains airborne when a granular material is dropped from a height. Exposure of any impregnated dust
to applicators must be avoided, especially with the newer generations of
pesticides which have higher mammalian toxicities.

The only practical test known to this author is one described in ASTM
D-547-41, "Dustiness of Coal" (35) and adapted by Engelhard (36) for evaluation of granular inerts and pesticidal products. The test utilizes a "dropping chamber" measuring 8 1/8 in. square and 32 in. long. The test sample is released from a platform on the top of the chamber and permitted to
fall through the chamber into a drawer receptacle at the bottom. Five sec-
onds after the sample is dropped, a polished chromium-plated dust collector tray is inserted into the column 25 in. below the point of release so
as to intercept and retain the falling dust particles that may be above that
point at the time of insertion. Thus the dust generated by the fall and impact at the base of the chamber can be collected and weighed.

3. Carrier/Pesticide Incompatibility

One of the most important properties to be considered when choosing an
inert for formulating dry pesticidal products is the compatibility of the
toxicant with the inert. Many of the toxicants available today are inherently
unstable to a degree when exposed to long term storage at some elevated

temperature that may normally be encountered in the field. Also many pesticidal chemicals are hydrolytically unstable to some degree, or have some tendency to suffer catalytic decomposition or degradation. Even if the inert were completely inactive chemically, merely spreading the toxicant over the large surface area of the inert could increase the rate of inherent degradation of the toxicant. But it has also been found that many inerts, especially the more sorptive carriers, have acidic, basic, and/or catalytically active sites on their surfaces, and can carry appreciable quantities of absorbed water. Many attempts have been made to study the roles of some of these carrier properties in the degradation of toxicants, but these studies have only revealed that carrier/pesticide incompatibility is very variable and complex. Thus, it would be impossible to discuss all the proposed theories concerning the causes of toxicant degradation in this chapter. An attempt will be made, however, to briefly discuss some of the more popular theories and the methods used today to avoid decomposition.

a. Chlorinated Hydrocarbons. The problem of the decomposition of certain chlorinated-hydrocarbon insecticides when formulated with mineral carriers was recognized as early as 1950. This decomposition was manifested when the pesticidal products were examined for toxicant content after a storage period, using either a chemical analysis or some type of biological assay. It was soon found that pesticides such as DDT, aldrin, dieldrin, endrin, chlordane, heptachlor, and toxaphene decomposed to varying degrees when impregnated on some inerts. It was also determined that merely measuring the pH of boiled water suspensions or extracts of inerts was not a true indication of the acidity or basicity.

D. S. McKinney et al. (37) discusses pH and extension of the acidity scale. Hammett and his co-workers (38-41) developed and reported the use of a series of pK_a indicators as early as 1930, and Walling (42) in 1950 described a method using these indicators to measure the surface acidity of a mineral carrier. A small sample (0.5 g) of the carrier is placed in a test tube containing 2 ml of indicator solution, and the appearance of the acid conjugate color indicates a surface acidity higher than the ionization constant of the indicator. Johnson (43) described a titration method by which the surface acidity could be quantitatively determined. Table 4 contains a list of indicators used in evaluation of the surface acidities of carriers and diluents.

The measurable acid sites on the mineral surfaces have been correlated with the decomposition of certain chlorinated organic insecticides by Benesi et al. (44), Fowkes et al. (45), Malina et al. (46), and Trademan et al. (47-49), and these workers have suggested "deactivators" to prevent

TABLE 4

Properties of Hammett Indicators

Eastman Cat. No.	Indicators[a]	pK_a	Basic color	Acid color
304	Anthraquinone	-8.2	Colorless	Yellow
1254	Chalcone	-5.6	Colorless	Yellow
2926	1, 9-Diphenyl-1, 3, 6, 8-nonatetraen-5-one	-3.0	Yellow	Red
3651	4, 4', 4"-Methylidynetris (N, N-dimethylaniline)	+0.8	Yellow	Blue
2494	N, N-Dimethyl-p-1-naphthyl-azoaniline	+1.2	Yellow	Blue
1714	4-Phenylazodiphenylamine	+1.5	Yellow	Purple
1210	4-O-tolylazo-o-toluidine	+2.0	Yellow	Red
1375	p-Phenylazoaniline	+2.8	Yellow	Orange
338	N, N-Dimethyl-p-phenyl-azoaniline	+3.3	Yellow	Red
5548	4-Phenylazo-1-naphthylamine	+4.3	Yellow	Red

[a]Indicator solutions are made up with 10 mg of indicator per 100 ml water-free benzene.

the decomposition. Organic chemicals such as ethylene glycol, diethylene glycol, triethylene glycol, propylene glycol, diacetone alcohol, isopropanol, urea, hexamethylenetetramine (HMT), monoethanolamine, and combinations of the glycols with sodium hydroxide have been found to have some value in "neutralizing" the active sites on carrier surfaces. The concentration of deactivator required to adequately stabilize a product depends on (1) the type and concentration of toxicant, (2) the deactivator to be used, and (3) the type of carrier used. Studies have been conducted to determine the extent of neutralization of active sites required for good stability with

a particular insecticide, and each manufacturer gives recommendations on
the concentration of deactivator to be used with a particular inert. The
"Handbook of Aldrin, Dieldrin, and Endrin Formulations" distributed by
Shell Chemical Co. (50) is typical of one such source of information on deactivator requirements. Shell reports that, for aldrin and dieldrin, most
carriers are deactivated with 1% urea (based on the weight of the carrier),
but that a few more active ones required as much as 3% urea; also for endrin, most carriers need less than 1% hexamethylenetetramine, but active
ones require as much as 5% HMT. Shell has determined the deactivator requirements by storing a series of field-strength dusts, made up with varying urea or HMT contents, at 130°F for two weeks, analyzing the stored
dusts for toxicant content, and ascertaining the minimum amount of deactivator required to minimize toxicant decomposition (51, 52).

A second source of deactivator requirements is Velsicol Chemical Co.,
for such products as chlordane, heptachlor, endrin, and Bandane. Velsicol
(16) takes the same approach as Shell in listing the deactivator requirements
for various carriers as determined by an accelerated, elevated-temperature
storage test, but also provides an additional test for the formulator based on
the surface acidity of the carrier. This test consists of placing a sample of
inert in the depression of a spot plate and adding a few drops of indicator
dye solution. If the dye turns color revealing catalytic activity, then further tests must be made by blending in one per cent increments of deactivator and testing in the same manner until no color changes can be observed in
the test plate. Table 5 contains information concerning the deactivator testing for the four toxicants mentioned above.

The surface activity for each type of mineral carrier varies, and variations have even been found for a given type of inert from different suppliers. The ranges of deactivator requirements for mineral types for the
four pesticides discussed above are given in Table 6. But Velsicol recommends testing of carriers by the individual formulators for each shipment received into the formulator's plant. It will be noted from the data
shown in Table 6 that the more sorptive carriers are more active and required a higher concentration of deactivator; diatomite is one exception to
this generality.

b. Organo-Phosphates. Much effort has been expended on the development of satisfactory dry formulations of phosphate-type insecticides since
the introduction of these toxicants. McPherson and Johnson (53) studied
the inherent instability of such toxicants as parathion, methyl parathion,
Chlorthion, and malathion, and reported data on rate of degradation of
these toxicants at various temperatures indicating poor stability. Yost,
Frederick, and Migrdichian (54) reported in 1955 the effect of carrier acid

TABLE 5

Surface Acidity Testing for Certain Pesticides

Pesticide	Type of deactivator	Desired pK$_a$ of carrier	Color indicator	Color change Base	Acid
Bandane	Deactivator H	+1.5	Benzene azodiphenyl- amine in isoctane	Yellow	Purple
Chlordane	Deactivator H	+1.5	Benzene azodiphenyl- amine in isoctane	Yellow	Purple
Endrin	Deactivator E or HMT	+3.3	Paradimethylamino azobenzene in isoctane	Yellow	Red
Heptachlor	Deactivator H	+3.3	Paradimethylamino- azobenzene in iso- octane	Yellow	Red

sites, basic sites, pH, and moisture content on the decomposition of malathion. These workers again reported data in 1959 on acceptable inerts and formulations of malathion (55).

Studies were conducted on the carrier properties that might cause decomposition of malathion, and the use of chemical deactivators to inhibit the decomposition by Polon and Sawyer (56–58). It was found that malathion was subject to alkaline hydrolysis and that alkaline materials such as urea, sodium hydroxide, hexamethylenetetramine, monoethanolamine, lime, sodium carbonate, and calcium carbonate, found to be effective for chlorinated hydrocarbon insecticides, can not be used for malathion. These workers proposed that basic sites on the surface of the clay caused the degradation and found that weak organic acids such as tall oil and rosin acids inhibited the degradation to a degree. Earlier, Trademan et al. (59) conducted work on the use of alcohols, ketones, glycols and ethers to stabilize dry compositions of parathion and methyl parathion.

Velsicol (60) suggests the use of a glycol-chemical Deactivator E, and recommends approximate concentrations for deactivating various carriers and diluents. Also Velsicol states that the threshold of stability for dry formulations of parathion and methyl parathion is between a pK$_a$ of 1 and 3.

TABLE 6

Deactivator Requirements for Certain Pesticides

Carrier or diluent type	Bandane Deacti- vator H[a]	Chlordane Deacti- vator H[a]	Endrin Deacti- vator E[a]	HMT[a]	Heptachlor Deacti- vator H[a]
Attapulgites	8-10%	6%	8-9%	5%	7-8%
Montmorillonites		15	8-12	4-9	
Kaolinites		2	3-5	1-2	3
Diatomites	0-2	1	1	1	3
Calcium carbonates	0-1	0	1	1	0
Talcs	0-1	0	1	1	0
Pyrophyllites	0-1	0	1	0.5	0

[a]Weight % based on weight of inert.

Thus, a formulator could determine the deactivator requirement for a particular carrier by use of the proper indicator solution (see Table 6).

Niagara (61) suggests the use of diethylene glycol, dipropylene glycol, or Deactivator E (Velsicol) for the stabilization of dry formulations of ethion. Diazinon is rapidly decomposed on all mineral carriers and has been formulated commercially on walnut shells. Schwint (62) has found a method of producing a low surface area granular attapulgite which suggests some promise as a compatible carrier for diazinon and other highly unstable phosphate pesticides. Oros et al. (63) have found that glycols are useful in stabilizing dry compositions of Thimet.

Phosdrin apparently is decomposed when in contact with carriers of high alkalinity and absorbed moisture. Shell (64) recommends the use of Pyrax ABB (pyrophyllite), Frianite M3X (volcanic dust), Emtal 43 (talc) and calcined gypsum as the only diluents that will give satisfactory toxicant stability under normal storage conditions. No deactivators are recommended by Shell.

Plapp and Casida (65) studied the degradation of ronnel and reported that this systemic insecticide was susceptible to alkaline hydrolysis at two sites on the molecule. Rosenfield and Van Valkenburg (66) studied the influence of various metallic ions and mineral pH's. They reported that nickel, iron, calcium, magnesium, aluminum, and zinc had little or no effect on the hydrolysis rate, while copper had a very pronounced effect on alkaline hydrolysis, and this was substantiated by Mortland and Raman (67). These workers tested various mineral carriers for pH and cation exchange capacity, and tried correlating these properties with degradation of ronnel. They reported that both acidic and basic pH minerals with cation exchange capacities over ten meq/100g caused decomposition, while those with CEC's of around six meq/100g gave fair stability. While the theories of these workers substantiated those proposed earlier by this author and co-workers (56-58), that basic sites on the mineral surface can contribute to decomposition of thiophosphates, the effects of basic materials either as the prime inerts, or as impurities are not considered.

c. Carbamates. Since their discovery in 1954, approximately 30 carbamates have been developed. Many of their chemical and biological properties have been discussed by Metcalf and Fukuto (68) in 1967. Probably the most well known carbamate is Sevin and formulation information on this insecticide has been published by Entley et al. (69), and by Union Carbide (70). It has been reported that decomposition of some carbamates is by alkaline hydrolysis, so that carriers and diluents with acidic pH's must be used if stable dry products are to be obtained. Fortunately not all carbamates suffer from degradation, and of those that do, the decomposition of Sevin is among the less serious. No deactivators have as yet been found for the carbamates.

d. Other Pesticides. There are a variety of pesticides that are incompatible with inerts, but only a few will be discussed here. Thiodan produced by Niagara (71) is slowly decomposed by the catalytic action of some clays and fillers. The kaolin type clays are especially active in this respect, while attapulgite clays are moderately active, and talcs and calcium carbonates are relatively inert. Niagara found that dipropylene glycol deactivates the carriers, enabling the formulation of stable dry products.

The first indications that Aramite was not stable with certain carriers became evident in 1953, and Smith, Gooden, and Taylor (72) reported in 1955 that their studies revealed that in a 24-hour test Aramite was highly effective with calcium carbonate, talc, or pyrophyllite in control of two-spotted spider mites, but that the efficiencies of preparations made with these carriers decreased when tested after one week. Naugatuck Chemical (73) studied the decomposition of Aramite on many carriers and diluents, and reported that materials such as most talcs, calcium carbonates, calcium silicates, pumices, and nut shell flour do not cause degradation, whereas diatomites, montmorillonites, attapulgites, pyrophyllites, and kaolins are catalytically active. Diluents from the kaolin group seem to be the most active. It was found that Maracarb N, a sodium lignosulfonate produced by the Marathion Division of American Can Co., could successfully deactivate active inerts when added in concentrations approximating 6% of the active carrier. It is believed by this author that this chemical deactivates by masking active sites on the surface of the inert. Urea can also be used at concentrations of 3-6% as a deactivator, and the role of neutralization of acid sites by this chemical is readily accepted.

III. DUST CONCENTRATES

Dust concentrates can be defined as a mixture of one or more toxicants with one or more carriers to form a dry, free-flowing powder containing from 10 to 50% total active ingredient. Dust concentrates are rarely used directly on the field, but are further diluted by blenders down to field strength.

A. Desired Properties

There are three properties that are important in dust concentrate products: (1) particle size distribution, (2) flowability, and (3) bulk density.

The particle size distribution of the product should be narrow to minimize settling, segregation, and layering of the different sizes, and also to minimize variation of toxicant concentrations on large and small particles. It is necessary for the average particle size of the product to be small enough to spread the toxicant evenly over the area to be controlled. Also, immediate toxicity has been found to be inversely proportional to particle size so that small particles are desirable. On the other hand, volatilization of toxicant will occur more rapidly with small particles, and very small particles are more apt to be blown off the contact control area. When considering all of these factors, it can be concluded that the choice of $40\,\mu$ average particle size with a range of perhaps 30 to $50\,\mu$ is desirable.

The flowability of the dust concentrate should not be any less than that of the original inert. Loss of flowability during impregnation indicates that the sorptive capacity of the inert is being exceeded, and the particles may tend to stick together and cake during storage prior to use. The importance of flowability of dust products (along with methods of testing) has been discussed earlier under the section on properties of dust inerts.

The bulk density properties have also been discussed above. It must be remembered that after impregnation of a toxicant, the bulk density of the product can increase considerably. The range of bulk densities of impregnated products can lie between 20 to 60 lb/ft^3, depending on the weight of the unimpregnated carrier, and the concentration and type of toxicants and solvents added. A product bulk density range of 30 to 50 lb/ft^3 seems to satisfy most commercial applications. Lower than 30 lb/ft^3 may be used for wettable powder concentrates or the home-owner market, whereas products heavier than 50 lb/ft^3 might be considered for application in water or from the airplane.

B. Methods of Preparation

The preparation of dust concentrates can be divided into three separate steps: (1) preblending, (2) pulverizing, and (3) postblending. These three steps are depicted in Fig. 1.

When developing a formula in the laboratory, there are some factors that must be considered before actually attempting to prepare a dust concentrate. The physical form of the toxicant must be first examined as follows: (1) For solid toxicants, is the toxicant easy to grind? If not, it may be necessary to dissolve it in a solvent. (2) For semisolid toxicants, could the toxicant be ground in the presence of a carrier? Could the toxicant be sprayed if heated? If not, it may be necessary to dissolve it in a solvent. (3) For liquid toxicants, is the viscosity of the toxicant low enough to be sprayed? If not, heating and/or dilution with a solvent may be necessary.

The next factor to consider is the solubility of the toxicant in impregnation solvents, if a solvent is necessary for impregnation. Normally, the type of solvent chosen must fulfill as many of the following requirements as possible: (1) high solvency for toxicant, (2) low volatility, (3) noninflammability, (4) high flash point if inflammable, (5) low phytotoxicity, (6) low viscosity, (7) high availability, and (8) low cost. Table 7 contains a list of solvents that have been used in formulation of pesticides. Normally, no attempt is made to recover the solvent, and it is assumed that the solvent will not volatilize from the product and is a part of the formula. Obviously, if a highly volatile solvent were used and left in the product, it would volatilize in storage and lower the package weight of the product, thereby increasing the percentage of toxicant in the product.

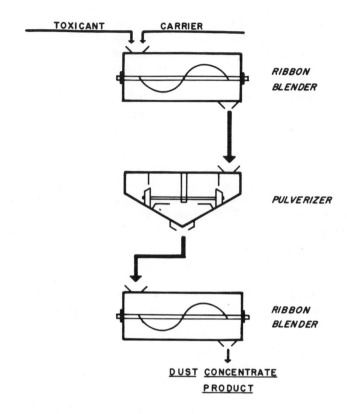

FIG. 1. Production of dust concentrates.

The next step is to determine if the desired concentration on the carrier can be attained. One chooses a particular carrier and determines its pesticide sorptive capacity following one of the methods described above under the section on carrier properties. If one of the special flowability/ sorptive capacity machines described earlier is not available, one can simply add increments of the toxicant (or toxicant/solvent system) to the carrier being mixed in a Waring Blendor, and visually examine each sample for loss of flowability. Once the approximate maximum concentration without loss of flowability has been determined, larger samples should be made. The Waring Blendor is an ideal piece of equipment for making samples of approximately 200 g. Solid toxicants or liquid solutions can be added using a syringe directly to the carrier. The blender will grind the solid toxicant or distribute the liquid homogeneously throughout the carrier. However, care must be taken not to overblend the mixture

TABLE 7

Solvents for Pesticidal Impregnations

Aliphatic Petroleum Solvents such as kerosenes, light refined mineral
      oils, range oils, and diesel oils.

Aromatic Petroleum Solvents

     a. Coal tar fractions such as xylene, toluene, and benzene.
     b. Light, medium and heavy aromatic naphthas (HAN).

Alcohols such as butanol, diethyl carbinol, ethanol, and isopropyl alcohol.

Cellosolves such as ethyl Cellosolve, and methyl Cellosolve.

Chlorinated Hydrocarbons such as carbon tetrachloride, chloroform,
      ethylene dichloride, methyl chloride, methylene chloride, trichloro-
      monofluoromethane.

Esters such as butyl acetate, dibutyl phthalate, di-2-ethylhexyl phthalate,
      ethyl acetate, and methyl acetate.

Ethers such as diethyl ether.

Ketones such as acetone, cyclohexanone, diacetone, methyl isobutyl
      ketone.

Vegetable Oils such as cottonseed, linseed, pine, and sesame oils.

---

because excessive grinding or heating will occur. Larger samples should
be made up using a small ribbon blender, followed by grinding with a laboratory pulverizer, and finally postblending in the ribbon blender. The
particle size distribution should be determined by one of the methods described earlier.

    Samples should be stored at ambient and elevated temperatures and
tested after periods of time for tendency to cake. If caking does occur,
the product must be reformulated with a lower active ingredient content,
or with a more sorptive carrier. The samples should also be checked
after the elevated temperature storage for active ingredient content by
chemical analysis and/or biological assay methods to determine if any

carrier/pesticide incompatibility exists. If any loss of active ingredient
does occur, studies will have to be repeated in evaluating various deactivators.

Production equipment is very similar to that described for preparation
of larger laboratory samples. The production equipment should include:

(1) Dust collecting system.

(2) Preliminary coarse grinder such as an ice crusher.

(3) Melting and blending kettles.

(4) Scales or metering devices for weighing or measuring ingredients.

(5) Ribbon preblender.

(6) Grinding equipment such as:

Raymond Imp Mill	Raymond Pulverizing Div., Chicago, Ill.
Mikro Pulverizer	Pulverizing Machinery Div., Summit, N.J.
Fitzpatrick Mill	W. J. Fitzpatrick Co., Chicago, Ill.

(7) Ribbon postblender.

(8) Packaging facilities.

Table 8 contains three typical formulas for dust concentrates, in which
solid, semisolid, and liquid toxicants are used.

IV. FIELD-STRENGTH DUSTS

Field-strength dusts can be defined as a mixture of one or more toxicants with one or more inerts to form a dry, free-flowing powder containing less than 10% total active ingredient. They are applied directly to the
field or other control areas without further dilution.

A. Desired Properties

The desired properties of field-strength dusts are the same as those of
the dust concentrates. The flowability of the product should be good, the

TABLE 8

Typical Formulas for Dust Concentrates

I. Solid Toxicant - 50% Sevin Dust Concentrate

Technical Sevin	50.50
Pikes Peak Clay (montmorillonite)	49.50
	Total 100.00 wt %

II. Semisolid Toxicant - 25% Aldrin Dust Concentrate

Technical aldrin	25.25
Attaclay (attapulgite)	74.00
Urea	0.75
	Total 100.00 wt %

III. Liquid Toxicant - 25% Malathion Dust Concentrate

Malathion LV concentrate 95% grade	27.5
Celite 209 (diatomaceous earth)	27.5
Barden Clay (kaolinite)	45.0
	Total 100.0 wt %

---

particle size range should be from 30 to 50 μ, and the bulk density should range from 20 to 60 lb/ft^3 depending on the final use.

B. Methods of Preparation

There are two methods of preparation of dilute or field-strength dusts: (1) direct impregnation method, and (2) dilution of dust concentrates. The methods and equipment used for direct impregnation would be the same as those used in the preparation of dust concentrates. The dilution of dust concentrates is a simple mixing operation that can be handled by blenders close to the point of consumption.

Figure 2 illustrates the blending operation for diluting dust concentrates to field-strength dusts. In the laboratory, a Waring Blendor or a ribbon blender can be used. Assuming that the dust concentrate has been properly prepared, it is necessary then to be concerned primarily with the

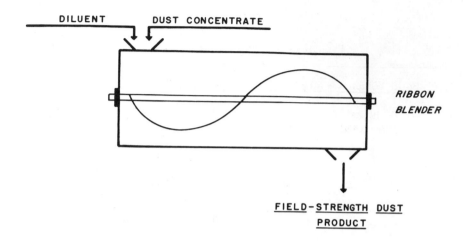

FIG. 2. Production of field-strength dusts.

properties of the carriers or diluents used to let down the concentrate. The
choice of the let-down inert will depend on the following factors: (1) desired
flowability of the final product, (2) desired particle size range of the final
product, (3) desired bulk density of the final product, (4) compatibility of
let-down inert with toxicant, and (5) cost of inert.

It is again assumed that the flowability of the dust concentrate is acceptable. However, some of the less costly diluents such as kaolin, pyrophyllites, or calcium carbonates do not have very good flowability, and it
may be necessary to use a portion of attapulgite in letting down to preserve
the same flowability as the dust concentrate. The kaolins are widely used
as let-down agents because of their low cost, and materials such as talcs,
pyrophyllites, and calcium carbonates are usually used in whole or part
when it is necessary to adjust the bulk density of the final products to higher ranges.

It is important that the particle size distribution of the let-down inert
resemble that of the dust concentrate. Big differences in particle size distributions will foster particle separation during handling, packaging, shipping, and possibly application.

The catalytic properties of the inerts used as let-down agents are no
less important than those used in making up the dust concentrate. The
pesticide/inert incompatibility must be tested for each of the diluents used
for let down. If deactivation is required, the proper concentration of

deactivator must be added for each diluent. Since the dilution of dust concentrates to field-strength dusts does not involve a grinding step, the distribution of the deactivator over the let-down inerts can readily become a problem. This problem is easy to overcome with the use of solid deactivators such as urea and HMT. A concentrate of the solid deactivator can be premade by grinding the let-down inert together with the solid deactivator at perhaps a high concentration such as 25% in a separate operation. Then the proper amount can be added during blending of the dust concentrate with the same inert. Most of the low sorptive inerts used for let-down are lower in catalytic activity than the high sorptive carriers used for production of the dust concentrate; these inerts in most cases require low concentrations of deactivator, and for the most part, practically the same concentration. It might be advantageous for dust-concentrate formulators to include an excess of the solid deactivator that will subsequently also deactivate the let-down inert.

Use of liquid deactivators presents a more serious problem. To obtain homogeneous distribution of the liquid deactivator on the active surfaces of the inert, it is mandatory that the additive be sprayed on the entire quantity of inert. This necessitates the installation of spraying equipment on the ribbon blender. Although it is not deemed as effective, a concentrate can be produced by spraying a large quantity of liquid deactivator onto the inert in a separate operation, and adding this concentrate during the blending operation. However, if the liquid deactivator is not mobile enough to redistribute itself homogeneously over the entire inert surface, this method may not provide good stability in the field-strength dust.

The production equipment used for manufacturing field-strength dusts by direct impregnation is the same as that necessary for making dust concentrates. Table 9 shows typical formulas for field-strength dusts by direct impregnation.

The primary equipment for letting down dust concentrates is a simple ribbon blender, with or without spraying nozzles for addition of liquid deactivators. Of course, other supporting equipment is necessary such as dust collectors, weighing and packaging equipment, etc. Table 10 shows typical formulas for field-strength dusts from dust concentrates. During the production of field-strength dusts either from the direct impregnation method or the dilution of dust concentrates methods, the deactivator is always added to the inerts and blended thoroughly prior to addition of the toxicant or dust concentrate. In some cases, if the deactivator is miscible, it may be preferable to mix it with the toxicant (and solvent if used) and spray the solution on the inert. Some workers believe that by using this method, the deactivator will at least be in close proximity to the carrier "hot spots" along with the toxicant, giving better stability.

TABLE 9

Typical Formulas for Field–Strength Dusts by Direct Impregnation[a]

I.   5% Sevin Field–Strength Dust

Technical Sevin	5.1
Pikes Peak Clay (montmorillonite)	14.9
Barden Clay (kaolinite)	80.0
Total	100.0 wt %

II.   2.5% Aldrin Field–Strength Dust

Technical aldrin	2.55
Attaclay X–250 (attapulgite)	12.91
CCC Diluent (calcium carbonate)	28.00
Barden AG Clay (kaolinite)	53.76
25% Urea concentrate (preground)	2.78
25% Urea	
15% Attaclay X–250	
60% Barden AG Clay	
Total	100.00 wt %

III.   4% Malathion Field–Strength Dust

Malathion LV concentrate 95% grade	4.21
Pyrax ABB (pyrophyllite)	95.79
Total	100.00 wt %

---

[a] Using basic steps of preblending, grinding, and postblending.

V.   WETTABLE POWDER CONCENTRATES

Wettable powders can be defined as finely divided pesticidal dust concentrates containing surface active agents that will allow the concentrate to be diluted to field strength to form stable sprayable suspensions.

A.   Desired Properties

There are many factors that have to be considered when formulating wettable powders. The product properties that are important are flowability,

TABLE 10

Typical Formulas for Field-Strength Dusts
by Dilution of Dust Concentrates

I. 5% Sevin Field-Strength Dust

	50% Sevin dust concentrate	10.2
	Pikes Peak Clay (montmorillonite)	9.8
	Barden AG Clay (kaolinite)	80.0
	Total	100.0 wt %

II. 5% Heptachlor Field-Strength Dust

	25% Heptachlor dust concentrate	20.2
	Emtal 43 (talc)	39.8
	Kao-X (kaolinite)	38.8
	Deactivator H	11.2
	Total	100.0 wt %

III. 2% Methyl Parathion Field-Strength Dust

	25% Methyl parathion dust concentrate	8.0
	Attaclay (attapulgite)	20.5
	CCC Diluent (calcium carbonate)	30.0
	Kao-X (kaolinite)	38.8
	Deactivator E	2.7
	Total	100.0 wt %

---

wettability, dispersibility, suspensibility, low foaming, and both physical and chemical storage stability. All of these properties are concerned with the ability of the product to be diluted with water to form a homogenous sprayable suspension.

One of the first properties of the concentrate that the consumer will encounter will be flowability of the dry concentrate--the ability to flow easily out of packages into hoppers and application sprayers. This property is influenced primarily by the flowability of the carrier, and the concentration of the toxicant impregnated. It can be measured by one of the methods described earlier for carriers and impregnated dusts. If the concentrate does not have good flowability, this deficiency will probably show up in the dispersibility testing.

Wettability is the next property to be encountered. This property is influenced by the type and concentration of toxicant in the product, and the proper choice of wetting agent to overcome surface tension. When a bag of the concentrate is dumped into the water in the spray tank, it should wet easily and sink to the bottom without carrying much entrained air. This property can be tested by carefully adding a quantity of the powder (equal to the field dilution ratio) into a beaker of water and measuring the time required for the powder to wet out and drop down into the water. Although the maximum wetting time for a product is an arbitrary choice, 90% dropping in 5 min can be used as a guide.

Dispersibility is the ability of the powdered concentrate to suspend in the water to form finely divided particles that will remain in suspension for a reasonable period of time. The dispersibility is related to the surface characteristics of the carrier and the toxicant, and dispersing agents are added to overcome attractive forces between particles.

Suspensibility is that ability of the dispersed particles to stay in suspension for an adequate period of time to allow all of the material to be sprayed homogeneously out of the application tank. The sedimentation of the particles in water obeys Stokes Law, thus the rate of sedimentation is directly proportional to the size and density of the particles. The density of the particles is dependent upon that of the toxicant and carrier (and any other components). Obviously the density of the toxicant is fixed, and in the high-concentration wettable powders, the small differences in the densities among the available sorptive carriers have only a nominal effect on the density of the final product. The particle size, on the other hand, can be varied, and it has been found that adequate suspensibility will result if the particles are from 1 to 3μ. This range is necessary when wettable powders are applied from spray equipment having no internal mixing provisions, such as knapsack sprayers, but most of the commercial equipment used in the United States does have internal mixers and larger particle size ranges can be tolerated. Obviously, during milling of a wettable powder, a wide particle size distribution will be obtained. There will even be a slight amount of plus-200-mesh (74μ) material, but this should be kept to a 2% maximum.

The particle size distribution can be determined by the methods described earlier for evaluation of carriers. Both the dispersibility and suspensibility can be determined simultaneously by methods described by Shell (74) for aldrin, dieldrin, and endrin, and by methods described by the World

Health Organization (WHO) (75) for evaluation of products used in their programs. Both of these methods are based on some original work by Folckemer et al. (76, 77). These methods involve dispersing the wettable powder in a graduated cylinder, allowing undisturbed settling to take place in the graduated cylinder in a controlled-temperature water bath for a period of time, withdrawing a sample of the suspension halfway down the graduate with a pipet, and finally determining the toxicant content in the aliquot. For a 30-min settling time, Shell states that minimum assays of 60% aldrin, 80% dieldrin, and 60% endrin (percentages of the original toxicant concentration added) should be obtained, whereas WHO specifies 50% minimums for the following pesticides: DDT, BHC, methoxychlor, chlordane, dieldrin, diazinon, malathion, and Chlorthion.

Foaming in a wettable powder dispersion is undesirable since air can be entrained in the spraying lines of the application rig and spraying will be sporadic and non-homogeneous. Also, with the larger commercial spraying rigs, recycling and mixing apparatus would cause the foam to rise out of the spray tank. The tests for tendency to foam are arbitrary and dependent on each manufacturer. One such test involves measuring the volume of foam after dispersing the wettable powder at field-use concentration in a 100-ml graduated cylinder; no more than 10 ml of foam should remain on top of the suspension after a 5-min waiting period. Foaming can be controlled by the proper choice and/or concentration of wetting agents.

Physical stability during storage is an important property of wettable powders. Because of the high concentration of toxicant in the product, individual particles can stick together and agglomerate during storage at ambient temperatures, resulting in a product exhibiting poor dispersibility and suspensibility. Both Shell (74) and WHO (75) describe methods by which the dry wettable powder concentrate is exposed to elevated temperature conditions to simulate storage in tropical climates, after which the test for dispersibility and suspensibility is conducted. Specifications for wettable powders for domestic use are not as rigid as those dictated by WHO for overseas use, and the tropical storage test is rarely applied to domestic products.

The problems of chemical stability of toxicants with carriers apply to wettable powders in the same manner as they apply to field-strength dusts and dust concentrates. While normally one would expect less degradation of the toxicant because of its higher concentration (and lower concentration of carrier) in the product, higher temperatures may be encountered in the tropical areas which would increase rate of degradation. All the normal precautions must be used in the choice of carriers and deactivators.

There is a trend today to add wettable powder concentrates to spray tanks already containing other wettable powders or emulsifiable concentrates. This is a dangerous practice not only because of the incompatibility of some toxicants, but also because of the incompatibility of the surface active agents used in formulating these products. Imprudent mixing of surfactants could cause flocculation and immobilization of the toxicants. Obviously, formulating a wettable powder that would be compatible with a half dozen other products is a task of large scope, but it has been done.

B. Methods of Preparation

The preparation of wettable powders can be divided into five steps: (1) preblending, (2) pulverizing, (3) reblending, (4) milling, and (5) postblending. These steps are depicted in Fig. 3.

In the laboratory formulation of a wettable powder, the first factor to consider is the compatibility of the carrier with the toxicant. More than likely some prior information will be available from formulation of other dry products. If not, a dust base can be prepared at low concentration, say 5 or 10% for stability testing. Chemical incompatibility usually is detected more easily at low toxicant loadings.

The next step is to determine the maximum toxicant loading without loss of flowability. This can be accomplished in the laboratory by using the Waring Blendor following methods described earlier in the section on laboratory preparation of dust concentrates. The maximum loading should be determined for all of the liquid components, i.e., the liquid toxicant, the liquid solvent if necessary for impregnation, the liquid dispersant and wetting agent, and the liquid deactivator.

Very often a mixture of carriers will be used. A primary carrier, such as a synthetic silica, can be used to provide the high sorptive capacity, and a second lower-cost, and high-flowability carrier such as attapulgite can be used in concentrations up to 30% of the total carrier weight. A unique innovation that has developed in the last five years is the use of up to 10% of colloidal attapulgite to improve suspensibility and prevent hard cake formation if sedimentation cannot be avoided in a spray tank by the user.

Once the maximum toxicant concentration has been established, the next step is the investigation of wetting and dispersing agents. Good wettability and dispersibility depend on the proper choice and concentrations of wetting and dispersing agents. When organic liquid toxicants are impregnated on the inert carriers, the normally hydrophilic surface of the inert is changed to a hydrophobic one. With solid and semisolid toxicants,

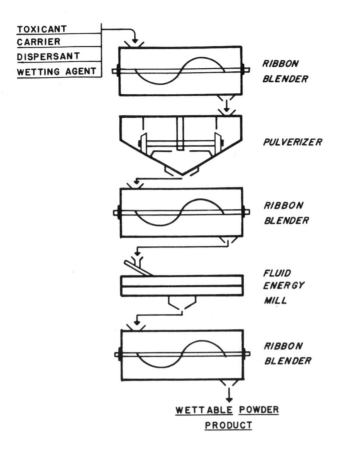

FIG. 3. Production of wettable powders.

the carrier is acting more like a conditioning agent to keep the solid particles from agglomerating, and the hydrophobic properties of the toxicant strongly influences the wettability of the concentrate. It is believed that choice of an acceptable wetting agent is the biggest hurdle in formulating wettable powders.

Table 11 contains a list of wetting agents that have successfully been used in the past, along with their pesticidal uses and approximate concentrations. In some cases information on surfactants used in preparation of emulsifiable concentrates may be helpful when commonly used wetting

TABLE 11

Wetting Agents for Wettable Powders

Wetting agents	Recommended concentration	Pesticide uses[a]	Supplier
Aerosol 22	1.5%	d	American Cyanamid Co. Wayne, N.J.
Agriwet 9086	2%	not	Nopco Chemical Co. Newark, N.J.
Duponol ME Dry	0.5-1.0%	afh	E.I. du Pont de Nemours & Co. Wilmington, Del.
Duponol WA Dry	0.25-1.0%	af	E.I. du Pont de Nemours & Co.
Igepal CA-630	4%	b	General Aniline & Film Corp. New York, N.Y.
Igepal CA-710	4%	b	General Aniline & Film Corp.
Igepal CO-630	0.1-0.5%	cq	General Aniline & Film Corp.
Igepon AC-78	0.5-0.75%	cv	General Aniline & Film Corp.
Igepon AP-78	1-2.5%	adefklmv	General Aniline & Film Corp.
Igepon TN-74	1.5%	b	General Aniline & Film Corp.
Igepon T-77	1-2%	acdefs	General Aniline & Film Corp.
Kyro EOB	1.5%	k	Procter & Gamble Co. Cincinnati, Ohio
Nekal BA-75	0.5-1%	rv	General Aniline & Film Corp. New York, N.Y.
Nekal BX-78	0.75-4.0%	bpuv	General Aniline & Film Corp.
Nekal WS-21	1%	iu	General Aniline & Film Corp.

TABLE 11 (continued)

Wetting agents	Recommended concentration	Pesticide uses[a]	Supplier
Nekal WS-25	0.25%	u	General Aniline & Film Corp.
NSAE Powder	1%	u	Onyx Chemical Co. Jersey City, N.J.
Renex 25	0.5-2%	dk	Atlas Chemical Industries, Inc. Wilmington, Del.
Renex 35	1.5%	dh	Atlas Chemical Industries, Inc.
Renex 648	1%	i	Atlas Chemical Industries, Inc.
Santomerse No. 1	1-3%	dk	Monsanto Co. St. Louis, Mo.
Santomerse No. 3	0.5-2%	k	Monsanto Co.
Sellogen HR	0.5-2%	cefghkrt	Nopco Chemical Co. Newark, N.J.
Sterox AJ	1.5%	dh	Monsanto Co. St. Louis, Mo.
Sterox CD	1.5%	d	Monsanto Co.
Tergitol NPX	0.2-2.0%	cde	Union Carbide Corp. New York, N.Y.
Tergitol NP-27	0.2-2.0%	cde	Union Carbide Corp.
Toximul MP	1%	in	Stepan Chemical Co. Maywood, N.J.
Toximul MP-8	1%	in	Stepan Chemical Co.
Triton X-100	0.3-2%	ek	Rohm & Haas Co. Philadelphia, Pa.

TABLE 11 (continued)

Wetting agents	Recommended concentration	Pesticide uses[a]	Supplier
Triton X-120	0. 7-4%	abcdefhkv	Rohm & Haas Co. Philadelphia, Pa.
Ultrawets DS, K, SK	0. 75-4%	bnu	ARCO Chemical Co. Philadelphia, Pa.
Vatsol OT-B	1%	a	American Cyanamid Co. Wayne, N. J.
Wetanol	0. 5-1%	bdf	Glyco Chemicals, Inc. New York, N. Y.

[a] Use Code:

a – aldrin	g – diphenamid	m – lindane	s – Strobane
b – Aramite	h – endrin	n – malathion	t – sulfur
c – BHC	i – ethion	o – parathion	u – Thiodan
d – chlordane	j – glyodin	p – Phygon	v – toxaphene
e – DDT	k – heptachlor	q – PMA	
f – dieldrin	l – IPC	r – Sevin	

agents are not effective. Also, sometimes wetting agents can be screened by measuring the interfacial tension between water and a liquid toxicant with the wetting agent added. In the case of solid or semisolid toxicants, the tests can be run with the toxicant and surfactant dissolved in a series of appropriate solvents. In this method, the concentration of toxicant in each solvent should be as high as possible. However, these last two methods of gathering information on suitable wetting agents are not always fruitful, and the researcher has to resort to trial and error methods.

The suitability of a dispersing agent depends not only on the characteristics of the toxicant, but perhaps just as much on the nature of the wetting agent used. Therefore, the dispersant should be chosen after the wetting agent. Also, the particle size distribution should be checked and found to

be in the proper range before a dispersing agent is rejected because of in-
effectiveness; poor dispersibility and suspendibility may be caused by too
large a particle size. Table 12 contains a list of dispersing agents along
with their successful uses with toxicants and concentration ranges.

The use of solid wetting and dispersing agents that have high water solubility, rather than liquid ones, are preferred for the following reasons:
(1) there is less liquid load on the carrier; (2) better distribution of the surfactant is obtained, rather than pockets absorbed on the carrier; (3) the
surfactant may be tightly absorbed in the carrier and not entirely released,
if added as a liquid; and (4) spraying equipment is not necessary with solid
surfactants. Surfactants, that are normally liquids, but marketed as dry
powders impregnated with the liquid surfactant, should definitely be avoided because the surfactant carrier may be incompatible with the toxicant,
and the surfactant may be held too tightly on its carrier.

In production, the surfactants should always be added after the toxicant
and carrier are blended together. If the toxicant is liquid, it will first be
absorbed into the inner pores of the carrier, and if followed by a liquid surfactant, the surfactant will be more readily accessible to the particle surface where it must function. If the toxicant is a solid, the liquid surfactant
will have a chance to coat the toxicant surface along with the carrier. If
the surfactants are solid, they can readily dissolve in the water when diluted to field strength, and can be chemically adsorbed on both the carrier
and toxicant surfaces.

In the laboratory testing of wetting and dispersing agents, all of the steps
in the production must be completed including the milling step as shown in
Fig. 3. Simply making unmilled blends of the carrier, toxicant, and surfactants may be used as a precursory screening step, but often milling to
a smaller particle size will increase the exterior particle surface area and
the surfactant requirement. There have been some cases where the wetting
or dispersing agent would not function at all after the milling step (78).

When checking the properties of a wettable powder, care must be taken
to use water of standard hardness, usually 342 ppm is used in laboratories,
calculated as calcium carbonate and prepared as follows:

Calcium chloride, anhydrous 0.304g
Magnesium chloride, hexahydrate..........0.139g
Distilled water, dilute to ....1 liter

Even though all of the screening tests should be done in hard water, promising formulations should also be checked in distilled and varying hardness

TABLE 12

Dispersants for Wettable Powders

Dispersant	Recommended concentration	Pesticide uses[a]	Supplier
Blancol	2–4%	adfkmv	General Aniline & Film Corp. Chattanooga, Tenn.
Blancol conc. N	1–2%	adfkv	General Aniline & Film Corp.
Darvan no. 1	1–2%	bdp	R. T. Vanderbilt Co., Inc. New York, N. Y.
Darvan no. 2	1–1.5%	dp	R. T. Vanderbilt Co., Inc.
Darvan no. 4	1–3%	nr	R. T. Vanderbilt Co., Inc.
Daxad 11	1–2%	bp	Dewey & Almy, W. R. Grace & Co. Cambridge, Mass.
Daxad 21	1–3%	nrs	Dewey & Almy, W. R. Grace & Co.
Daxad 23	1%	p	Dewey & Almy, W. R. Grace & Co.
Daxad 27	3%	v	Dewey & Almy, W. R. Grace & Co.
Indulin C	1–3%	e	West Virginia Pulp & Paper Co. North Charleston, S. C.
Marasperse C-21	5–6%	afj	American Can Co., Marathon Div. New York, N. Y.
Marasperse CB	5–6%	af	American Can Co., Marathon Div.
Marasperse N-22	1–4%	abcdeklnpuv	American Can Co., Marathon Div.
Lomar PW	3%	g	Nopco Chemical Co. Newark, N. J.
Orzan P	1.5%	d	Crown-Zellerbach Corp. Camas, Wash.

TABLE 12 (continued)

Dispersant	Recom- mended concen- tration	Pesticide uses[a]	Supplier
Polyfon F	1.5-3.0%	den	West Virginia Pulp & Paper Co. North Charleston, S.C.
Polyfon H	1.5-4.0%	abcdefknov	West Virginia Pulp & Paper Co.
Polyfon O	1.0-1.5%	gu	West Virginia Pulp & Paper Co.
Polyfon T	1.5-3.0%	dnv	West Virginia Pulp & Paper Co.
Tamol N	1.0-2.0%	abcdefhkpv	Rohm & Haas Co. Philadelphia, Pa.
Tamol 731	2%	h	Rohm & Haas Co.
Tamol 850	1.5%	b	Rohm & Haas Co.

[a]See footnote to Table 11 for a listing of the Use Code.

water. There have been some rare cases where the wettable powder func-
tioned well in hard water, but failed in soft water.

In the laboratory, a Waring Blendor can be used for the initial impregnation, blending, and pulverizing of the components. Laboratory-sized as
well as production fluid energy mills are available from the following manufacturers: (1) Micronizer, from Sturtevant Mill Co., Boston, Mass. (2)
Jet-O-Mizer, from Fluid Energy Processing & Equipment Co., Philadelphia, Pa. (3) Reductionizer, from Reduction Engineering Corp., Newark,
N.J., and (4) Micron-Master, from The Jet Pulverizer Co., Palmyra,
N.J. For the laboratory-sized mills, it is preferable to have the mill
rubberlined to minimize wear and facilitate cleaning. Care must also be
taken to use oil-free, cold, dry air, and the mill should be operated in a
manner such that the product will not heat up. Often with solid toxicants
with low melting points, crushed dry ice is passed through the mill along
with the toxicant blend to prevent melting. After milling, the material

must be postblended, and the Waring Blendor can also be used for this operation in the laboratory.

In regular production, all the equipment described above for the manufacture of dust concentrates is necessary, along with the fluid energy mill. In production, it is preferable to have a separate blender for the postblending operation so that chances of contamination of the final product with unmilled components are eliminated. When high suspensibility products are not necessary, the use of fluid energy mills may be replaced with mills capable of grinding down to 5μ. The list of pulverizing mills given under the section on production of dust concentrates may be suitable for production of wettable powders for some domestic uses.

After formulation in the laboratory, samples should be tested for particle size distribution, wettability, dispersibility, and suspensibility, and rechecking these properties after some storage period should not be neglected. Table 13 contains typical formulas for wettable powders.

VI. GRANULAR PRODUCTS

As described earlier, a granular product is comprised of distinct particles within the range 4 to 80 mesh. Usually, there is approximately a 2:1 ratio of the largest particle to the smallest particle. Granular products have several advantages. Particle weight reduces drift, granules are easy to apply with accurate control of rate and placement, and granules are safer to applicators and other desirable forms of life. Since their inception in the early fifties, this form of pesticide has probably become the most widely used and most versatile of the available products.

A. Desired Properties and Types of Products

The desired properties of granular pesticidal products are the same as described above for granular carriers. They must be attrition resistant, flowable, and dust-free. The granular carrier must be capable of carrying the toxicant to the site of control without any losses, and then release the toxicant for control. Much work has been conducted on the methods of production of granules, but little has been done on altering the nature of the granular products and their action at the site of control.

The concentrations of active ingredients in granular products range from 2% to as much as 40%, with the most common concentrations within the range of 5 to 20%. When concentrations are 5% or lower, the economics of use of such a large proportion of carrier in the product must be

TABLE 13

Typical Formulas for Wettable Powders

I.   75% DDT Wettable Powder

DDT (technical)		76.5
Hi–Sil 233 (hydrated silicon dioxide)		20.5
Igepon T–77 (wetting agent)		1.5
Marasperse N (dispersing agent)		1.5
	Total	100.0 wt %

II.   40% Aldrin Wettable Powder

Aldrin (technical)		40.4
Attaclay (attapulgite)		50.9
Urea (deactivator)		2.7
Duponol WA Dry (wetting agent)		1.0
Marasperse CB (dispersing agent)		5.0
	Total	100.0 wt %

III.   50% Sevin Wettable Powder

Sevin (technical)		50.5
Pikes Peak (montmorillonite)		48.0
Nekal BA–75 (wetting agent)		0.5
Daxad 21 (dispersing agent)		1.0
	Total	100.0 wt %

IV.   50% Malathion Wettable Powder

Malathion (technical)		55.0
Micro-Cel E (hydrous calcium silicate)		42.0
Polyfon H (dispersing agent)		3.0
	Total	100.0 wt %

considered. When concentrations are higher than 20%, adequate distribution of the active ingredient might be hindered. Choice of concentration depends on many factors including (1) the nature of the pest to be controlled, (2) the application rate of the active ingredient necessary to give control, (3) the ability of the available equipment to accurately apply the granular product, and (4) the type and economics of competitive products used for the same control. As described earlier, the size of the particles also depends on the end use. The rate of granular disintegration can be controlled to a limited extent by the choice of granular carrier.

The types of granular products available on the market today are limited. Regular granular products include (1) granular carriers impregnated by conventional methods, (2) agglomerated granules, and (3) granular carriers with toxicant plastered to the outside. Pelleted products ranging in diameters from 1/4 in. to 1/32 in. must also be included among granular products.

B. Methods of Preparation

The most common method of producing granular pesticidal products is by the direct impregnation method. This technique which has been in use for almost 20 years is simple and processing costs are comparatively low. In the last 10 years efforts have been made by toxicant manufacturers to develop more effective, but less toxic pesticides. Many of these new pesticides, however, have different physical and chemical properties--properties that do not allow the use of the direct impregnation methods. The problem most often encountered is that of insolubility of the pesticide in commonly used, low-cost, impregnation solvents. Efforts have therefore been made to develop new formulation techniques, some of which have become commercial. Some of these available methods will be discussed below.

1. Direct Impregnation Technique

Granular carriers and diluents and their properties were described above. These preformed inerts are impregnated with liquid toxicants (or a solvent solution if the toxicant is a solid) by simply spraying the liquid onto the granules. During the first few years of granular production, ribbon blenders were used because of the unavailability of other suitable equipment. To minimize attrition of the granules by the blender, it was recommended that there be a minimum clearance of 1/4 in. between the blender walls and blades, and that the blender blades have a maximum rotational speed of about 40 rpm. Subsequently more acceptable equipment was developed and became available to the industry. This equipment, depicted in Fig. 4, closely resembles a cement mixer. Spray nozzles, lifting flights, and speed of rotation has been designed to give homogeneous impregnation with minimum attrition of the granules. Since this type of equipment has become available, the use of ribbon blenders for impregnation should be avoided.

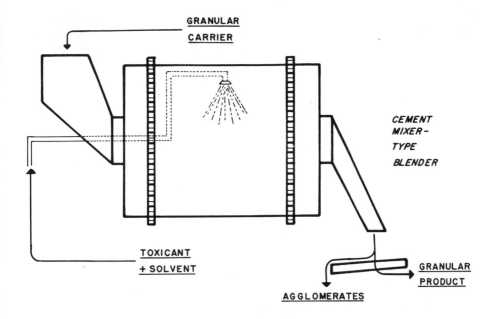

FIG. 4. Solvent impregnation process.

Since only liquids can be impregnated onto granular carriers, semi-solid and solid toxicants must be dissolved in suitable solvents. Many of the same solvents listed in Table 7 and used for making dust concentrates, dilute dusts, and wettable powders, can be used for impregnation of granules. As a general guide, a minimum of one gallon of liquid for every 100 lb of carrier should be impregnated to obtain adequate distribution of the toxicant over the granules. Thus, even for the impregnation of normally liquid toxicants at low concentrations, e.g., 2-10%, these toxicants should be diluted with a solvent. The amount of solvent used, however, is dictated by the solubility of the toxicant and the low viscosity necessary for the solution to be sprayed. Quite often the toxicant solution is heated to increase the solubility of the toxicant in the solvent, and/or to lower the viscosity of the solution. The upper limit on the volume of solution that can be impregnated depends on the sorptive capacity of the carrier. The excessive use of solvent, on the other hand, should be avoided to keep from increasing the density excessively and lowering the particles per pound in the final product. Whenever possible, deactivators should be dissolved in the toxicant solution and sprayed simultaneously on the carrier. With solid deactivators such as urea, a water solution should be prepared and impregnated before the toxicant solution. The addition of solid deactivators to the granular carrier should be avoided since the ability of the deactivator to reach active sites inside the granules is questionable. The spray from the nozzles should be directed into the falling curtain of granules, where the turnover of the material is the greatest. Impingement of the spray on the bare walls of the blenders should be avoided. Loose dust tends to be picked up by the larger granules during the spraying. Some agglomerates may form, and not totally broken up in the postblending process; therefore, the product should be passed through a scalping screen to remove these agglomerates.

The time necessary for the spraying of solution onto the granules should be kept below ten minutes. If the volume of solution is such that this time would be exceeded, additional spray nozzles should be installed. The impregnated product is usually postblended to allow distribution of the toxicant solution on the granules and to break up agglomerates. This postblending time should not exceed 15 min to minimize attrition of the granules.

Table 14 contains some typical formulas which are examples of the various solutions which are sprayed onto the granules. Formula I uses no solvent since the 2, 4-D is liquid and has a low enough viscosity to be sprayed. In Formula II, the water solution of the sodium salt of DNOSB is liquid and low enough in viscosity, but water is added to this toxicant to obtain adequate spreading over the granules. Formula III is typical of a high concentration granular product in which the toxicant is a solid and must be dissolved in an organic solvent. This formula also contains a water solution of a deactivator which must be sprayed on the carrier before the toxicant solution. Formula IV uses an organic solvent-soluble deactivator; the Bandane, HAN solvent, and Deactivator H are mixed together and sprayed on the granules as one solution. Formula V is typical of a case where the toxicant is a liquid, but a small amount of solvent is necessary to obtain a viscosity low enough for spraying. The DDT in Formula VI is a solid and must be dissolved in a solvent to be impregnated.

2. Suspension Impregnation Technique

For those pesticides that are insoluble in water or the low-cost, low-phytotoxic solvents that are normally used for direct solution impregnation methods, a suspension impregnation method may be useful. This method which has been successfully used for some products consists of first preparing a suspension of the finely divided powdered toxicant in an organic solvent, and spraying the suspension on a granular carrier. A high concentrate wettable powder, if available, might be used in preparing a water suspension. As the solvent is absorbed, the solid toxicant is drawn

TABLE 14

Typical Formulas for Impregnated Granular Pesticides

I.   10% 2,4-D granules

2,4-D solution (50% acid equivalent)		20.5
25/50 A LVM Attaclay (attapulgite)		79.5
	Total	100.0 wt %

II.   4% Dinitro granules

Dinitro sodium salt solution (40% in water)		10.5
25/50 A LVM Attaclay (attapulgite)		89.5
	Total	100.0 wt %

III.   20% Aldrin granules

Aldrin (75% solution in diesel oil)		26.9
Urea (50% solution in water)		1.5
30/60 AA RVM Florex (attapulgite)		71.6
	Total	100.0 wt %

IV.   7.5% Bandane granules

Bandane (technical)		7.5
Heavy aromatic naphtha (HAN)		10.0
Deactivator H		6.1
16/30 AA RVM Attaclay (attapulgite)		76.4
	Total	100.0 wt %

V.   15% Bandane granules

Bandane (technical)		15.0
Heavy aromatic naphtha		1.0
Ground corn cobs, 16-60 mesh		84.0
	Total	100.0 wt %

VI.   5% DDT granules

DDT (technical, 60% solution in xylene)		8.5
Vermiculite #4 (mica)		91.5
	Total	100.0 wt %

into pockets on the surface of the granules. Obviously, this technique is limited to low concentrations of active ingredient--about a maximum of 5%. Smaller size granules should be used for this method since they offer a greater exterior surface per unit weight than the larger mesh sizes. The toxicant in products prepared by this method also has a tendency to slough off the granules. Impregnation and postblending times should be kept short to minimize flaking off of the toxicant from the granules. Table 15 shows some typical formulas for toxicants applied as an organic solvent suspension, as well as an aqueous suspension.

3. Sticking Techniques

An extension of the suspension impregnation technique is the sticking technique, which was developed in an attempt to obtain higher concentrations of active ingredient in the product. This method has the same disadvantages as the suspension technique, having to plaster active ingredient only on the exterior surface of the granules. This technique is rarely used where sloughing off of the powdered toxicant is not a problem.

The variations in the sticking methods may be described as follows: (1) spraying of a suspension of powdered toxicant on a granular carrier followed by a postspraying of binding agent solution, (2) prespraying of the granular carrier with a binder solution, then adding the powdered toxicant to the tumbler to be picked up by the sticky granules, and (3) as in (2) above but followed by a postspray of binding agent solution.

Table 16 contains some typical formulas utilizing sticking techniques. Example I shows how a postspray of a binder can be used to minimize flaking off of a powdered toxicant impregnated by the suspension technique. Example II shows how the sorptive capacity of the clay carrier can be satisfied with an oil in water emulsion, while at the same time coating the granules with a light mineral oil that can be used to adhere the pesticide to the carrier. Example III utilizes an organic binder solution to prepare the clay carrier to pick up the powdered toxicant and also to bond the toxicant to the granules.

The carriers that are used when utilizing the sticking techniques need not be high sorptive carriers; low sorptive materials can be used as the base. However, bases within the desired bulk density, with properties that are consistent are not readily available. Thus, it is less troublesome and more reliable to use standard sorptive carriers, satisfying their sorptive capacities by first impregnating them with water or an oil in water emulsion, so that most of the binder will remain near the exterior of the granule where it will be effective in picking up the powdered toxicant.

TABLE 15

Typical Formulas for Suspension Impregnation Technique

I. 4% Pesticide "A" granular product

Step A. Blend the following:

Pesticide "A" (finely powdered)	4. 0 parts/wt
White oil	16. 0 parts/wt

Step B. Impregnate suspension from Step A on:

30/60 AA RVM Attaclay (attapulgite)	80. 0 parts/wt
Total	100. 0 parts/wt

II. 2.5% Pesticide "B" granular product

Step A. Blend the following:

Pesticide "B" - 50% wettable powder	5. 0 parts/wt
Water	25. 0 parts/wt

Step B. Impregnate suspension from Step A on:

25/50 A LVM Attaclay (attapulgite)	70. 0 parts/wt
Total	100. 0 parts/wt

There are many types of binders that can be used: the natural adhesive bases including animal glues, fish glues, caseins, blood albumen, starches, dextrins, natural gums and latexes, wood rosin and shellacs; mineral bases including silicates and asphalts; and the synthetic adhesive bases which are numerous and more costly than the natural and mineral adhesives. The rate of release of the toxicant from the granules can be controlled to a degree by the choice of the binding agent.

4. Post-Drying Technique

Many toxicants have been encountered that are insoluble in organic solvents. If the pesticide is slightly water soluble, but impregnation to the desired concentration of the solution would exceed the sorptive capacity of the granular carrier, the impregnation can be followed by a drying step to remove the excess water. During the impregnation, the sorptive capacity of the carrier can be exceeded only slightly, perhaps resulting in a slightly moist, low-flowable, granular material. Then this blend is dried to remove the excess moisture. If the desired concentration of active ingredient

TABLE 16

Typical Formulas for Granular Products Prepared
by Sticking Techniques

I.   2.5% Pesticide "C" granular product

    Step A. Blend the following:
             Pesticide "C" - 50% wettable powder   5.0 parts/wt
             Water   25.0 parts/wt

    Step B. Impregnate suspension from Step A on:
             25/50 A LVM Attaclay (attapulgite)   65.0 parts/wt

    Step C. Postspray mix with:
             Raw sugar solution (20% in water)   5.0 parts/wt
                                                        ──────
                                         Total   100.0 parts/wt

II. 10% Pesticide "D" granular product

    Step A. Prepare the following emulsion:
             Water   14.8 parts/wt
             Triton X-100 (emulsifier)   0.4 parts/wt
             Atlantic 85B (light mineral oil)   14.8 parts/wt

    Step B. Impregnate emulsion from Step A on:
             16/30 AA RVM Attaclay (attapulgite)   57.5 parts/wt

    Step C. Add the following to the blend:
             Pesticide "D" - 80% wettable powder   12.5 parts/wt
                                                        ──────
                                         Total   100.0 parts/wt

III. 10% Pesticide "E" granular product

    Step A. Prepare the following solution:
             Acintol R (rosin)   3.0 parts/wt
             Atlantic 85B (light mineral oil)   12.0 parts/wt

    Step B. Impregnate solution from Step A on:
             30/60 AA RVM Florex (attapulgite)   70.0 parts/wt

    Step C. Add the following to the blend:
             Pesticide "E" (finely powdered)   10.0 parts/wt

    Step D. Post-spray with the following solution:
             Acintol R (rosin)   1.0 parts/wt
             Atlantic 85B (light mineral oil)   4.0 parts/wt
                                                        ──────
                                         Total   100.0 parts/wt

cannot be obtained in one impregnation/drying step, multiple steps can be used. Figure 5 illustrates the steps involved in two impregnation/drying steps.

The temperature used in the drying steps must be high enough to remove the moisture in a short period of time in order to minimize attrition of the granules. On the other hand, the bed temperature must be lower than the decomposition temperature of the toxicant. Temperature degradation studies should be conducted while the toxicant is impregnated on the carrier to determine the appropriate temperature.

The amount of residual water left in the final product should be carefully selected to minimize gain or loss of moisture in retail packages. Studies should be conducted to determine the moisture equilibrium concentration when samples are exposed to ambient temperatures and humidities that may be encountered in storage. The moisture equilibrium concentration will vary with the type of carrier used for the impregnation and with the concentration of active ingredient impregnated. To further minimize changes in moisture content of the packaged product, low moisture exchange packages should be used. Three-ply craft bags with polyethylene or asphalt laminated moisture barriers are effective in reducing moisture exchange in many cases.

Table 17 contains some typical formulas and the description of the steps involved in one and two step impregnation/drying techniques.

5. Vacuum Evaporation Technique

The vacuum evaporation technique is used when the pesticide is not soluble, or has limited solubility, in those solvents commonly used for direct impregnation, but may be soluble in costly, phytotoxic, and/or volatile solvents. The exotic solvent can be removed by application of heat and/or vacuum to recover the solvent for reuse. Single units are available in which the carrier can be impregnated and then the solvent can be removed by applying heat and/or vacuum. The operation of one such unit, manufactured by the Patterson-Kelly Co., Inc. of East Stroudsburg, Pa., is depicted in Fig. 6. This unit is available with a steam jacket around the blender shell to aid in the evaporation and recovery of the solvent. Pesticide impregnation can be carried out using single or multiple impregnation/solvent recovery steps.

There are two main points, besides pesticide/solvent solubility, to be considered in using this technique. The first is the heat stability of the toxicant if heat is to be applied to facilitate removal of the solvent. The

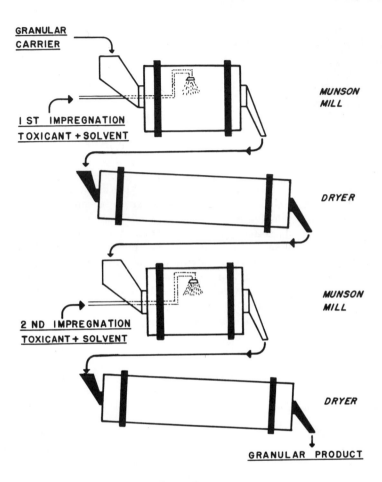

FIG. 5.   Post-drying process.

second is the vapor pressures of the solvent and the toxicant and their solutions.   Care must be taken not to remove the toxicant from the granules during the solvent recovery step.

Table 18 gives two examples of use of the solvent recovery technique. Example I is a case where the Pesticide "H" is fairly soluble in acetone, but it is preferable to remove and recover the acetone because of its high volatility.   Example II is a case where the Pesticide "I" has limited solubility in xylene, and the impregnation is carried out in two impregnation/ recovery steps.   There have been products on the market where xylene

TABLE 17

Typical Formulas for Formulation of Pesticides by Post-Drying Technique

I. Single impregnation/drying step

10% Pesticide "F" granular product

Step A. Impregnate the following:
           Pesticide "F" (28% water solution)   36.0 parts/wt
           16/30 A LVM Attaclay (attapulgite)   82.0 parts/wt

Step B. Rotary dry the impregnated product to re-
           move 18 parts/wt of moisture. The follow-
           ing formula will result:
           Pesticide "F"   10.0 wt %
           Water (residual)   8.0 wt %
           16/30 A LVM Attaclay (attapulgite)   82.0 wt %
                                           _____
                                    Total   100.0 wt %

II. Double impregnation/drying steps

10% Pesticide "G" granular product

Step A. Impregnate the following:
           Pesticide "G" (15% water solution)   40.0 parts/wt
           25/50 A LVM Attaclay (attapulgite)   82.0 parts/wt

Step B. Rotary dry the impregnated product to
           remove 30 parts/wt of moisture.

Step C. Impregnate the following:
           Dried product from Step B   92.0 parts/wt
           Pesticide "G" (15% water solution)   26.7 parts/wt

Step D. Rotary dry the impregnated product to
           remove 18.7 parts/wt of moisture. The
           following formula will result:
           Pesticide "G"   10.0 wt %
           Water (residual)   8.0 wt %
           16/30 A LVM Attaclay (attapulgite)   82.0 wt %
                                           _____
                                    Total   100.0 wt %

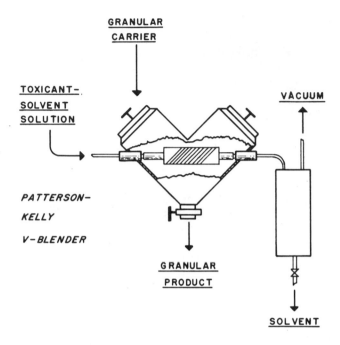

FIG. 6.   Vacuum evaporation process.

used as the solvent for direct impregnation was left in the product; however, since solvent recovery equipment is available, it is preferable to remove this solvent from the product as shown in Example II.

6.   Melt-On Technique

Again for use with pesticides that are insoluble in aqueous and organic systems, the melt-on technique may be considered.   Pesticides with low melting points, low volatility, and good chemical stability with respect to temperature can be used in this process.   The pesticide and granular carrier are simultaneously fed into a rotary dryer, and heated to a bed temperature above the melting point of the pesticide.   The pesticide becomes liquid upon melting and is absorbed by the granular carrier.   This operation can be a batch operation or continuous depending on the type of drying equipment chosen.   Figure 7 illustrates the type of equipment that can be used in this process.

TABLE 18

Typical Formulas for Formulation Pesticides by
Vacuum Evaporation Process

I. Single Impregnation/Solvent Recovery Step

10% Pesticide "H" granular product

Step A. Impregnate the following:
         Pesticide "H" (25% solution in acetone)   40. 0 parts/wt
         30/60 AA RVM Florex (attapulgite)   90. 0 parts/wt

Step B. Remove 30 parts/wt of acetone by vacuum
         evaporation, and recover for reuse

II. Double Impregnation/Solvent Recovery Step

10% Pesticide "I" granular product

Step A. Impregnate the following:
         Pesticide "I" (20% solution in xylene)   40. 0 parts/wt
         18/35 A LVM Attaclay (attapulgite)   90. 0 parts/wt

Step B. Remove 32 parts/wt of xylene by steam
         heating and vacuum evaporation, and
         recover for reuse.

Step C. Impregnate the following:
         Product from Step B   98. 0 parts/wt
         Pesticide "I" (20% solution in xylene)   10. 0 parts/wt

Step D. Remove 8 parts/wt of xylene by steam
         heating and vacuum evaporation, and
         recover for reuse.

---

7. Extrusion Technique

    This technique is used with insoluble pesticides, or where a pelleted
pesticide is desired. The pesticide is pugged with clay and water, extruded, and dried. If a pelleted product is desired, the extrudate is cut
to the desired length (usually equal to the diameter); if a granular product
is desired, the dried extrudate is ground and screened for the proper

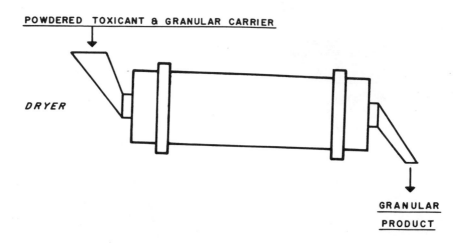

FIG. 7. Melt-on process.

granular size. Fines must be fed back into the pugger if the process is to
be economically sound. This process is illustrated in Fig. 8.

Many times the toxicant can act as a binder, but if dry pellet strength
is not obtained, a separate binder must be added. Normally an attapulgite
clay or inert of lower sorptivity is used as the primary carrier. It has
been found by Marples and Sawyer (79) that colloidal attapulgite clay can
be used as a binder in pesticidal mixtures. The hardness and rate of water disintegration of the pelleted or granular product can be controlled to
a degree by the ratio of colloidal to calcined attapulgite clay. Water is
added to the blend to make it just slightly wet and plastic. If a pelleted
product is desired, the material is usually extruded through holes ranging
from 1/4 to 1/32 in., and the extrudate is cut during extrusion. Larger diameters can be extruded, and cutting is not necessary of the material will
later be ground to obtain a granular product.

The material is then dried by some means to remove the excess moisture and heat set the extrudate. The quantity of moisture removed will depend upon the type and concentrations of toxicant, binder, and carrier used
in the manufacture. The moisture equilibrium conditions of the product
when exposed to ambient temperatures and humidities should be determined, and the material should be dried to this equilibrium percentage to
minimize exchange of moisture in packaging. Obviously the temperature
used in the drying should be lower than the temperature at which the toxicant decomposes in the presence of the carriers and binders.

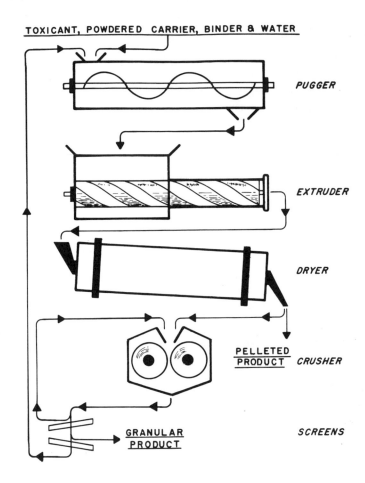

FIG. 8. Extrusion process.

The pelleted product can be ground using corrugated roll mills, and then screened to obtain the proper mesh distribution. If the product is to conform to the NACA recommendation of 90% by weight of granules lying between the designated product screens, and if attempts are made to match the granular products made by direct impregnation techniques, it will be found that the grinding yield for the desired fraction is usually less than 50%. Because the fines contain the same concentration of active ingredient as the product, the fines cannot be discarded and some means must be found to recycle them to the pugger, and adjustments must be made in the binder requirements to reagglomerate the fines. The greatest problem with continual recycling of fines is the ability of the toxicant to withstand the repeated drying steps without degradation in activity.

Instead of using the auger-screw extruder to obtain the extrudate, either a pellet mill such as the California pellet mill manufactured by California Pellet Mill Co., Crawfordsville, Indiana, or a compressed cylindrical pill making machine such as that manufactured by Stokes Co., Div. of Pennsalt Chem. Corp., Philadelphia, Pa. can be utilized. With the pellet mill, less pressure is applied and the pellets tend to be not as hard as those produced by an auger extruder. Pill making machines usually apply very high pressures, and can usually agglomerate dry blends of the toxicant, carriers, and binders. Products made by both of these types of machines are not usually milled to obtain granular products.

Table 19 shows the steps involved in producing a granular 20% Pesticide "J" product. It will be noted that an excess of water is added to the pugger to render the mix plastic and extrudable. Also, all of the water is not removed; 8% residual water remains which was found to be the equilibrium moisture content of this product.

8. Agglomeration Techniques

These methods are also used to manufacture granular products from insoluble pesticides. One such granulation method is depicted in Fig. 9. The pan granulator made by Dravo Corp., Neville Island, Pittsburgh, Pa., has been used to granulate fertilizer products and iron ore for many years. The powdered toxicant and carrier are preblended and proportioned onto the granulating pan while being sprayed with water or an aqueous binder solution. The particles tend to roll and build up, and when they attain the desired size, are automatically discharged out over the rim of the pan. The size of the granules depend on many operational factors such as rate of addition of the dry blend and liquid, the positions at which these materials are added to the pan, the speed of rotation of the pan, and the tilt of the pan. A wide distribution of sizes are obtained from the discharge of the pan and must be screened for the desired mesh size. The undersize granules can be returned to the blender to act as seeds, but the oversize must first be crushed before returning to the blender. The granular product must be dried and probably rescreened after drying.

Another machine with a different configuration but giving the same type of product as the pan granulator is the rotary drum granulator. This machine does not have as many variables and does not have the ability to discharge balls when they have grown to a desired size. The products made

TABLE 19

Typical Material Balance in Extrusion Process for Production
of 20% Pesticide "J" 30/60 Mesh Product

Step A.	Add the following to the pugger:	
	Pesticide "J"	20 parts/wt
	Attaclay (calcined attapulgite carrier)	60 parts/wt
	Attagel 150 (colloidal attapulgite binder)	12 parts/wt
	Water	38 parts/wt
Step B.	Extrude through 1/2-in. holes.	
Step C.	Dry in rotary dryer at 250°F bed temperature.	
Step D.	Mill for maximum 30/60 mesh yield.	
	Return fines to pugger.	
	The 30/60 mesh product should have the following composition:	
	Pesticide "J"	20
	Attaclay (attapulgite)	60
	Attagel 150 (binder)	12
	Water (residual)	8
	Total	100 wt %

by both of these machines tend to be lightweight and soft. However, Diamond Chemicals and A. L. Galloway (80) have overcome the hardness deficiency by use of dehydrated calcium sulfate (plaster of Paris) as the
prime inert in the production. The plaster of Paris is mixed with the
pesticide and water, formed into balls by one of the machines above, dried,
crushed, and screened to obtain the desired granular size. The plaster of
Paris when dried and dehydrated sets the material into attrition-resistant
balls.

A second type of equipment that can be used for granulation is the Patterson-Kelly V-Blender manufactured by the Patterson-Kelly Co., Inc.,
East Stroudsburg, Pa. A patent that has been obtained by Polon (81) on a
modification of the machine to improve the granulation efficiency teaches
the process for granulation of pesticidal products. In this process the
carrier and pesticide are charged to the V-blender, water or water/binder

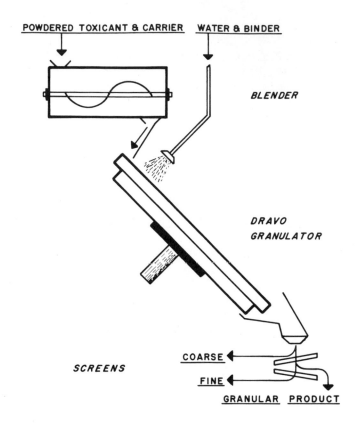

FIG. 9.   Granulation process.

solution is introduced through a dispersion bar, and the machine is run until spherodizing of the material takes place. As in the pan or drum granulators, the spherical yield of a specific mesh grade is 50% or less, and off-size material must be returned to the beginning of the process after adjustments. The spherical product must be dried to equilibrium moisture conditions. Thus far, this process has only been successful on a batch basis.

   A third type of equipment that can be used for agglomeration is the pressure roll maching. Three manufacturers of such machines are: (1) Allis-Chalmers, Milwaukee 1, Wisconsin, (2) Sprout, Waldron & Co., Inc., Muncy, Pa., and (3) Komarek-Greaves and Co., Rosemont, Ill. Usually a preblender is used to mix the toxicant, carrier, binder, and water, and the mixture is passed between the pressure rolls. If smooth rolls are used, the roll discharge will come out as flakes, which may or may not require drying depending on the formula and its components. Then the flakes must be crushed and screened to obtain the desired granular mesh distribution. Briqueting rolls to give particles as small as 1/8 in. are available from most companies. In the briqueting operation, little if any off-size material is obtained, and it is rarely followed by a drying step.

VII. POTENTIAL RESEARCH AREAS

The potential research areas in formulation of dry pesticidal products are many. Many millions of dollars and years of research are spent to develop a chemical with pesticidal properties, selectivity, and low mammalian toxicity, and its successful testing and use in the field depends upon how well the pesticide is formulated in the laboratory. And yet the same importance and monetary expenditures are not given to techniques of formulating. In fact, many of the pesticide companies do not have a formulating chemist, and depend on the large carrier companies, few of which have extensive research facilities, to formulate and prepare samples for field testing. It is generally agreed that it is getting more difficult to develop toxicants with the desired pesticidal properties, which also have the physical and chemical properties that will allow the use of simple and low-cost formulating methods, such as the direct impregnation method for granular products. More money and research must be spent on formulating techniques in the future. Some of those discussed above for granular products are not commercial, but they undoubtedly have merit, even though the processes are more complex, and the equipment and production are more costly.

Carrier/pesticide incompatibility has been encountered and known to exist for 20 years, and yet little has been done to determine the potential causes and modes of decomposition. The prime approaches taken thus far have been to find successful deactivators or to switch to more compatible carriers. Little work has been conducted thus far by the carrier producers in altering the carrier surfaces to eliminate the catalytic activity. However, because of the vast number and variety of pesticides, the cost for this work cannot be borne alone by the carrier producers, especially since the monetary value of the carrier in a product is low in comparison to the pesticide and other additives.

Some work is being conducted in the field of toxicant release, primarily because it is believed that the same control can be obtained with a fraction of the concentrations of active ingredients presently being used. If the

active ingredient could be released from granules in two steps, less active
ingredients should be required to obtain better results. The first step
would involve the release of a high concentration for a very short period
of time to obtain kill, and then followed by release of the chemical at a
very low rate over a long period of time to continue control. The present
granular products do not have such mechanisms. Some work has been conducted in attempts to increase or decrease the rate of release by use of
various impregnating solvents and/or surfactants. However, little has
been done in coating the final granular product to control the release.
Also, it has been found that, for a particular pesticide, the rate of release can vary depending on the carrier used. It is anticipated that granular products of the future will be comprised of two or more carrier bases,
or blends of granules that have been postsprayed with different coatings
that will dissolve at different rates.

It will be noted that only the rudiments of formulating were discussed
above. Where exact formulas were given as examples, these were taken
from published and readily available data, and do not disclose any proprietary information. In the discussion of various techniques for formulation of granular products, care was taken not to use any specific names,
since there are some newer products made by nonstandard techniques. It
is hoped that the above discussion will serve as a guide to the novice in
formulating the simple products, and perhaps to incite the experienced
formulating chemists to develop and utilize new formulating techniques.

REFERENCES

1. T. C. Watkins and L. B. Norton, Handbook of Insecticide Dust Di-
    luents and Carriers, Dorland Books, Caldwell, New Jersey, 1947.

2. D. D. Weidhaas and J. L. Brann, Jr., Handbook of Insecticide Dust
    Diluents and Carriers, 2nd ed., Dorland Books, Caldwell, New Jersey,
    1955.

3. ASTM D-281-31, Oil Absorption of Pigments by Spatula Rub-Out, Part
    20, American Society for Testing and Materials, Philadelphia, Pa.,
    1966, p. 161.

4. ASTM D-1483-60, Oil Absorption of Pigments by Gardner-Coleman
    Method, Part 20, American Society for Testing and Materials,
    Philadelphia, Pa., 1966, pp. 670-1.

5. H. A. Gardner and G. G. Sward, Physical and Chemical Examination
    Paints, Varnishes, Lacquers and Colors, 12th ed., Gardner Labora-
    tory, Inc.

6. Floridin, Technical Data and Product Specifications - Fullers Earth and Activated Bauxite, Floridin Company, Tallahassee, Fla., pp. 27-28.

7. M. A. Malina, "Flowability of Dust Formulations," Agri. Chem., September 1960.

8. Engelhard Minerals & Chemicals Corp., Flowability of Dry and Impregnated Dusts, Diluents, and Carriers, Standard Method No. 4252, Menlo Park, Edison, N.J.

9. T. D. Oulton, Encyclopedia of Industrial Chemical Analysis, Vol. 4, Wiley, New York, 1967, pp. 440-441.

10. K. T. Whitby, "The Mechanics of Fine Sieving," Am. Soc. Testing Mater. Spec. Tech. Publ. 234, 3-25 (1959).

11. Ref. 9, pp. 441-2.

12. A. H. M. Andreason, Angew. Chem., 20, 283 (1935).

13. E. H. Amstein and B. A. Scott, J. Appl. Chem. (London), 1, 510 (1951).

14. G. Herder, Small Particle Statistics, Elsevier, Amsterdam, 1953.

15. ASTM D-422-63, Method for Grain Size of Soils, Part 11, American Society for Testing and Materials, Philadelphia, Pa., 1964, pp. 194-205.

16. TAPPI T649 sm-54, Technical Association of Pulp and Paper Industry, New York, N.Y.

17. J. F. Lesveaux, H. West, and F. S. Black, J. Ag. Food Chem., 2(20), 1022-4 (1954).

18. P. S. Roller, "Metal Powder Size Distribution with the Roller Analyzer," Am. Soc. Testing Mater. Spec. Tech. Publ. 140, 54 (1950).

19. E. L. Gooden and C. M. Smith, Ind. Eng. Chem. (Anal. Ed.), 12, 497 (1940).

20. R. P. Loveland, "Methods of Particle-Size Analysis," Am. Soc. Testing Mater. Spec. Tech. Publ. 234, 57-88 (1959).

21.   R. R. Irani and C. F. Callis, Particle Size: Measurement, Interpretation, and Application, Wiley, New York, 1963, pp. 93-106.

22.   Ref. 20, pp. 71-3.

23.   Ref. 9, p. 443.

24.   Ref. 5, p. 210.

25.   Engelhard Minerals and Chemicals Corp., Determination of the Tamped Volume Weight or Bulk Density of Granular and Powdered Materials, Standard Method No. 3341, Menlo Park, Edison, N.J.

26.   NACA Granular Pesticide Committee Bulletin Reference M-3, National Agricultural Chemicals Association, Washington, D.C. p. 188 (March 19, 1962).

27.   J. A. Polon, Method for Determining the Flowability and Pesticide Sorptive Capacity of Granular Inerts, Standard Method No. 4254, Engelhard Minerals and Chemicals Corporation, Menlo Park, Edison, N.J.

28.   J. A. Polon, Method for Determining Saturation Sorptive Capacity of Granular Inerts, Standard Method No. 4251, Engelhard Minerals and Chemicals Corporation, Menlo Park, Edison, N.J.

29.   ASTM E-11-61, Specification for Sieves for Testing Purposes (Wire Cloth Sieves, Round-Hole and Square-Hole Plate Screens or Sieves), American Society for Testing and Materials, Philadelphia, Pa., 1965.

30.   Technical Information Bulletin No. 153, Agricultural Products for Granular Formulations, Engelhard Minerals & Chemicals Corporation, Menlo Park, Edison, N.J.

31.   H. M. Gwyn, Jr., "Determining the Particle Count Per Pound of Granular Pesticides," Agri. Chem., June (1964).

32.   Interim Federal Specifications O-I-00528 (AGR-ARS) February 6, 1958 (revised March 7, 1961).

33.   Engelhard Minerals & Chemicals Corp., Determination of Water Disintegration of Granular Carriers and Diluents, Standard Method No. 3132.3, Menlo Park, Edison, N.J.

34. Engelhard Minerals & Chemicals Corp., Determination of Hardness
     of Granular Fullers Earth and Bauxite, Standard Method No. 3111,
     Menlo Park, Edison, N.J.

35. ASTM D-547-41, Test for Index of Dustiness of Coal and Coke,
     Part 8, American Society for Testing and Materials, Philadelphia,
     Pa., 1961, pp. 1299-1302.

36. Engelhard Minerals & Chemicals Corp., Determination of Dustiness
     of Granular Materials, Standard Method No. 3131, Menlo Park,
     Edison, N.J.

37. D. S. McKinney, P. Fugassi, and J. C. Warner, ASTM Technical
     Publication No. 73, American Society for Testing Materials,
     Philadelphia, Pa., 1946.

38. L. P. Hammett et al., J. Am. Chem. Soc., 52, 4795 (1930).

39. L. P. Hammett et al., J. Am. Chem. Soc., 54, 2721 (1932).

40. L. P. Hammett et al., J. Am. Chem. Soc., 54, 4239 (1932).

41. L. P. Hammett, Physical Organic Chemistry, McGraw-Hill, New
     York, 1940, pp. 273-8.

42. C. J. Walling, J. Am. Chem. Soc., 72, 1164 (1950).

43. O. Johnson, J. Phys. Chem., 59, 827 (1955).

44. H. A. Benesi, Y. P. Sun, E. S. Loeffler, and K. D. Detling (to
     Shell Development Co.), U.S. Pat. 2,868,688 (January 13, 1959).

45. F. M. Fowkes, H. A. Benesi, L. B. Ryland, W. M. Sawyer,
     K. D. Detling, E. S. Loeffler, F. B. Folckemer, M. R. Johnson
     and Y. P. Sun, J. Agri. Food Chem., 8, 203 (1960).

46. M. A. Malina, A. Goldman, L. Trademan and P. B. Polen, J. Agr.
     Food Chem., 4, 1038 (1956).

47. L. Trademan, M. A. Malina, and L. P. Wilks (to Velsicol Chemical
     Corp.), U.S. Pat. 2,875,119 (February 24, 1959).

48. L. Trademan, M. A. Malina, and L. P. Wilks (to Velsicol Chemical
     Corp., U.S. Pat. 2,875,120 (February 24, 1959).

49.   L. Trademan, M. A. Malina, and L. P. Wilks (to Velsicol Chemical Corp.), U.S. Pat. 2,875,121 (February 24, 1959).

50.   Shell, Handbook of Aldrin, Dieldrin, and Endrin Formulations, 2nd ed., Shell Chemical Corp., New York, N.Y., June, 1959.

51.   Shell, Appendix, Shell Method Series 571/58, Determination of Urea Requirements for the Deactivation of Solid Carriers of Aldrin and Dieldrin.

52.   Shell, Appendix, Shell Method Series 565/58, Determination of HMT Requirements for the Deactivation of Solid Carriers of Endrin.

53.   J. B. McPherson, Jr., and G. A. Johnson, J. Agr. Food Chem., 4(1), 1956.

54.   J. F. Yost, I. B. Frederick, and V. Migrdichian, Agr. Chem., 10(9), 43-5, 127, 139; (10) 42-4, 105, 107 (1955).

55.   J. F. Yost and I. B. Frederick, Farm Chem., 122, 64 (1959).

56.   E. W. Sawyer, Jr. (to Minerals & Chemicals Philipp Corp.), U.S. Pat. 2,962,418 (November 29, 1960).

57.   E. W. Sawyer, Jr. and J. A. Polon (to Minerals and Chemicals Philipp Corp.), U.S. Pat. 2,967,127 (January 3, 1961).

58.   J. A. Polon and E. W. Sawyer, Jr., J. Agr. Food Chem., 10(3), 244 (1962).

59.   L. Trademan, M. A. Malina, and L. P. Wilks (to Velsicol Chemical Corp.), U.S. Pat. 2,927,882 (March 8, 1960).

60.   Formulation Manual No. 514-1, Parathion (Technical) and Methyl Parathion, Velsicol Chemical Corp., Chicago, Ill., December, 1957.

61.   Niagara, Formulation Information for Ethion, FMC Corp., Niagara Chemical Division, Middleport, New York., February 15, 1965.

62.   I. A. Schwint (to Minerals & Chemicals Philipp Corp.), U.S. Pat. 3,232,831 (February 1, 1966).

63.   N. R. Oros and R. D. Vartanian (to American Cyanamid Co.), U.S. Pat. 2,970,080 (January 31, 1961).

64. Shell, Handbook of Phosdrin Insecticide Formulations, No. SC:57-28, Shell Chemical Corp., New York, N.Y., June, 1957.

65. F. W. Plapp and J. E. Casida, J. Agr. Food Chem., 6, 662 (1958).

66. C. Rosenfield and W. Van Valkenburg, J. Agr. Food Chem., 13(1), 68-72 (1965).

67. M. M. Mortland and K. V. Raman, J. Agr. Food Chem., 15(1), 163-7 (1967).

68. R. L. Metcalf and T. R. Fukuto, J. Agr. Food Chem., 15(6), 1022-9 (1967).

69. W. J. Entley, D. C. Blue, and H. A. Stansbury, Jr., Farm Chem., pp. 52, 53, 56, 58, 60 (June, 1965).

70. Union Carbide, Sevin Carbaryl Insecticide Manual for Formulators, No. F-41-54, Union Carbide Corp., New York, N.Y., February, 1964.

71. Niagara, Formulation Information for Thiodan, FMC Corp., Niagara Chemical Division, Middleport, New York, February 15, 1965.

72. F. F. Smith, E. L. Gooden, and E. A. Taylor, J. Econ. Ent., 48(6), 762-3 (1955).

73. Aramite Technical Formulators Handbook, Naugatuck Chemical, Division of United States Rubber Co., February, 1954.

74. Ref. 50, Appendix, Shell Method Series 586/58, Determination of Suspendibility of Wettable Powders.

75. WHO, Specifications for Pesticides, 2nd ed., World Health Organization, Geneva, Switzerland, 1961, pp. 85-138.

76. F. B. Folckemer and R. G. Dimartini, "Particle Size Distribution of High-Suspendibility Dieldrin Wettable Powders," Paper No. 926, Shell Chemical Corp., New York, June 4, 1959.

77. F. B. Folckemer and A. Miller, "Some Factors Affecting The Storage Stability of Dieldrin Wettable Powders," Paper No. 1026, Shell Chemical Corp., New York, September, 1960.

78.   J. W. Miles and M. B. Goette, J. Agr. Food Chem., 16(4), 635-8
      (1968).

79.   J. O. Marples and E. W. Sawyer, Jr. (to Minerals & Chemicals
      Philipp Corp.), U.S. Pat. 3,062,637 (November 6, 1962).

80.   A. L. Galloway (to Diamond Alkali Co.), U.S. Pat. 3,056,723
      (October 2, 1962).

81.   J. A. Polon (to Minerals and Chemicals Philipp Corp.), U.S. Pat.
      3,192,290 (June 29, 1965).

APPENDIX A

List of Pesticide Carriers and Suppliers

Attapulgites

Pulgite	Dresser Minerals, Div. Dresser Industries, Dallas, Texas
Granulex	Dresser Minerals, Div. Dresser Industries
Attaclay	Engelhard Minerals & Chemicals Corp., Menlo Park, Edison, N.J.
Attaclay X-250	Engelhard Minerals & Chemicals Corp.
Granular Attaclays	Engelhard Minerals & Chemicals Corp.
Attasorbs RVM and LVM	Engelhard Minerals & Chemicals Corp.
Attacote	Engelhard Minerals & Chemicals Corp.
Diluex	Floridin Co., Pittsburgh, Pa.
Diluex A	Floridin Co.
Florex	Floridin Co.
Micro-Sorb RVM and LVM	Floridin Co.
Micro-Cote Micromesh Grade	Floridin Co.
Megsite	Waverly Petroleum Products Co., Philadephia, Pa.
#1304 Clay	Whittaker, Clark, & Daniels, Inc., New York, N.Y.

APPENDIX A (continued)

Diatomites

Aquafil	Aquafil Co., Los Altos, Calif.
Frianite M3X	California Industrial Minerals Co., Friant, Calif.
Frianite TP	California Industrial Minerals Co.
Aquafil	E.I. du Pont de Nemours & Co., Wilmington, Del.
Celatom MN-35	Eagle-Picher, Cincinnati, Ohio
Celatom MN-39	Eagle-Picher
Celatom MP-61	Eagle-Picher
Celatom MP-78	Eagle-Picher
Dicalite IG-3	Great Lakes Carbon Corp., New York, N.Y.
Dicalite IG-5	Great Lakes Carbon Corp.
Dicalite IG-33	Great Lakes Carbon Corp.
Dicalite SA-5	Great Lakes Carbon Corp.
Dicalite 109-3	Great Lakes Carbon Corp.
Dicalite 2065	Great Lakes Carbon Corp.
Celite 209	Johns-Manville Products Corp., New York, N.Y.
Celite 400	Johns-Manville Products Corp.
Celite 394 EE	Johns-Manville Products Corp.
Celite SSC	Johns-Manville Products Corp.

APPENDIX A (continued)

Celkate	Johns-Manville Products Corp.
Kenite 75	Kenite Corp., Scarsdale, N.Y.
Kenite 102	Kenite Corp.
Kenite 2200	Kenite Corp.
#68 Multicell	Tamms Industries Inc., Chicago, Ill.
Tamms Diatomaceous Silica	Tamms Industries Inc.
Quaker Diatomite	Charles A. Wagner Co., Philadelphia, Pa.
Hysorb Clay	Wharton-Jackson Co., Los Angeles, Calif.

Fullers Earth

Emathlite VMP 38	Mid-Florida Mining Co., Lowell, Fla.
Emathlite VMP 150	Mid-Florida Mining Co.
Emathlite VMP 3000	Mid-Florida Mining Co.
Emathlite (granules)	Mid-Florida Mining Co.

Montmorillonites

Pikes Peak Clays Types 9401, D6, G6, H6, J6, S6, and T6	General Reduction Co., Chicago, Ill.
Wilkinite	Harry Haze, Inc., Chicago, Ill.
PRC Dust	Pearl River Clay Co., McNeill, Miss.

APPENDIX A (continued)

Montmorillonites (cont)

    Fesco-Jel   F. E. Schundler & Co.,
                                                   Joliet, Ill.

    Aero Dusts "F" and "G"   Southern Clay Products Inc.,
                                                   Gonzales, Texas

    Creek-O-Nite #6   Star Enterprises, Inc.
                                                   Cassapolis, Mich.

    Creek-O-Nite Granules   Star Enterprises, Inc.

    Adsorbol F-100   United Sierra Div., Cyprus Mines Corp.,
                                                   Trenton, N.J.

    Argosite Clay   United Sierra Div., Cyprus Mines Corp.

    Inca Clay   United Sierra Div., Cyprus Mines Corp.

    Clay DM   United Sierra Div., Cyprus Mines Corp.

    Olancha Clay Types BW & MO   United Sierra Div., Cyprus Mines Corp.

Vermiculites

    Vermiculite   American Vermiculite Corp.,
                                                   Roan Mountain, Tenn.

    Vermiculite (Exfoliated)   C. R. Graybeal and Sons,
                                                   Roan Mountain, Tenn.

    #2626 325 Mesh Expanded   Whittaker, Clark, & Daniels,
       Vermiculite   New York, N.Y.

    #2629 Delaminated Vermicu-   Whittaker, Clark, & Daniels
       lite Fines

    Vermiculite No. 4   Zonolite Div., W.R. Grace & Co.,
                                                   Chicago, Ill.

    Terralite   Zonolite Div., W.R. Grace & Co.

APPENDIX A (continued)

Synthetics

Flo-Float	Alabama Calcium Products, Co., Gantts Quarry, Ala.
Cab-O-Sil	Cabot Corporation, Boston, Mass.
CCC Diluent	Calcium Carbonate Co., Chicago, Ill.
Hi-Sil C	Columbia Chemical Div., Pittsburgh Plate Glass Co., Pittsburgh, Pa.
Hi-Sil 101	Columbia Chemical Div., Pittsburgh Plate Glass Co.
Hi-Sil 202	Columbia Chemical Div., Pittsburgh Plate Glass Co.
Hi-Sil 233	Columbia Chemical Div., Pittsburgh Plate Glass Co.
Hi-Sil 266X	Columbia Chemical Div., Pittsburgh Plate Glass Co.
Calcene NC	Columbia Chemical Div., Pittsburgh Plate Glass Co.
Silene EF	Columbia Chemical Div., Pittsburgh Plate Glass Co.
Millical Brand $CaCO_3$	Diamond Alkali Co., Cleveland, Ohio
Non-Fer-A1 Brand	Diamond Alkali Co.
Zeolex 7A	J. M. Huber Corp., New York, N.Y.
Zeosyl 10	J. M. Huber Corp.,

APPENDIX A (continued)

Synthetics (cont)

    Zeothix 60   J. M. Huber Corp.,
                                        New York, N.Y.

    Micro-Cel   Johns-Manville Products Corp.,
                                         New York, N.Y.

    Vansil   R. T. Vanderbilt Co.,
                                         New York, N.Y.

APPENDIX B

List of Pesticide Diluents and Suppliers

Aluminum Silicates

    Sorbex   Roger G. Brown Co.,
                                           Macon, Ga.

    Bleaching Clay #140   Industrial Minerals & Chemical Co.,
                                           Berkeley, Calif.

    Ser-Fil   Industrial Minerals Inc.,
                                           York, S.C.

    Rev Dust   Milwhite Co.,
                                           Houston, Texas

Apatite

    Tricalcium Phosphate-   Monsanto Chemical Co.,
      Conditioner Grade   Inorganic Div.,
                                           St. Louis, Mo.

Bentonite

    Panther Creek Bentonite   American Colloid Co.,
                                           Skokie, Ill.

    Standard Powdered Volclay   American Colloid Co.
      (SPV) 200 Mesh

    Volclay KWK - Granular   American Colloid Co.

    Volclay No. 90 - Granular   American Colloid Co.

    Wyo Jel   Archer Daniels Midland Co.,
                                           Minneapolis, Minn.

APPENDIX B (continued)

Bentonite (cont)

Wyo Jel–C	Archer Daniels Midland Co., Minneapolis, Minn.
Wyo Bond	Archer Daniels Midland Co.
Bentonite	Baroid Chemical, Inc., Houston, Texas
Thixo–Jel	Georgia Kaolin Co., Elizabeth, N.J.
Black Hills Bentonite	Eastern Clay Products Dept., International Minerals & Chemicals Corp., Skokie, Ill.
Dixie Bond	Eastern Clay Products Dept., International Minerals & Chemicals Corp.
Bentonite	Magnet Cove Barium Corp., Houston, Texas
T.A.T. Bentonite	Tamms Industries, Inc., Chicago, Ill.
#49 Bentonite – 200 Mesh S.P.V.	Whittaker, Clark, & Daniels, Inc., New York, N.Y.
#475 Bentonite – Panther Creek	Whittaker, Clark, & Daniels, Inc.
#890 Bentonite – Volclay S.P.V.	Whittaker, Clark, & Daniels, Inc.
#891 Bentonite – Volclay KWK	Whittaker, Clark, & Daniels, Inc.
#2245 Insecticide – KY–1 Clay	Whittaker, Clark, & Daniels, Inc.
Wyo–Bond Bentonite	The Wyodak Chemical Div., Federal Foundry Supply Co., Cleveland, Ohio

APPENDIX B (continued)

Limestones and Lime

Calcium Carbonate	Allied Chemical Corp., New York, N.Y.
Calcium Carbonate	Austin White Lime Co., Austin, Texas
Carbex	Roger G. Brown Co., Macon, Ga.
VEE Dust Filler	California Chemical Co., Richmond, Calif.
Grade 98V	Columbia Quarry Co., St. Louis, Mo.
Pulverized Limestone	Columbia Quarry Co.
Commercial Minerals Whiting	Commercial Mineral Co., San Francisco, Calif.
Corson's Actomag	G. & W.H. Corson, Plymouth Meeting, Pa.
Corson's Agricultural Hydrated Lime	G. & W. H. Corson
Corson's Ground Burnt Lime	G. & W.H. Corson
Corson's High Magnesium Spray Lime	G. & W.H. Corson
Corson's Pulverized Limestone	G. & W.H. Corson
AERO Limestone Grade FF	Cyanamid of Canada, Ltd., Montreal, P.Q., Canada
Calcium Carbonate	Diamond Alkali Co., Cleveland, Ohio
Dilcard (#10 White)	The Georgia Marble Co., Tate, Ga.

APPENDIX B (continued)

Limestones and Lime (cont)

Georgia Marble No. 10	The Georgia Marble Co., Tate, Ga.
No. 10 T	The Georgia Marble Co.
Talcarb	The Georgia Marble Co.
Spray Hydrated Lime	M. J. Grove Lime Co., Div. of Flint Kote, Lime Kiln, Md.
Chemical Pebble Lime	M. J. Grove Lime Co., Div. of Flint Kote
Dolomite	James River Hydrate & Supply Co., Buchanan, Va.
Lime Crest I. F. No. 100	Limestone Products Corp. of America, Newton, N.J.
Lime Crest Calcite Superior Hydrated Lime	Limestone Products Corp. of America,
Calcium Carbonate	Marblehead Lime Co., Chicago, Ill.
Airfloated Pulverized Limestone	H. E. Millard Lime and Stone Co., Annville, Pa.
Millard's Modern Chemical Hydrated Lime	H. E. Millard Lime and Stone Co.
Mississippi Dust Diluent	Mississippi Lime Co., Alton, Ill.
Gold Bond Spraying & Dusting Hydrated Lime	National Gypsum Co., Buffalo, N.Y.
Calcium Carbonate	New England Lime Co., Adams, Mass.

APPENDIX B (continued)

Ohio Superspray Hydrated Line	The Ohio Hydrate & Supply Co., Woodville, Ohio
Oyster Shell	Oyster Shell Products, New Rochelle, N.Y.
VLR Diluent	Charles Pfizer & Co., Easton, Pa.
Magnesium Carbonate	Selco Supply Co., Eton, Colorado
Tamms Calcium Carbonate	Tamms Industries, Inc., Chicago, Ill.
Kaluent #1	Thompson, Weinman & Co., New York, N.Y.
Kaluent #2	Thompson, Weinman & Co.
Bell-Mine High Calcium Spray Hydrated Lime	Warner Co., Philadelphia, Pa.
Bell-Mine Rotary Kiln Pulverized Lime	Warner Co.
#629 Whiting	Whittaker, Clark, & Daniels, Inc. New York, N.Y.
#1701 Norris Filler	Whittaker, Clark, & Daniels, Inc.
VLR Diluent (CCC)	C. F. Williams Co., Los Angeles, Calif.
Purecal O	Wyandotte Chemical Corp., Wyandotte, Mich.
Calwhite	Wyrough & Loser, Trenton, N.J.
Calwhite T	Wyrough & Loser
No. 10 White	Wyrough & Loser

APPENDIX B (continued)

Limestones and Lime (cont)

    Talcarb   Wyrough & Loser,
                                              Trenton, N.J.

    Gamatalcarb   Wyrough & Loser

Calcium Sulfate

    Snow White Filler   U.S. Gypsum,
                                              Chicago, Ill.

    Ben Franklin Dusting Gypsum   U.S. Gypsum

Kaolinites

    Clay P-3 S   Albion Kaolin Div.,
                                              Interchemical Corp.,
                                              New York, N.Y.

    Prill Grade Kaolin   Bell Kaolin,
                                              Batesburg, S.C.

    Mileestee Clay (kaolin #16)   Roger G. Brown Co.,
                                              Macon, Ga.

    Mileestee Clay (kaolin #20)   Roger G. Brown Co.

    Mileestee Clay (kaolin #22)   Roger G. Brown Co.

    Sax Clay   Charles B. Chrystal Co.,
                                              New York, N.Y.

    Dilu-Dust   Dresser Industries, Inc.,
                                              Dallas, Texas

    Kao-X   Engelhard Minerals & Chemicals Corp.
                                              Menlo Park, Edison, N.J.

    Kaolin   Freeport Sulphur Co.,
                                              New York, N.Y.

APPENDIX B (continued)

Standard Air Float	Georgia Kaolin Co., Elizabeth, N.J.
Airfloat Kaolin PD 181	Georgia Kaolin Co.
Pioneer Air Float	Georgia Kaolin Co.
Inert "C"	Hammill & Gillespie, Inc., New York, N.Y.
Barden Clay	J. M. Huber Corp., New York, N.Y.
Barden AG Clay	J. M. Huber Corp.
Hi white R Clay	J. M. Huber Corp.
Nulflo	J. M. Huber Corp.
Suprex Clay	J. M. Huber Corp.
Suprex T	J. M. Huber Corp.
Suprex LG Clay	J. M. Huber Corp.
B-5-6 Clay	Industrial Minerals & Chemical Co., Berkeley, Calif.
Palmetto Clay	International Clay Co., Graniteville, S.C.
Xact Clay 811	Magnet Cove Barium Corp., Houston, Texas
Wilco Brand Clay	M & M Clays, Inc., McIntyre, Ga.
Narvon IF2	Narvon Mines Ltd., Lancaster, Pa.
Narvon Agri-F2	Narvon Mines Ltd.
Narvon Agri-F3	Narvon Mines Ltd.

APPENDIX B (continued)

Kaolinites (cont)

Natka Clay	National Kaolin Products, Aiken, S. C.
Natka Clay-D	National Kaolin Products
M Clay	F. E. Schundler & Co., Joliet, Ill.
Velvex Clay	Southeastern Clay Co., Aiken, S.C.
Kaocote	Southeastern Clay Co.
Type 41	Southeastern Clay Co.
Spinks Extender Clay	H. C. Spinks Clay Co., Cincinnati, Ohio
#2 Tamfloss Clay	Tamms Industries, Inc., Chicago, Ill.
Tako Airfloated Colloidal Kaolin	Thomas Alabama Kaolin Co., Baltimore, Md.
Kaolloid Clay	Thompson, Weinman & Co., New York, N.Y.
Bancroft Clay	United Sierra Div., Cyprus Mines Corp., Trenton, N.J.
Barnet Clay	United Sierra Div., Cyprus Mines Corp.
Brunswick Clay	United Sierra Div., Cyprus Mines Corp.
Dilex	United Sierra Div., Cyprus Mines Corp.
Extendite "S" Clay	United Sierra Div., Cyprus Mines Corp.
Franklin Clay	United Sierra Div., Cyprus Mines Corp.
Franklin WW Clay	United Sierra Div., Cyprus Mines Corp.

APPENDIX B (continued)

Homer Clay   United Sierra Div., Cyprus Mines Corp.

Perry Clay   United Sierra Div., Cyprus Mines Corp.

Saxon Clay   United Sierra Div., Cyprus Mines Corp.

Continental Clay   R. T. Vanderbilt Co.,
                                        New York, N.Y.

#1431 Clay (Run of the Mine)   Whittaker, Clark, & Daniels, Inc.,
                                        New York, N.Y.

#1433 (Kaolloid) Clay   Whittaker, Clark, & Daniels, Inc.

#2451 Clay - Standard Air   Whittaker, Clark, & Daniels, Inc.
    Floated

Micas

Mineralite Mica 3X, 4X,   Mineralite Sales Corp.,
    and 5X   New York, N.Y.

Ser-X Clay   Summit Mining Corp.,
                                        Carlisle, Pa.

H 15   Summit Mining Corp.

Triple A Grade Mica   Thompson, Weinman & Co.,
                                        New York, N.Y.

Perlites

Panacalite   Combined Metals Reduction Co.,
                                        Salt Lake City, Utah

Ryolex Aggregate   Silbrico Corp.,
                                        Chicago, Ill.

Tenn Flo   Tennessee Products and Chemical
                                        Corp.,
                                        Nashville, Tenn.

APPENDIX B (continued)

Phosphate Rock

Phosphodust   American Agricultural Chemical Co.,
                                     New York, N.Y.

Pumice

Frianites 4AR, 4ARS, 16X,   California Industrial Minerals Co.,
   25X, and TP   Friant, Calif.

Vol Dust   Valley Brick and Tile Co.,
                                     Mission, Texas

Pyrophyllites

Air Float #1 Pyrophyllite   A. C. Drury & Co.,
                                     Chicago, Ill.

Glendon Pyrophyllite   General Minerals Co.,
                                     Greensboro, N.C.

Pyrolite   Huntly Industrial Minerals,
                                     Bishop, Calif.

Pyrophyllite (Kennedy)   Kennedy Minerals Co., Inc.,
                                     Los Angeles, Calif.

Pyrophyllite   Los Angeles Chemical Co.,
                                     South Gate, Calif.

Nuclay   Charles Pfizer & Co.,
                                     Easton, Pa.

Pyrax B   R. T. Vanderbilt,
                                     New York, N.Y.

Pyrax ABB   R. T. Vanderbilt

Pyrax Granules   R. T. Vanderbilt

Diluex B #261   Whittaker, Clark, & Daniels, Inc.,
                                     New York, N.Y.

APPENDIX B (continued)

#29 Pyrophyllite   Whittaker, Clark, & Daniels, Inc.

#261 Pyrophyllite   Whittaker, Clark, & Daniels, Inc.

Pyrolite   Wharton-Jackson Co.,
                                          Los Angeles, Calif.

Silica

Amorphous Silica   Illinois Minerals Co.,
                                          Cairo, Ill.

Sulfur

Electric Brand Dusting   Stauffer Chemical Co.,
  Sulfur (Code 3-1)   New York, N.Y.

Owl Brand Dusting Sulfur   Stauffer Chemical Co.,
  (Code 1-1)

Perfection Brand Dusting   Stauffer Chemical Co.
  Sulfur (Code 2-1)

Swan Brand Dusting Sulfur   Stauffer Chemical Co.
  (Code 4-1)

Talcs

Talc No. 202   American Mineral Co.,
                                          Los Angeles, Calif.

Clatal   American Talc Co.,
                                          Chatsworth, Ga.

Highwater Talc   Baker Talc Ltd.,
                                          Highwater, Quebec, Canada

Blue Soapstone   Blue Ridge Talc Co., Inc.,
                                          Henry, Va.

Talc   Broughton Soapstone and Quarry Co.
                                          Broughton Station, Quebec, Canada

APPENDIX B (continued)

Talcs **(cont)**

Asbestol Regular	Carbola Chemical Co., Natural Bridge, N.Y.
Asbestol Superfine	Carbola Chemical Co.
Micro Velva A	Carbola Chemical Co.
No. 270 Talc	Clinchfield Sand and Feldspar Corp., Townson, Md.
Cohutta Talc	Cohutta Talc Co., Dalton, Ga.
Talc	Commercial Mineral Co., San Francisco, Calif.
Emtal 43	Engelhard Minerals & Chemicals Corp., Menlo Park, Edison, N.J.
Talc	Georgia Kaolin Co., Elizabeth, N.J.
A - White Talc #901	Georgia Talc Co., Chatsworth, Ga.
Soapstone	Industrial Minerals and Chemical Corp., Berkeley, Calif.
Talc P-28	Kennedy Mineral Co., Inc., Los Angeles, Calif.
Ken Talc	Kennedy Mineral Co., Inc.
Soapstone	Liberty Talc Mines, Inc., Sykesville, Md.
Loomkill Talc	W. H. Loomis Talc Corp., Gouverneur, N.Y.
Soapstone	Marquette Products, Ltd., Quebec, P.Q., Canada
Mefford Talc (velvetone)	Mefford Chemical Co., Los Angeles, Calif.
Clear Lake Talc	Carl F. Miller Co., Seattle, Wash.

APPENDIX B (continued)

Mil-Slip N.W.	Milwhite Co., Inc., Houston, Texas
Mil-Slip Regular	Milwhite Co., Inc.
Milwhite Soapstone	Milwhite Co., Inc.
Code #961	Southern Talc Co., Chatsworth, Ga.
Llano Talc	Southwestern Talc Corp., Llano, Texas
Grade A-250 Talc	Southwestern Talc Corp.
Roofing Talc (30 Mesh to 325)	Summit Industries, Carlisle, Pa.
6 J Gray Talc	Tamms Industries, Inc., Chicago, Ill.
Mistron T-076	United Sierra Div., Cyprus Mines Corp., Trenton, N.J.
Mistron Talc 139	United Sierra Div., Cyprus Mines Corp.
Pyrotalc	United Sierra Div., Cyprus Mines Corp.
Sierra White IR	United Sierra Div., Cyprus Mines Corp.
Sierra Cloud Talc	United Sierra Div., Cyprus Mines Corp.
Nytal Talc Types 100 & 200	R. T. Vanderbilt Co., New York, N.Y.
Talc	Vermont Talc Co., Chester, Vt.
De Sal Talc	Wharton Jackson Co., Los Angeles, Calif.
Sawed Talc (Saline Valley Talc)	Wharton Jackson Co.
Velvetone Talc	Wharton Jackson Co.
Clatal	Whittaker, Clark, & Daniels, Inc., New York, N.Y.
#5 Canadian Talc	Whittaker, Clark, & Daniels, Inc.

APPENDIX B (continued)

Talcs (cont)

#150 Vermont Talc   Whittaker, Clark, & Daniels, Inc.,
                                     New York, N.Y.

#367 Georgia Talc   Whittaker, Clark, & Daniels, Inc.

#1094 Talc   Whittaker, Clark, & Daniels, Inc.

#1367 Georgia Talc   Whittaker, Clark, & Daniels, Inc.

#2367 Talc   Whittaker, Clark, & Daniels, Inc.

#2952 Clatal   Whittaker, Clark, & Daniels, Inc.

#2953 Clatal   Whittaker, Clark, & Daniels, Inc.

Tripolite

RD-1 Ground Silica   Illinois Minerals Co.,
                                     Cairo, Ill.

RD-2 Ground Silica   Illinois Minerals Co.

Double Ground Rose Tripoli   Tamms Industries, Inc.,
                                     Chicago, Ill.

#85 Tripoli   Whittaker, Clark, & Daniels, Inc.,
                                     New York, N.Y.

#2 Double Ground Cream   Whittaker, Clark, & Daniels, Inc.

Unidentified Minerals

Amcolite No. 1   American Mineral Co.,
                                     Los Angeles, Calif.

Amcolite No. 5AA   American Mineral Co.

Pigment No. 28   Kennedy Minerals Co., Inc.,
                                     Los Angeles, Calif.

APPENDIX B (continued)

Francite   Ol' Rebel Minerals, Inc.,
                                        San Francisco, Calif.

Gamaco   Wyrough & Loser,
                                        Trenton, N.J.

Botanicals

  Vegetable Shell Flours   Agrashell, Inc.,
    WF-3, WF-5 & WF-7   Los Angeles, Calif.

  Corn Cob Granules   Anderson Cob Mill, Inc.,
                                        Maumee, Ohio

  Kaysoy 200 (Soybean Flour)   Archer Daniels Midland Co.,
                                        Minneapolis, Minn.

  Solka Floc (BW-40)   Brown Co.,
                                        New York, N.Y.

  Solka Floc (BNB-40)   Brown Co.

  Solka Floc (fil B)   Brown Co.

  Ground Tobacco Stems   Chemical Formulators, Inc.,
                                        Nitro, W. Va.

  Parmite no. 5   Composition Materials Co., Inc.,
                                        New York, N.Y.

  Corncob Grits   The C. P. Hall Co.,
                                        Akron, Ohio

  Corn Cob Grits, Meal,   Kobrite Div., John N. Bos Sand
    and Flour   Company
                                        Chicago, Ill.

  Lo-Fat Soy Flour   A. E. Staley Mfg. Co.,
                                        Decatur, Ill.

  Tobacco Stems   Virginia Carolina Chemical Co.,
                                        Richmond, Va.

APPENDIX B (continued)

Botanicals (cont)

    Sterilized Tobacco Diluent   Virginia Carolina Chemical Co.,
                                     Richmond, Va.

    Gold Leaf Tobacco Dust   Virginia Carolina Chemical Co.

Synthetic Organic

    Furafil M   Quaker Oats Co.,
                                     Chicago, Ill.

    Furafil C   Quaker Oats Co.

    Furafil 100   Quaker Oats Co.

Chapter 6

PLANT FOR THE FORMULATIONS OF INSECTICIDES

J. F. Vine

C. Eng. M. I. Mech. E.
Chigwell, Essex, England

I. INTRODUCTION

The production of insecticides, herbicides, and allied products in the forms used by agriculture and horticulturists involves both the art of the chemist to produce toxic technical ingredients of good insecticidal killing properties and the skills of the mechanical engineer to convert these often difficult-to-handle and process materials into products that comply with known and/or established international formulation specifications. Ideally, the roles of the formulation chemist and of the engineer in the selection and processing of test formulations are integrated, with the arts and technique of both combined to produce a satisfactory end product.

This chapter deals with some of the many aspects and types of equipment used in the production of formulated insecticide powders in their various forms. The three principal groups into which powdered products fall
are: (1) dust concentrates, which contain the minimum of filler carrying
agents compatible with satisfactory mill operations, (2) field strength dusts,
which are dust concentrates diluted with a suitable diluent material to toxicity levels suitable for field application, and (3) wettable powders, which
contain a maximum of toxicant compatible with satisfactory milling and
mill operation and with the physiochemical properties of the formulation.
These three categories broadly form the principal groupings of products and aspects of manufacturing processes. In each of the categories,
the end product is in the form of a combination of either solid or liquid
technical ingredients formulated with suitable fillers or diluents and additives, compounded into an homogeneous mixture.

The extent, size, and degree of sophistication of an insecticide formulation plant, like a great number of other process plants, is related to output, product range and other economic factors which will not be referred to
here. The chapter therefore deals first with the preparation of filler and
diluent ingredients.

II. INTAKE AND HANDLING

It is an accepted fact that for the purposes of field application, insecticide formulations can contain as little as 1% active ingredient and up to 98%
or more of inert filler-ingredients such as the many types of kaoline inter-
nationally available. On the other hand, high concentrates, dust concentrates, and wettable powders can contain as much as 75% w/w toxicant with
15-20% filler, the remaining 5-10% being made up of additives, etc. In
instances of this order, it is often necessary to use synthetic fillers that
have sorptive capacity values in excess of normal mineral fillers.

In plants where the products are to be based on the use of indigenous
mineral fillers, it is often necessary and desirable to incorporate a fillers
processing plant, alongside the formulation plant, for controlled supply of
fillers and diluents, in order to render bulk filler ingredients into a proper
condition. To obtain the desired condition, agglomerations and large
lumps of raw feed material are usually broken prior to further processing.
Initial breaking ensures that the material will have a known piece size,
thus rendering them suitable for drying prior to final reduction.

In Fig. 1 a flow sheet of a typical installation, using an orthodox rotary type dryer is shown. The arrangement and type of equipment forming
such an installation can vary a great deal depending on the characteristics

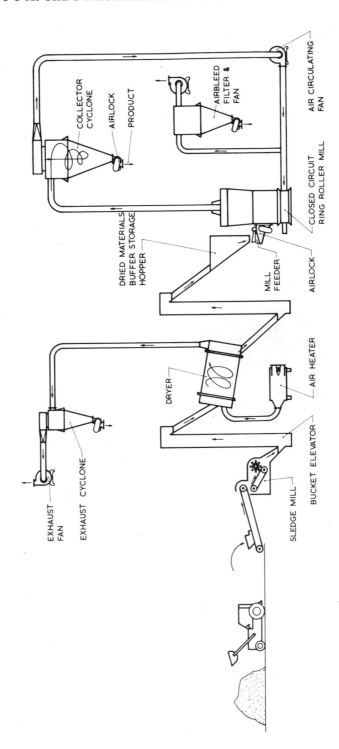

FIG. 1. Flow diagram of a typical intake and preparation plant.

of the raw feed material, its delivered plant feed size, moisture content, etc. (Where moisture level is sufficiently low, circuit milling and drying can be utilized as shown in Fig. 3.) Primary crushers incorporated in the type of plant shown in Fig. 1 must handle materials that have a wide range of moisture levels, so that consideration must therefore be given to the maximum level of moisture which can be expected at any one time. A point to note on the type of primary crusher shown in Fig. 2 is the provision of a constantly moving breaking surface, often described as "a traveling breaker plate." Crushers with moving breaker plates of this type have the advantage of preventing the "pasting-up" of material in the crusher, as would occur on a crusher having a fixed breaker plate. It is also good practice for the traveling breaker plate to have its own separate drive arrangement.

FIG. 2. Swing sledge mill with breaker plate.

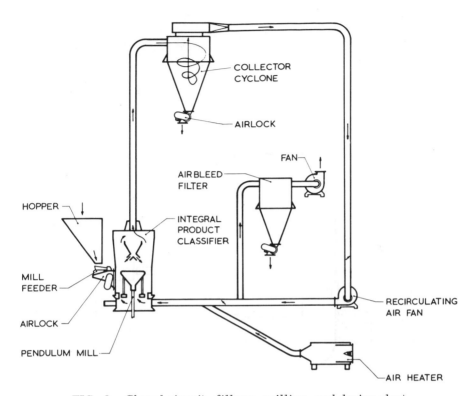

FIG. 3. Closed circuit, fillers, milling, and drying plant.

For the final milling of the dried raw feed, the air swept ring roll mill is probably one of the most useful units, particularly when milling to finenesses in excess of 76 micron. A normal arrangement of such a mill is shown in Fig. 3. Mills of this type need not incorporate an air heater when milling pre-dried materials or materials of low moisture content.

On completion of the drying and milling process, the ingredients can be fed into the formulation plant. How this is achieved is a factor of the location of each plant and the quantity of material to be handled from one plant to the other. Where the quantities are of an amount to merit some form of conveying, a pneumatic conveying system can be utilized. In systems of this type the conveying media, air, can be routed conveniently with the material through suitably sized trunking from one plant to the other. Figure 4 illustrates a simple installation arranged to convey to two separate points from a common feed point. Installations such as this form totally enclosed

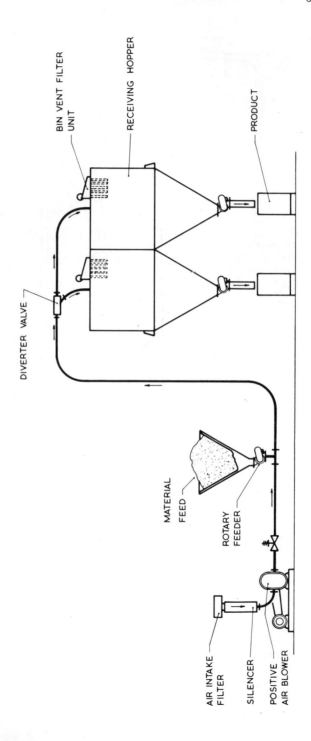

FIG. 4. Pneumatic conveying system.

systems, which can be expanded by the addition of single or multiple diverter control valves and additional trunking lines.

III. FORMULATION PLANT

The requirements of formulation plants involve criteria laid down by the production controller, plant budget controller, and the chemist who determines the product quality requirements, from which the engineer proceeds with his designs. The basic functions of a formulation plant can be broken down as shown in Fig. 5. The layout of equipment and the degree

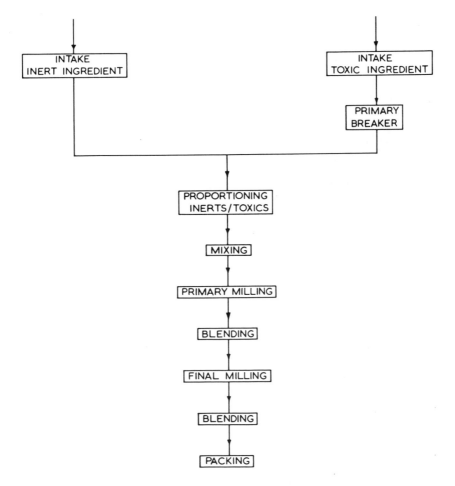

FIG. 5. Basic functions of a formulation plant.

of automation and control are each predicated on individual manufacturers needs and budget, as well as the degree of incorporation with existing plant and facilities. The handling and preparation of filler and/or diluent ingredients can be accomplished as already described. In instances where the manufacturer purchases such ingredients in prepared form, the only requirement is to feed these ingredients into the formulating plant. The means by which this is achieved must be related to the quantities to be handled, distances involved, and the physical characteristics of the materials.

A. Intake and Toxics

The intake of toxic ingredients to the plant presents problems of hygiene for those persons responsible for the handling and emptying of the bags or drums in which they are contained. The harmful effects of the ingredients, both from contact and inhalation, demand care and consideration to ensure maximum safety for the operatives. Good housekeeping is essential at all points where bags or drums are manually opened and discharged. Spillage of materials and dust nuisances must be kept to an absolute minimum.

A useful machine for the opening and discharge of bags is that shown in Fig. 6. This machine automatically opens the bag, allows for the discharge of the contents at point A and the removal of the bag remnants at point B. It only requires placing unopened bags on the feed mechanism. The machine can be vented by means of a central exhaust system or alternatively be provided with its own integral exhaust unit arranged to discharge collected dusts through the same outlet as the principal materials outlet A. From this point the dusts can be collected or conveyed. Should the materials be agglomerated and need pre-breaking, a suitable breaker can be located at the outlet for this purpose. Inclusion of special items of this type in a plant, apart from the benefits of hygiene, and operator safety, also considerably reduces the operative time in feeding materials into the plant. Bag handling rates can be as high as 200-300 per hour.

B. Flow Plan

The basic flow plan of a plant capable of producing each of the three basic groups, dust concentrates, field strength dusts, and wettable powders, is shown in Fig. 7. The flow plan illustrates each functional section of plant necessary to make up a flexible production unit. The measure and degree of automatic control, and also the size of the plant, can vary from a basic but comprehensive plant (Fig. 7) to a fully automatic plant (Fig. 8) or a complete complex (Fig. 9). The fundamental process requirements are similar in each case. The controlling aspects of design include range of products and capacity.

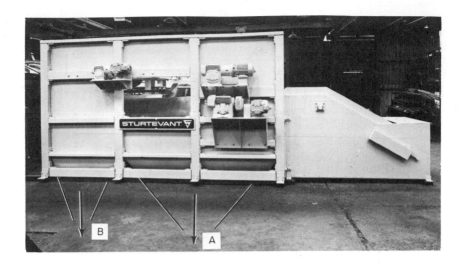

FIG. 6. Automatic bag slitter.

C. Weighing and Proportioning

Following the processing of fillers and diluents to meet the requirements of dust concentrates and field strength dusts and their charging into
the formulation plant, the next operation is to proportion the correct quantities of the technical pesticide and filler materials and any other additives,
such as stabilizers, into known and weighed batches. To achieve these
aims an arrangement of small storage hoppers having bottom discharge
outlets is necessary to hold the materials at a common central point, under which variable-speed controlled feeders can be located for individual
control and feeding of the ingredients into a centrally located weighing machine. Irrespective of the number of hoppers, such a system can be operated by one of three different means: (1) arranged for manual push button start and stop of each feeder, with the operator responsible for feeding
into the weigher the correct amount of each ingredient; (2) as a system
where each feeder is manually started by push button, but having automatic
weight control stop; or (3) as a fully automatic capsule or punch card control system, having selective formulation features and programmed time
control of weighing and blending programs, plus automatic recording and
print out of weights and batch numbers.

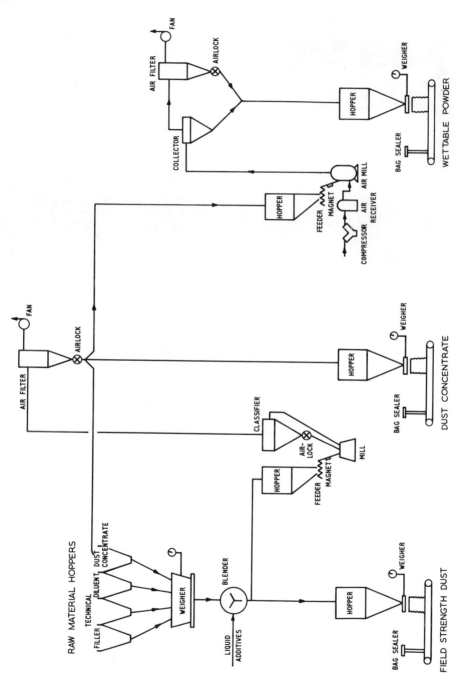

FIG. 7. Basic flow diagram of a plant producing dust concentrates, field strength dusts, and wettable powders.

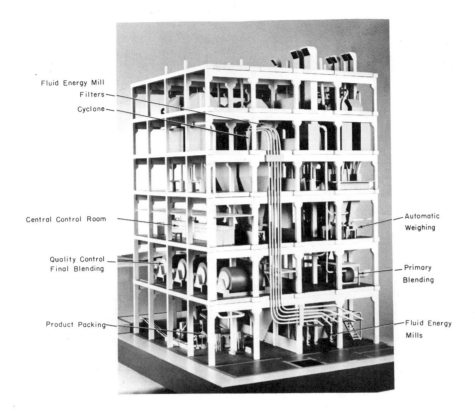

Fluid Energy Mill
Filters
Cyclone

Central Control Room

Quality Control
Final Blending

Product Packing

Automatic
Weighing

Primary
Blending

Fluid Energy
Mills

FIG. 8. Automatic plant.

The essential element of all good weighing systems is constant and reliable flow of materials from the storage hopper onto the controling discharge feeders. Without good flow of materials, the metering effect of the feeders and integrated weighing controls are nullified. The basic hopper requirements include maximum valley angle, maximum discharge openings which can be accommodated by the discharge feeders, and the best shape and form to assist material flow. Where possible the hopper should preferably have one or more vertical sides, to assist internal shearing of materials. In the determination of hopper design it is essential to consider the angle of repose of the materials to be stored along with their flow ability and other characteristics.

Delay and stoppages in the flow of material from hoppers is invariably the effects of arching occurring at the throat outlet of the hopper. Many devices are available such as bolt-on vibrators, aeration pads, pulsating

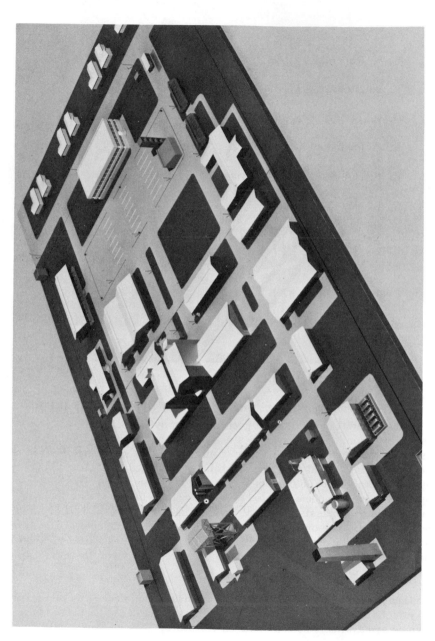

FIG. 9. Insecticide plant.

rubber diaphragms arranged to bellow inwards against the materials in the hopper when the diaphragm is supplied with air, and activated live hopper/ bin outlet of sections. Their selection and their use demand caution in certain instances. Hopper sections at throat outlets should be free of internal projections, rough welding, and bolt heads. Where possible the hopper bottom section should preferably be a fabricated one-piece unit, having the corners (valley angle) folded to form radiused corners or else filletted. Loosening hoppers which are subsidiary discharge hoppers lend themselves to greater precision to manufacture on these points. They can also be provided with inspection doors, and generally arranged to accommodate discharge gates and feeder equipment in a better manner. Ventilation of hoppers subjected to high entry rates of materials is often necessary and desirable for removal of displaced air, and prevention of pressurization of the hopper and the probable ensuing of dust laden air streams from various points not properly sealed or badly jointed.

Feeders for the feeding of materials from the hopper into the weighing machine should preferably have two speeds, fast and slow, the fast speed being used for high feed rates, the slow speed for final monitoring of the ingredient to the required total amount in the weighing machine. Automatic control of the switch from fast to slow can be achieved by suitable contacts operated by the weigher dial head, set to switch from fast to slow at a preset weighment figure slightly in advance of the total weight required.

The selection of feeders resolve basically into four categories or a combination of types, depending on the materials being handled: electric vibrating tray feeders, screw feeders having live vibrating screws and feed cones, screw feeders having multiple ribbon worms mounted in the feed inlet, and rotary feeders. Of course, feeder capacity should be sufficient to handle the desired quantity of material to be weighed.

D. Mixing

Following the weighing and proportioning of the formulation batch, the ingredients are mixed into an homogeneous mass. For the dispersion and mixing of a solid technical pesticide and filler prior to milling, perhaps the most suitable machine is the common ribbon or paddle type of mixer. These machines are straightforward in design, easily cleaned, and by virtue of the disposition of their flights or paddles, the material flow within the mixer is actuated in a nonuniform flow pattern. This causes the advantageous shearing of technical agglomerates, as well as the kneading of filler into the solid technical particles, particularly with technical pesticides which are waxy, low melting point solids.

Impregnation of fillers and miscellaneous formulations can also be accomplished during the mixing phase. Metering pumps or metering valves incorporated in a pump system may be utilized to apply liquid toxicants or solutions of solid toxicants. Solutions are pumped and sprayed under pressure onto and into the solids being mixed in the mixer trough.

E. Milling Dusts

The milling of mixed ingredients depends on the plant arrangement and equipment used, but in the main, it is considered advantageous to have two separate categories: (1) premilling of mixed ingredients prior to final milling in fluid energy type mills, as with high grade wettable powders and seed dressings, and (2) milling of mixed ingredients to produce dust concentrates and field strength dusts. The opening paragraph of this chapter refers to the integrated roles of the formulation chemist and the engineer in the selection and processing of test formulations. It is in this area of the process probably more than any other that the joint efforts of each should be concentrated. It is no secret that the selection of balanced formulations using filler ingredients with suitable absorbency factors can result in mixtures having tolerable milling characteristics or otherwise. The formulation with respect to the milling depends in no small way on the type and quality of the inert filler used as a grinding/carrying media.

Balanced formulations and good selection of filler media inevitably lead to longer production runs and lesser cleaning time, often with the added bonus of a product having good flow characteristics. The fundamental factor can be termed compatability, and the controling factors are softening temperature of the technical and absorbency of the inert filler.

For milling to specifications in the order of 200 mesh, a range of conventional mills can be used. Some of these utilize screens or grates, hammers, or beaters. Others utilize disks in the vertical or horizontal plane. A further category utilizes separately mounted pneumatic classifiers.

Figure 10 shows a Mikro Pulverizer, high speed hammer pulverizer, fitted with stirrup type hammers and patterned mesh grates. Material is fed by means of screws directly into the path of the rotating hammers. Particle reduction is achieved by impact and forcing of the material through the mesh forming the bottom section of the grinding chamber. Mill ventilation is achieved by an air vent seen on the side of the mill.

Pin mills in the main take the form of two disks, one static and the other driven. Each has an arranged pattern of pins to ensure that those

FIG. 10. Mikro pulverizer.

on the driven disk will move through the spaces between the pins fixed to
the static disk (grinding disk). The configuration of disks can be in the
horizontal plane (Fig. 11) or alternatively in the vertical plane (Fig. 12).
The machines usually operate on a single pass of material.

Another type of mill is the screenless Victoria Hammer Mill (Fig. 13),
utilizing a separate pneumatic classifier and operating in conjunction with
a product filter collector. This type of mill is fed by a variable speed
screw feeder, but is not fitted with grates or screens to control product
size. The grinding chamber is free of obstructions. Particle reduction
is dependent on hammer impact and pneumatic evacuation of a size band of
product. The particles are later separated in a classifier arranged to discharge oversize particles through an airlock and return them back for further milling. Product particles of the requisite size are conveyed by the
outgoing air stream leaving the classifier, and eventually collected in a
filter collector, which is exhausted to atmosphere by means of a fan. This
arrangement forms a closed circuit milling system, having the advantage
of induced air entry with the benefits of lower milling temperatures, plus
the freedom to locate the filter collector at a point where the product is
needed.

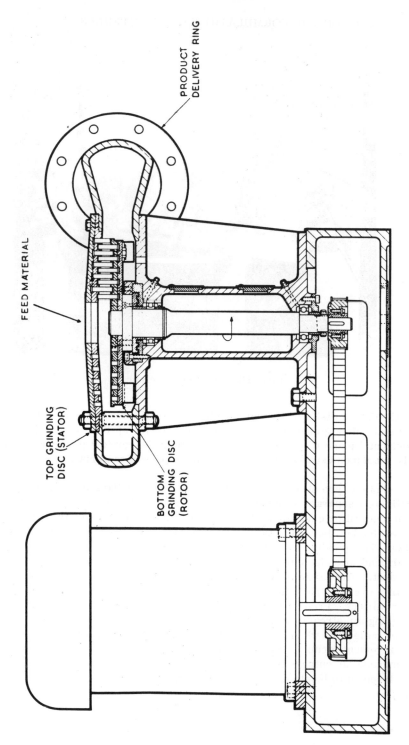

PRODUCT DELIVERY RING

FEED MATERIAL

TOP GRINDING DISC (STATOR)

BOTTOM GRINDING DISC (ROTOR)

FIG. 11. Horizontal pin disk mill.

250

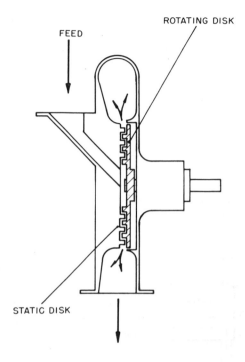

FIG. 12.   Vertical-type pin disk mill.

Figure 14 shows an aspirated beater type of mill. The mill incorporates a form of integral classifier and exhaust fan which can be adjusted to control the internal air current velocities, thus controlling particle size leaving the mill. Mills following the same basic principles but arranged on an horizontal axis are also available.

F. Wettable Powders

The most wellknown mills for the final milling of products to sizes below 10 micron, those described as fluid energy mills, utilize compressed air as the grinding energy media. In the main the principles of operation of the various types of mills falling under this category are similar. That is, particle reduction is achieved by the effect of jets or streams of high pressure, of high velocity air, on the material particles fed into the mill. Each relies on the fluid energy of compressed air, obtained from a suitable compressor unit.

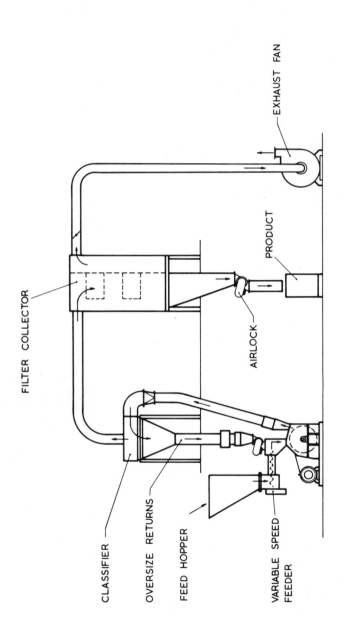

FIG. 13. Victoria hammer mill.

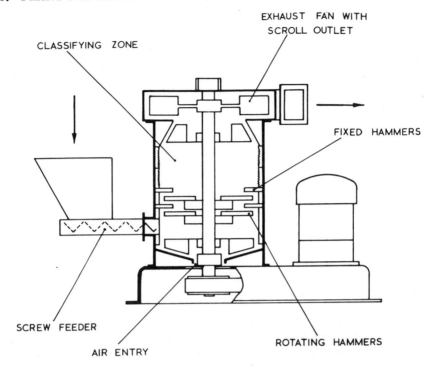

FIG. 14. Aspirated beater-type mill.

Each type of mill operates with a filter collector (1) for the collection of product, and (2) for filtration of energy spent air passing to atmosphere. As precollectors in advance of the final filter collectors, high efficiency cyclones can be incorporated in series. Their use depends to a large measure on layout, preference, and filter loading and media. The use of a cyclone precollector, however, does have the effect of reducing the dust burden passing to the final filter collector.

A built-in advantage in the use of fluid energy type mills is that the air expansion characteristics on entering the mill are adiabatic. The air exit temperature is usually therefore no greater than when entering the mill after having done the work of grinding. It also follows that there is little to no gain in the temperature of material at the point of entry and leaving the mill.

Mills of this type are capable of reducing materials down to the smallest of micron size, from feed sizes of thirty mesh and greater. They can be operated with steam or air, but in the art of milling insecticide formulations it is essential that air is used. The arrangements of air supply can be from standard compressors working in conjunction with suitably sized air receivers, intermediate air coolers, and controls. Where it is necessary that the air supply is free of oil, oil-free compressors can be utilized. Also, when grinding temperature-sensitive ingredients, refrigerated-type chilling units can be used for final cooling of the air supply from the compressor air receiver. Figure 15 illustrates a compressor unit incorporating chilled water units for the final air cooler.

Examples of these mills (Figs. 16 and 17) illustrate two distinct types. The micronizer fluid energy mill comprises a circular grinding chamber, surrounded by a pressurized manifold, arranged to pass compressed air through a number of tangentially placed jets into the grinding chamber. During operation materials are fed into the hollow grinding chamber through an aspirated venturi feeder which, when the mill is operating, has a negative condition at the point of material entry into the venturi. The number and size of jets fitted to the mill is relative to the size and duty of the mill, and is usually supplied by the manufacturer to suit the duty specified. It is possible, however, for plant operators to change jet sizes. The mills are usually supplied having top or bottom discharge, and are arranged for cleaning by splitting along the horizontal axis of the mill. The upper section of the mill is usually fixed by a series of clamps to the lower section. Outlet air pressure from the mill is usually slightly above atmospheric, the velocity being sufficient to pneumatically convey the finely ground outgoing particles to a locally placed collector.

A second example of this type of mill is the Jet-O-Mizer fluid energy mill (Fig. 17). In this instance, the mill configuration is dissimilar to the micronizer type, being arranged to operate in the vertical plane. The mill consists of three basic sections: (1) the lower section comprising grinding chamber, (2) vertical lengths (stacks), and (3) the upper classifying section. The sections are hollow and constructed in a varying elongated torus, the grinding chamber being in the form of an inverted trapezoidal cross section. Each of the sections is fixed to the other by means of flanged and bolted connections. Cleaning can be carried out by removal of a stack section. Feed to the mill is through a venturi connected to the grinding chamber manifold, which also has a negative condition at the entry point when in operation. Placed tangentially to the curvature of the grinding chamber are removable jets, located and sized to permit entry of high velocity air onto and into the incoming feed material. The placement and direction of the nozzles causes the entering jets of air to flow in the desired direction around the mill. Grinding is achieved by the velocity energy effect of the pressurized air entering the grinding chamber and the

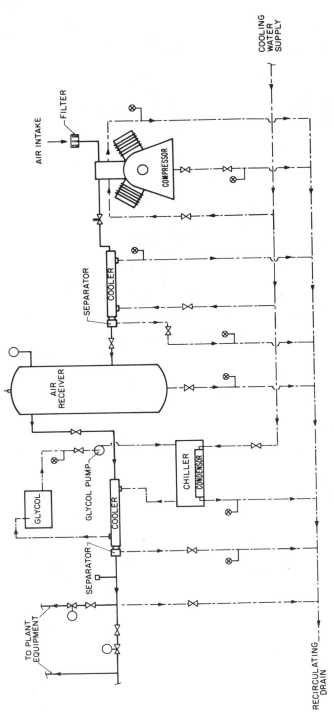

FIG. 15.   Flow diagram for compressed air supply.

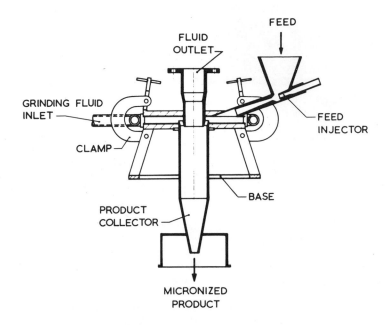

FIG. 16. Micronizer mill.

entrainments of the feed material with the air, causing the particles to break by pressure impact and attrition against each other. Materials circulate around the mill during the reduction period, the heavier and larger particles centrifuging toward the outer surfaces of the mill; the lighter and smaller particles circulate within the inner confines of the mill until of a particle size whereby they are removed from the main rotating flow path of the mill by the viscous drag of the outgoing air stream.

A characteristic problem in grinding insecticide formulations is their tendency to adhere in the form of a film to both stationary and moving parts of milling equipment. Films of this type develop into a quite hard and solid build-up, causing the equipment to choke and block. In the case of high-speed hammer mills, it can also cause them to run out of balance. Once films of this type commence to form, their build-up is accompanied by additional and progressive increases of frictional heat and additional grinding work as a result of reduced mill chamber area.

Attempts to reduce this problem have been made by the addition and inclusion of quantities of "dry ice" to the mixtures to be milled, i.e., "freeze grinding." Also liquid nitrogen may be used before or during the passage

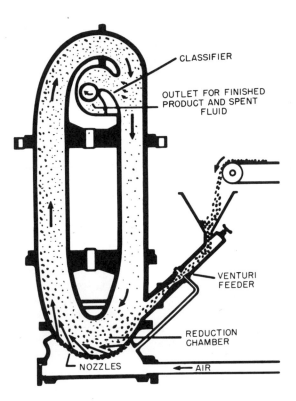

FIG. 17.   Fluid energy Jet-O-Mizer mill.

of materials through the mill. In the cases of air attrition mills of the fluid
energy micronizer type and special types of hammer mills, further attempts
have been made to reduce the problem by lining the mills with a rubber
having a suitable schore hardness.

It is, however, an unfortunate factor that the stopping of milling equipment for removal of build-up and cleaning must take place from time to
time. The operating period before cleaning is necessary varies according to the make-up and compatibility of the formulation and the type of inert used, not only with respect to its absorbency value, but also to its
scouring nature. In addition, where plants are installed in climatically
hot countries, there is the additional problem of high ambient temperatures and humidity.

Mill running times vary a great deal, from as short as thirty minutes
under adverse conditions to as long as complete shifts, even days, before

stoppage for cleaning becomes necessary. A practical solution to minimize mill down time is to establish by record the length of time a mill will run before blockage or cleaning becomes necessary. Then deduce from the observations the point at which the time necessary for cleaning represents a balance of maximum milling time coupled to minimum cleaning time. Such observations in one plant led to a program of three hours running coupled to scheduled fifteen minute stoppages for cleaning. In a shift of nine hours, down time amounted to forty five minutes representing some 10% of the total shift's period. The previous practice had been the running of mills on an unscheduled program, resulting in runs up to four hours and up to thirty minutes down time for cleaning, equivalent to one hours down time per shift and a down time factor for cleaning of 13-1/2% of the total shift's work period.

G. Blending

The effective dispersion of active and inert ingredients to form a properly blended and homogeneous product meeting toxicant variations allowed in the W.H.O. specification is extremely important. For this purpose ribbon, drum, or cone blenders are the most widely used. The action of these machines during the period of blending tends to increase the free flowing properties of finely ground aerated materials.

Figure 18 is a cross-sectional view of a drum blender, illustrating in the left-hand diagram the machine in the blend position, while the right-hand diagram shows the machine in the discharge position. The advantages of this type of machine is that it is continuously running and does not require stopping for charging and discharging. Control of the blend discharge gate can be automatic from a system program controller linked to an automatic weighing system. Blending times in this type of machine can be as short as three minutes, and less.

Figure 19 shows two 300-cubic-feet capacity blenders handling an explosive formulation and arranged to operate with the drum contents blanketed with inert gas. The small trolley-like machine in the foregound is a device for pneumatic removal of samples from the blender drum during the blending cycle.

By virtue of the variations that are possible with respect to the arrangement of flights or paddles for the movement of the materials contained in the mixer trough, ribbon mixers lend themselves to a wide number of applications. They are adaptable to the mixing of materials having low densities, to materials having high levels of density, and to those requiring positive and mechanical effort to be discharged from the trough.

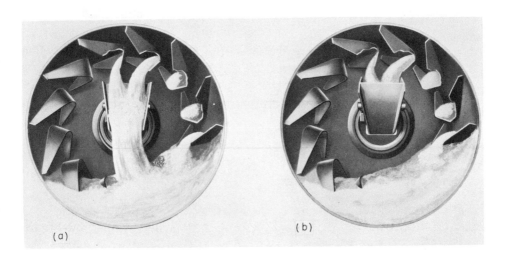

FIG. 18. Cross section of a drum blender.

The selection of mixing or blending machinery must be considered in relation to plant requirements and layout. Where the outputs do not warrant duplicate items of mixing equipment, and where mixing times are such that a single unit can be incorporated in the plant, the ribbon mixer can fill the role of both primary and final mixer. Ribbon mixers are capable of primary mixing most formulations in times less than 15 minutes. The mixing of finished fine products can be carried out in correspondingly less time.

A further consideration in the selection of mixing and blending equipment is the frequency that formulations are going to be changed. To avoid contamination between batches, cleaning of equipment must be thorough and effective. To ensure that this is achieved, mixing and blending equipment should have maximum accessibility to the various parts in contact with the materials being handled. The alternative to repeated cleaning is the use of several mixers or blenders designed to operate with small batches and arranged to overcome the necessity of frequent cleaning.

IV. EXPLOSION HAZARDS

Throughout a great number of process plants the hazard of explosion exists. Where hazardous materials are handled it is essential that adequate and proper care is taken to minimize the potential dangers. Such

FIG. 19. Two 300-cubic-foot blenders operating with inert gas.

precautions do not of necessity mean the inclusion of flame-proof explosion
equipment but rather a practical approach to relieve explosion pressures
in the event of an explosion coupled with safeguards to prevent and suppress
explosions.

Most explosions start by ignition at a single point, and in themselves
can often be small. Small explosions often create, by virtue of their turbulence, the conditions whereby a secondary and more violent explosion
can occur. It is to safeguard against these dangers that the use of automatic explosion detection and suppression equipment becomes important,
both from the primary and secondary conditions. Their incorporation into
milling, blending, and dust collection systems ensures a good measure of
protection. A typical milling protection system is shown in Figs. 20 and 21.

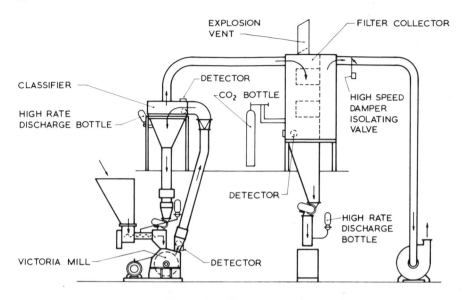

FIG. 20. Typical milling protection system.

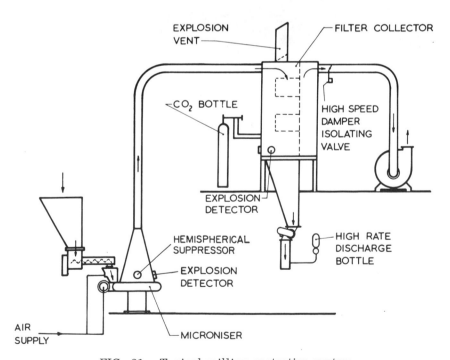

FIG. 21. Typical milling protection system

The essentials of an explosion protection system are to detect an incipient explosion and actuate devices that suppress, vent, or initiate other actions designed to control and prevent the spread and effect of the explosion. Key units making up a system are described in the following paragraphs.

A. Detectors

Figure 22 shows an explosion detector set to operate when a predetermined increase in the estimated system pressure is exceeded.

FIG. 22. Explosion detector.

B. Suppressors

Figure 23 illustrates a hemispherical-type suppressor which is designed to allow the spherical container to be housed within the system ducting or equipment. It disperses its contents when activated. The suppressor is explosively operated when activated electrically by a system pressure detector.

C. Suppressant Bottles

Alternative to the use of hemispherical suppressors, high rate discharge bottles (Fig. 24) having single or dual outlets can be used. These are particularly useful where it is not possible or practical to fit hemispherical suppressors. Where blanketing of a high order is considered necessary, larger bottles of CO_2 can be utilized. Figure 25 illustrates a milling unit fitted with a detector and hemispherical suppressor.

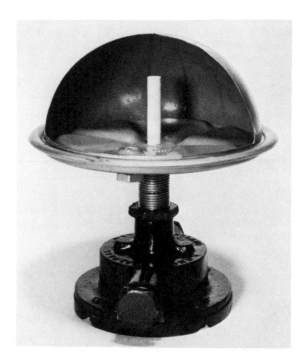

FIG. 23. Hemispherical suppressor.

Isolation of interconnecting items of plant equipment and shutdown of equipment in that area of plant in which the explosion occurs must also be related to initial explosion protection. Where fans, exhausters, etc., form part of the equipment it is not only necessary to shut down the motors driving these units, but also to isolate them immediately when the explosion occurs. Fans and exhausters often take minutes to come to a standstill and these minutes are critical during the initial explosion pressure rise. Their immediate isolation prevents them from spreading combusted matter. Isolator control mechanisms are designed to close immediately by cartridge explosion activated by an electrical signal from a system detector. They are particularly suitable for isolation of fans and exhausters, filters and cyclones, and to retain within the latter units the inerting concentration while a fan or exhauster is coming to rest.

The degree of explosion protection built into the plant must be related to the explosive nature of the materials to be processed. The range of materials which are combustible and which can explode when in a sufficiently

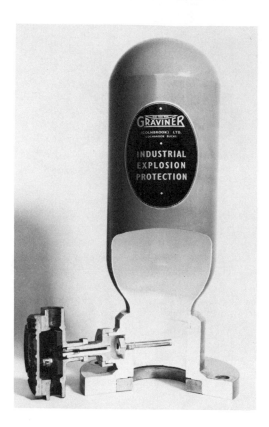

FIG. 24. High-rate discharge bottle.

divided state in atmosphere are numerous. The protection system design
must take these facts into account.

In a number of installations simple relief by venting is quite often sufficient. Care must be taken that, where blow off panels, etc., are used,
these are secured to a safe anchor point by chain to prevent their being
blown elsewhere and perhaps causing further damage or injury. Relief
panels should be fitted on all cyclones, hoppers, bag filters, etc., which
are relatively weak with respect to withstanding sudden explosive pressures. It is desirable that they are located at points of ignition or where
pressure waves enter.

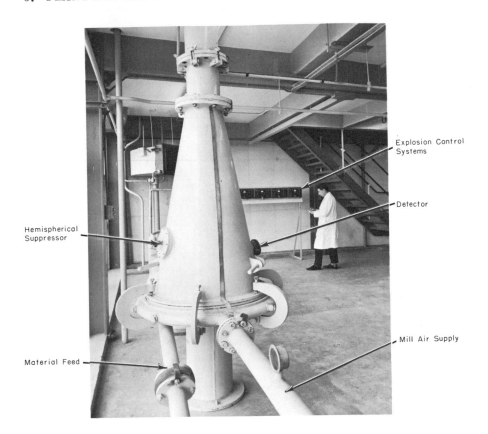

FIG. 25. Micronizer mill with explosion protection.

V. PACKING

All products have to be packed whether in drums, open mouth bags, valved bags or containers, etc. A wide range of filling and weighing machines are available to meet these needs, i.e., drop weight fillers, auger filling machines, etc.

Filling and packing difficulties may be encountered when the product is micron in size, and in particular when linked with a low bulk density value. It is not uncommon to allow the containers or bags to be in excess of the required size to enable the materials with their inflated bulk densities at the time of feeding into the bag or containers to be more easily accommodated. After initial filling, settling of the product in the bag or container gradually takes place. The inevitable result is that after settling the space occupied by the material can quite often be only 60% of the occupied space at the instance of filling.

A practice indulged in by some operators it to top up the bags or containers by hand after a suitable settling period. This method is slow and unhygienic from the operative's aspect. For this class of filling, vacuum type packing or densifying machines are desirable. These have the advantage of reducing the bulk density of material at the instant of filling to a value equal to a figure after a normal period of settling, or to a figure which can only be achieved by artificial compaction. The economics regarding the use of sophisticated machinery of this kind must be related to savings in operator time in topping up bags. Consider also improved packing conditions, savings in quantities of containers or bags used, superior packed and presented bagged product, and total enclosure during the filling of the bag or drum.

Figures 26 and 27 show a pure vacuum filling machine. In the first instance the machine is shown in the open position with a bag in the fill

FIG. 26. Automatic vacuum filling machine with bag in fill position.

position, and in the second the machine is closed and in the process of filling a bag. This type of machine uses vacuum techniques and utilizes a weight indicator dial. The machine has the advantage of being able to fill paper and vapor barrier bags and drums or fiber packs with plastic bag inserts. A pallet of bags stacked ten high indicates the measure of compaction and trueness of bag form after filling (Fig. 28). In the packing of D. D. T. wettable powders gains in densities of 60% per container (normal packing 21.9 lb/cu ft, vacuum packing 35 lb/cu ft) have been obtained. The measure of compaction and final bulk density of the packed product may be adjusted by controls on the machine. Furthermore, the entire operation is carried out with the machine door closed, thus isolating the operator from dust nuisances.

A further technique in vacuum packing is the use of machines for de-aerating product prior to the point of entry into the bag or container filling machine. Figure 29 illustrates the Gerivac machine. The technique adopted in this type of unit is the evacuation of air from the product during the period of time the product is passing through a regulated vacuum zone surrounding part of the feeder body. Providing there is no delay from the

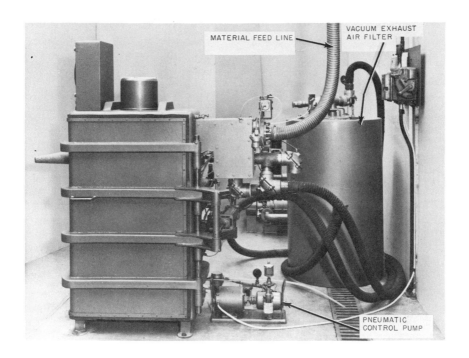

FIG. 27. Vacuum packing machine during process of filling bag.

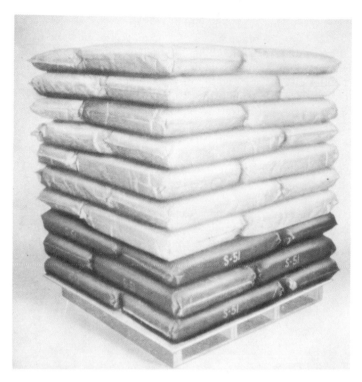

FIG. 28. Above is a stack of bags ten high. Note there is no loss of space or listing of the stack, which makes it possible to stack three and four pallets without danger of toppling over. Also, considerable space is saved in both warehousing and shipping, due to greater density packing over conventional methods, thus using smaller bags for the same weights.

instant of discharge from the Gerivac densifier to the packing into the final container or bag there will be no noticeable increase in bulk density. The adjustable controls on the unit are the speed of auger, which can be set to suit varying conditions by means of the variable speed driving gear, and the degree of vacuum on the densifying zone by adjustment of vacuum pump controls.

The aim of both types of machine is the removal of relatively large volumes of air locked in the many voids contained in the mass of the finely ground powders, with a resultant increase in bulk density without the disadvantages of the formation of agglomerated particles. The removal of air also has the advantage of taking away a source of moisture. Experience

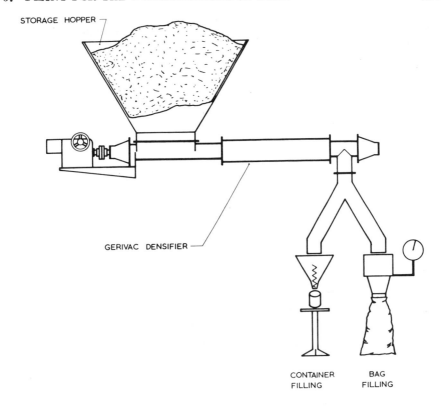

FIG. 29. Gerivac machine.

has shown that in both types of machine there are benefits in the increase of bulk density and in the flow characteristics on immediate discharge of the materials from the container or bag into which they have been packed.

VI. GOOD HOUSEKEEPING

The primary requisite of all plants handling dusty materials, toxic or otherwise, is to ensure a clean atmosphere to practical limits. The governments of many countries have laid down values of threshold limits for various obnoxious and toxic ingredients, which when airborne in dust form constitute a health and explosive hazard to operators. It is necessary, therefore, to ensure within the limits of practibility that plant, and in particular plant and equipment in operator areas, will operate in as clean and dust-free condition as possible. To this end an essential element of all plants handling dusts or fine powders is a properly designed dust exhaust and collection system.

The extent and design of such systems depends on the disposition of the equipment and plant layout generally. It is to be considered whether it is better and desirable to install separate systems, each designed to work in conjunction with a section of the process/formulation plant or whether a central plant arranged to accommodate all points of dust nuisances would be more efficient. In either arrangement, the system should allow both for the exhaust of dusts at each point where dust nuisances can occur, i.e., charging of raw material, packing stations, etc., and for the conveying of collected dusts to a central point where they can be removed and collected from the exhaust air, prior to the air passing to the atmosphere.

To ensure exhaust conditions at such points, hoods or canopies should be provided and arranged to envelope, as far as possible, each offending point. Secondly, the volume of air entering the system at each point must be sufficient to satisfactorily collect airborne dusts and have a duct speed to prevent "drop out" of particles when traveling through the duct. The controlling factors affecting volume of air to be handled are hood area, speed of air at face of hood, and air duct speed.

Because of the low particle size of the dusts being handled, the selection of collection and air filtration equipment invariably brings into favor the filter type of bag collector. Filters of this type fall into several categories, but in each case, the filtering media is a bag or pad, made from materials such as wool felt or combinations of man-made fibers. The method for the removal of collected dusts, whether they are on the inside or outside of the bags or wallets, depends on the filter design. Invariably, the dust is removed by a mechanical shaking mechanism, a reverse flow of exhaust air by damper positioning or subsidiary fan, or by short sharp jets of air at pressures up to 100 psi arranged to blow into the filtering media. Examples of this latter type are the Dust Control Engineering reverse jet filter having an arrangement of pads constructed in the form of envelopes (Fig. 30), and the Mikro Pulsair Filter (Fig. 31) having an arrangement of circular bags connected to a venturi at the top outlet of the bag, and through which a jet of compressed air enters the bag when signalled by an electrical timing device to do so. Both types of filters have collection efficiencies in the order of 99.9% and greater. Their virtues are high air/ dust burden ratio, automatic operation, and absence of mechanical shaking devices.

For the collection of heavy dusts and where efficiency of collection and degree of dust exhaust to atmosphere is not a prime factor, cyclones are

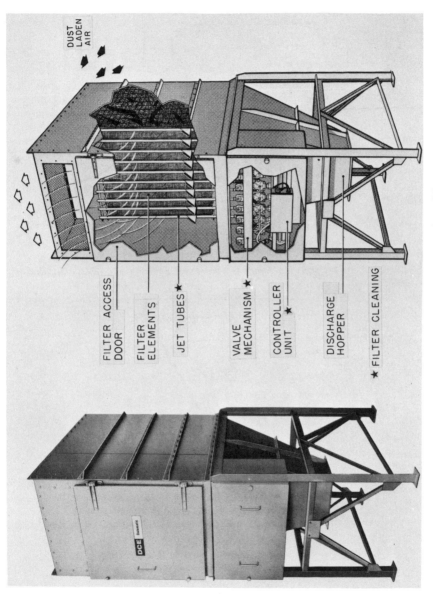

DUST LADEN AIR

FILTER ACCESS DOOR

FILTER ELEMENTS

JET TUBES ★

VALVE MECHANISM ★

CONTROLLER UNIT ★

DISCHARGE HOPPER

★ FILTER CLEANING

FIG. 30.   DCE Dalamatic automatic dust filter.

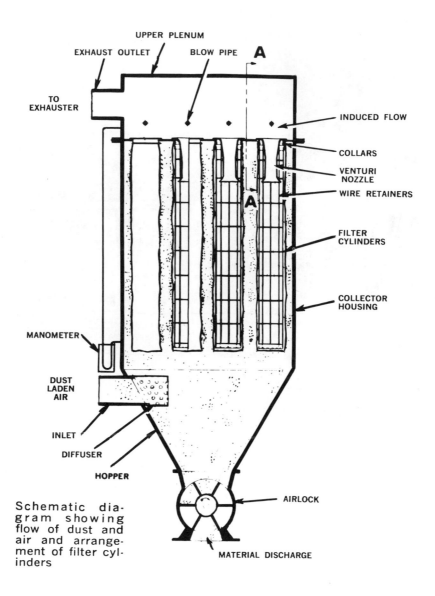

FIG. 31.   Mikro Pulsair Filter.

convenient units.   They are also useful as precollectors in advance of filters for reduction of the final dust loading on the filtering media.

For the cleanliness of floor areas, it is good practice to build into the plant a central vacuum cleaning system where the plant size warrants such a system.   An arrangement of this type can take the form of a centrally located collector-receiver unit exhausted by means of a suitable turbo exhauster.   From the collector there can run through the plant a main line into which flexible hand hoses can be connected at suitable points for the cleaning of floors and wall.   Removal of dusts from floor areas, walls, and ledges are contributions to plant cleanliness and general hygiene and preventative with respect to breakdowns caused by dusts.   The central vacuum cleaning is as illustrated in Fig. 32, which shows a typical arrangement of collector and exhauster.

FIG. 32.   Central vacuum cleaning plant.

ACKNOWLEDGMENT

Acknowledgment is accorded to Sturtevant Engineering Co. Ltd. (London); K. E. K. Limited; H. E. Kingston & Co. Limited; Fluid Energy Processing Co. Ltd; Dust Control Engineering Ltd; Mikropul Limited; Graviner Limited; and Gericke EmbH.

Chapter 7

REDUCING PESTICIDE CHEMICAL DRIFT

Wesley E. Yates and Norman B. Akesson

University of California
Davis, California

I. INTRODUCTION

Pesticide chemicals have been used by man to increase production of world food and fibers and to reduce insect and other vector-borne disease, and thus to increase the standard of living of mankind throughout much of the world. In the industrialized countries in particular the use of pesticides has contributed to increased food quality as well as quantity and also to reduced farm labor demand, which frees labor to industry and assists in raising overall productivity. However, the extensive use of these chemicals has resulted in widespread pollution of the air, water, and land environment (1); has posed a threat to human health (2); has through biological magnification in the food chain seriously threatened certain species of wild-life (3); and particularly in relation to economic insect control, has created serious ecological imbalances, destroying predators, parasites, and crop pollenators and generating resistant strains of economic insects no longer controllable by the chemical pesticides (4). Thus, in spite of their tremendous value to mankind increasing limitations have had to be placed on pesticides to reduce the use of the low degradability chemicals such as DDT, to limit the use of the proven carcinogenic chemicals, and overall to increase the efficiency of chemical control through better timing, more careful applications, and reduced dosages. Alternative means to chemical control, such as the use of biological and chemosensory materials, sterile male techniques, and cultural or environmental practices are being investigated (5) but the nonpesticidal chemical control techniques will not likely have significant effect for many years (6). In the meantime as many as possible of the nonpesticidal means must be brought to bear on pest problems, and chemicals will continue to be used with greater care and under increasingly restrictive conditions. This consideration of all the ecological factors associated with crop production and protection is a form of systems analysis, or integrated control by entomological defination (7). In certain crops such as cotton, the ecological imbalances created by excessive chemical use has resulted in insect resistance and resurgence and consequent large crop losses in spite of increased chemical use. In many cotton-producing areas the only avenue remaining open to successful crop culture is the integrated control concept (8).

The total United States production of synthetic pesticide chemicals (including export) was estimated at one billion pounds for 1966 (9), and indications are that this has increased at a rate of about 15% per year through 1969 (6). The sale of $1 billion per year of synthetic pesticides for domestic and export use was reached in 1965 (10) and it is projected that this will rise to $3 billion by 1975. A grand total of U.S. sales of all pesticide materials reached $12 billion in 1969 (6).

During this period (1964–1968) of sharp increase of chemical use in the United States, the Food and Drug Administration's market surveillance program sampled over 85,000 raw and ready-to-eat agricultural products and analyzed these for as many as 54 different chemicals. Approximately one-half of these samples have shown detectable residues, with the yearly average remaining about the same (11). The food supply on the U.S. market, while continuing to show low levels of pesticide residues, has seldom shown amounts above the tolerances established for these chemicals on foods (12).

Certain wildlife and fish species have not fared as well, however, and reports have continued to appear on biological magnification problems associated primarily with fish-eating and predator birds (13-15). The greatest share of the responsibility for damage to wildlife, fish, and mollusks rests with the chlorinated hydrocarbon chemicals: DDT (and metabolites), aldrin, dieldrin, endrin, chlordane, heptachlor, and toxaphene (6). Degradation under certain conditions such as in waterway mud and in soils can be very slow (16) and air and water transport to distant parts of the world, including the polar ice caps seems to have occurred (17, 18).

Research on application techniques and machines has been increasingly directed toward the overall increase in pesticide chemical efficacy in relation to control of specific pests and toward reduction of the hazards posed by pesticide chemicals being transported from the field of application. This movement occurs several ways.

One means of movement is transport out of the field on the crops harvested (which may result in seizures of the crop if there is an over-tolerance residue) and usual disposal of the chemical along with the processing wastes. The contaminated waste water finds its way into sewer and drainage systems.

A second means of movement is transport by ground and surface water, after the chemicals have been deposited on the soil, or washed from plants to the soil. Chemicals are transported into the drainage areas (as from irrigation water run-off) into streams and lakes and finally into the oceans. Since most pesticide chemicals are not very soluble in water there is a tendency for these to be bound to the mud, silt, and plant debris normally occurring in the waterways. At this point chemicals may enter the biological food chain as fish, plankton, or crustaceans feed in the contaminated waterways.

The third means of movement is transport by air, which occurs primarily as a direct drift of chemical at the time of application but secondarily as the chemical may be removed by wind from treated plants, trees, and shrubs, or as the chemical deposited on soil particles is picked up as dust and may be transported by winds from a few miles to many thousands of miles across oceans and polar regions. However, this paper is concerned specifically with more objective measurements of primary pesticide drift from a source or application site to a few miles downwind.

Thus, the drift of pesticide chemicals is but one of several means of entry into the general environment (19). The release of large quantities of chemicals into the air at the time of application by ground or aircraft type machines constitutes a uniquely hazardous period of exposure in the total application and control operation. While the machine can exert a significant influence on the drift potential by the height above ground and manner of discharge, it is quite apparent that after release the extent of aerial transport is going to be primarily a function of the ambient weather at the application site.

It is possible to make use of the drift characteristics of fine sprays and favorable weather conditions to cover very broad swaths in large area spraying, thus reducing the per acre cost and time spent for the application. Control programs such as for the desert locust (20, 21) and tsetse fly in Africa (22), and for grasshoppers in Canada (23) and the United States (24, 25) have been used successfully for many years, but are permissable today only with rapidly degradable materials applied at low rates per acre. Forest insect control applications have also used the large scale technique associated with drift spraying (26, 27), and aerosols or fog (very fine airborne spray) applications have long been used in mosquito adulticiding (28). In more recent time the use of undiluted, finely-atomized sprays has been shown promising for treating large rice acreages for larval control in standing water (29).

However, the heavier rates (2 to 10 times the dosage per acre) and mixed cropping generally associated with agricultural applications has resulted in significant drift problems. For example, in California as early as 1945 the movement of calcium arsenate dust used for worm control on tomatoes resulted in residues on nearby alfalfa fields of sufficient levels to kill 75 dairy cows (30). The subsequent development and use of the phenoxyacetic acid herbicide, 2,4-D, for use on broadleaf weed control resulted in extensive damage to many susceptible crops such as cotton, grapes, beans, and tomatoes (31). Symptoms of 2,4-D can be observed on sensitive grape leaves at levels as low as 0.0001 μg (32). Although 2,4-D was banned as a dust form for agricultural use by aircraft (Civil Aeronautics Ad. 1948) the use of sprays continued and lawsuits involving drift damage resulted in more than 21 such cases reaching the Appellate Courts by 1953 (33). The damage from herbicides has continued into recent years and in 1968 the herbicide, propanil, was banned from use on rice in Northern California because of injury resulting from drift of this material to prune orchards, some at distances of 3 to 5 miles from the nearest application (34). More subtle, however, is the over-tolerance residue on food and feed crops resulting from heavy applications of insecticides to crops such as cotton and orchards, and drift of these chemicals to nearby human food or animal feed crops. Residue tolerances on raw agricultural commodities are set at the Federal Government level by the Office of Pesticides Programs, a division of the Environmental Protection Agency. State governments generally follow the federal laws and accept tolerances prescribed by them. Thus, if food and feed residue analyses disclose above-tolerance levels, the shipment can be held and ordered destroyed if no means (such as removing outside leaves) are found possible to remove or reduce the residue. Drift suits continue to be filed and damage suits claiming drift from 3 to 20 miles distance have entered Appellate

Courts (35). Drift probably constitutes the most serious element of finan-
cial risk for commercial pesticide applicators (36).

II. THE BASIC DRIFT PARAMETERS

The basic parameters affecting aerial transport of finely dispersed
pesticide chemicals may be separated into gravitational forces relating to
settling rates of discharged chemicals, electrostatic forces relating to application equipment and plant deposit, and the various forces of meteorology relating to transport and deposit of the chemicals. Field studies have
provided reproducible data on amounts and extents of drifting materials,
and studies of machines and meteorology have indicated means for reducing and predicting downwind drift residue levels.

A. Gravitational Forces

The gravitational force is one of several important factors that the applicator can utilize to reduce drift. The movement of a particle in air is
a function of the resultant of the gravitational and aerodynamic drag forces.
The gravitational force acts straight downward and is simply the volume of
the particle multiplied by the difference of the particle density and the air
density. The aerodynamic force on a rigid particle is related to the velocity of air relative to the particle, its size and shape, and the density and
viscosity of the air. In addition, for liquid particles the surface tension
and viscosity of the liquid may also affect the drag force. Whenever the
forces are unbalanced, the particle will accelerate in the direction of the
resultant force at a rate defined by Newton's second law of motion: force =
mass x acceleration. Thus a particle falling from rest into still air will
accelerate until the gravitational force is counterbalanced by the drag force.
When the forces are balanced, the particles will fall at a constant terminal
velocity, v_t. For water drops falling in air, particles less than 100 μ in
size will approach their terminal velocity in less than 1 in. The distance
required to achieve 95% of the terminal velocity increases to approximately 2 ft for a 500 μ particle and 15 ft for a 2000 μ particle. Thus, for
particles larger than a few hundred microns it may be desirable to consider the particle displacement during the period of acceleration. Lapple
and Sheppard (37) describe a technique to calculate the trajectory of ac-
celerating particles. However, under most drift conditions the terminal
velocity of a particle can be used to identify the effect of the gravitational
forces on the drift of various types of pesticide particles.

The terminal velocity of a rigid sphere can be calculated by equating
the weight (W) to the drag force as follows:

$$W = \frac{4}{3} \pi r^3 (\rho_p - \rho_a) g = \frac{1}{2} \rho_a v^2 \pi r^2 C_D \qquad (1)$$

where ρ_a is the air density, g is the gravitational acceleration, particle of radius r and density ρ_p, ν terminal velocity, and C_D the drag coefficient. The drag coefficient can be related empirically to the Reynolds number, Re, and is available in many fluid dynamic texts or handbooks (38). The terminal velocities that are of most interest in pesticide applications may be classified into three different flow regimes: turbulent, intermediate, and streamline. For Reynolds numbers between 1,000 and 200,000 the drag coefficient for rigid spheres is substantially constant at approximately 0.44. Thus for the turbulent region the terminal velocity reduces to:

$$\text{(Turbulent)} \quad \nu_t = 1.74 \sqrt{dg \, (\rho_p - \rho_a)/\rho_a} \qquad\qquad (2)$$

For the intermediate range, Re between 2 and 1000, (37), the drag can be approximated by $C_D = 18.5 \, Re^{0.6}$, and the terminal velocity is given by:

$$\text{(Intermediate)} \quad \nu_t = 0.153 \, g^{0.714} d^{1.142} (\rho_p - \rho_a)^{0.714} / \rho_a^{0.286} \mu_a^{0.428}$$

where μ_a is the viscosity of the air.

For streamline flow, with Re less than 0.1 and a good approximation up to 2, the $C_D = 24/Re$ results in the following, commonly known as Stokes Law:

$$\text{(Streamline)} \quad \nu_t = gd^2 \, (\rho_p - \rho_a)/18\mu_a$$

It is interesting to note that changes in diameter and particle density do not affect the terminal velocity as much under turbulent conditions as under laminar conditions. During turbulent conditions the terminal velocity changes approximately as the square root of the diameter and density whereas under the laminar region the terminal velocity changes as the square of the diameter and approximately proportional to the density.

When the particle size approaches the same order of magnitude as the mean free path of the molecules in the air, the Cunningham correction (39) should be applied, but it is not significant for particles larger than 3 μ. In addition, the bombardment by the molecules in the air will cause displacement of small particles in a random manner which is known as Brownian movement. For spherical particles falling in air the Brownian motion is not significant unless the particle diameter is smaller than 1 μ.

Data on the terminal velocities of irregularly shaped particles is very limited and is complicated by the difficulty of expressing the particle size. Some irregularly shaped particles with a Reynolds number below 50 appear to have fall velocities within ±20% of the value obtained from Stokes Law (39) while data (40) on dusts of coal, china clay, and quartz indicate terminal velocities from 36 to 55% less than the fall velocity of a sphere based on microscopic mean diameter measurements. Orr presented information (41) from several sources on the drag coefficient for variously shaped particles over a wide range of Reynold's numbers. The shape factor, called sphericity, was used and defined as the surface area of an equivalent sphere divided by the surface area of the particle. With a sphericity of 0.1, the drag coefficient was approximately 100 times as great as a smooth sphere in the turbulent region and about 3 times as great in the laminar region.

The terminal velocity of liquid droplets may vary from the rigid sphere data due to deformation of the particle resulting from the aerodynamic forces (42, 43) as well as internal circulation within the droplet (44). The terminal velocity of water drops falling in air was accurately determined by Gunn and Kinzer (45). Their results indicate that the terminal velocity for drops below 80 μ approached the terminal velocity calculated by Stokes Law and for larger drops (up to 5800 μ) velocities were not as high as equivalent rigid spheres. Table 1 illustrates the terminal velocities of rigid spheres as well as water drops. Buzzard and Nedderman (44) reported the separate effects of surface tension and viscosity on the drag coefficients of droplets. Results show an increase in drag coefficient at high Reynolds numbers as the surface tension decreased from 71 to 18.5 dyn/cm. At a Reynolds number of 1000 a high liquid viscosity (50 cP, centipoise) produced approximately 1.5 times greater drag coefficients compared to liquids with a low viscosity (1 cP). The reduction in drag coefficient for low viscosity was attributed to the greater internal circulation which would retard the point of boundary layer separation. The effect was reduced at lower Reynolds numbers with a very small difference in drag coefficient with a Reynolds number from 400 to 500.

Table 1 shows the terminal velocities for a typical range of physical factors. The data clearly indicates that the particle size is by far the most important property affecting the fall velocity. Physical factors of secondary importance on fall velocities are the density, shape, surface tension, and viscosity of the particle. To minimize drift, the particle should be large. However, for a given application rate the number of particles available varies inversely with the cube of the diameter. Table 1 also illustrates the theoretical number of uniformly sized drops per square inch of flat surface area for a 10 gal/acre application. This hypothetical case was included to illustrate the relative effect of particle size on the coverage and distribution aspects. It should be remembered that the plant canopy is three dimensional and the surface area of the plant that

TABLE 1

Terminal Velocities of Particles in Air and Number of Drops/in.[2]

Diameter (μ)	Rigid sphere[a]			Water droplet[b]	No. of drops based on 10 gal/acre (drops/in.[2])
	Sp.G. = 0.8 V_t (ft/sec)	Sp.G. = 1.0 V_t (ft/sec)	Sp.G. = 2.5 V_t (ft/sec)	Sp.G. = 1.0 V_t (ft/sec)	
1	8.8×10^{-5}	0.00011	0.00028	0.0001[a]	1.15×10^{10}
10	0.008	0.01	0.025	0.01[a]	1.15×10^7
50	0.20	0.25	0.63	0.25[a]	92,195.00
100	0.68	0.85	1.8	0.89	11,524.00
200	1.9	2.4	4.6	2.4	1,441.00
300	3.2	3.9	7.5	3.8	427.00
400	4.6	5.3	10.0	5.3	180.00
500	5.6	6.8	12.5	6.8	92.00
1000	11.0	13.3	23.0	13.2	11.5
2000	19.0	22.0	37.0	21.0	1.4
3000	25.0	29.0	46.0	26.0	0.43
4000	30.0	34.0	54.0	29.0	0.18
5000	34.0	38.0	60.0	30.0	0.092

[a]Values obtained from Ref. 38, p. 1021 for settling in air at 70° F (0.0749 lb/ft³ and 0.0181 cP).

[b]Values obtained from Ref. 44, p. 246 for settling in air at 20°C, 760 mm, 50% R.H.

requires coverage by the pesticide is many times larger than the surface ground area it occupies. Determination of the optimum particle size or particle size distribution is extremely complex and is related to the type of pest, pesticide, location of pest in crop canopy, application equipment, meteorological conditions, and drift hazards. Although the theoretical number of particles continues to increase with a reduction in size, it should be noted that the gravitational settling velocity decreases and the resultant deposition at the desired location may reach a peak and then drop off rapidly with a further reduction of particle size. Aerodynamic catch also plays a part in the deposit of small particles which below 25 μ increasingly tend to be directed around an object rather than impacting (30).

For maximum control of drift the largest particle size should be used which is consistent with acceptable pest control and reasonable volumes to be applied per acre. For a given particle size the fall velocity can be enhanced by selecting the following physical properties: high density of carrier (if solid, a smooth nearly spherical particle is desired) and high surface tension which will minimize drop deformation and a low viscosity which will enhance internal circulation and increase fall velocity.

1. Dust Formulations

The uses of agricultural dusts are well known, and very effective formulations have emerged from several decades of experience with various toxicants, carriers, and particle sizes. However, the physical properties will dramatically point out the extremely low terminal velocities and consequently the inherently high drift hazard associated with conventional applications of agricultural dusts.

Dusts are generally formulated in one of two ways. If the toxicant is solid under normal conditions, it may be ground to the specified size and mixed with an inert powder. To avoid separation the density and particle size of the toxicant and inert material should be similar. An alternate procedure, the toxicant may be dissolved or if it is a liquid under normal conditions it may be adsorbed onto an inert powder to form the agricultural dust. In most agricultural areas the application rate for the diluted material is from 20 to 50 lb/acre with toxicant concentration ranging from 1 to 50%. Sulfur is one of the few chemicals applied without dilution.

The particle size, shape, and density of the dust is important in relationship to penetration and deposition at desired locations within different crop canopies. Weidhaas and Brann (46) revised and edited a second edition of a handbook listing some of the physical and chemical properties of an extensive number of commercially available dust diluents and carriers. The handbook includes information on a variety of botanical flours and numerous types of minerals including oxides of silicon, calcium, carbonates, sulfates, silicates, phosphates, and indeterminate minerals. The specific gravity of the materials ranges from 1.3 to 3.0. Most of the powders have a number median diameter ranging from 1 to 10 μ and can be controlled within certain limitations by the processor. Table 2 illustrates the particle size distributions and specific gravity of a few common dust diluents and carriers (46). The Asbestol Regular and Micro Velva A talcs are examples of two particle size ranges available with a number median diameter of 6.0 μ and 1.5 μ, respectively.

Theoretically, a 10-μ dust particle with a specific gravity of 2.5 would require over 100 sec to settle 3 ft, and a 1-μ particle would require over 3 hr. Thus, it is apparent that the application equipment must carry the dust to the target site. In addition, impingement or deposition is very low with this particle size, and a high percentage may remain in the air and drift with the air currents for long periods of time. As indicated by the above theoretical considerations, the sedimentation and impingement would be extremely poor if all of the dust were dispersed thoroughly in the air. However, observations have revealed that dust clouds consist of mixtures of individual particles and agglomerates of from 25 to 300 particles (47). Therefore, the particle size of the processed dust is not an accurate indication of the particle size distribution of material released in the air. Thus, to select materials for maximum fall rate, it may be necessary to determine the agglomeration characteristics produced by the combination of the type of dust and the application equipment.

Most commercial dusts are graded to have 80-90% of particles less than 30-μ diameter. Consequently, the deposition by sedimentation and impingement is low and a major portion has been shown to drift out of the treatment area (48).

2. Granular Formations

Granular materials offer many potential advantages for certain aircraft and ground applications. Granulars are generally formulated in a manner similar to dusts and may utilize impregnation, dry mixing, or adhesive binding procedures (49). The active ingredient generally ranges from 2 to 25% (by weight) of the commercial granular formation. Some of the inert materials that have been used as granular carriers are attapulgite, bentonite, celite, mica, perlite, and tobacco. Many of the materials that are used for dust carriers are also used for granules, the basic difference being the particle size. Granules generally refer to material that is considerably larger in size and accurately screened to a rather narrow range of sizes. Table 3 lists the typical size ranges available. The 15/30 mesh

TABLE 2

Physical Properties of Typical Agricultural Dust Diluents and Carriers[a]

Classification, code	Specific gravity	Particle size (μ)		
		Av (number)	Av (air permeation)	Distribution
Talc				
Asbestol Regular	2.6	6 μ		16.5% < 1 μ
				46% < 5 μ
				62% < 10 μ
				80% < 20 μ
				93% < 30 μ
Talc				
Micro Velva A	2.6	1.5 μ		40% < 1 μ
				58% < 2 μ
				87% < 5 μ
				97% < 7 μ
				99.98% < 10 μ
Pyrophyllite				
Airfloat #1	2.7		3.3 μ	11% < 2.5 μ
				27% < 5.0 μ
				50% < 10.0 μ
				99% < 74.0 μ
Attapulgite				
Attaclay	2.3–2.6		1.2 μ	12% < 5 μ
				27% < 10 μ
				42% < 15 μ
				56% < 20 μ
				69% < 25 μ
				80% < 30 μ
				88% < 35 μ
				94.5% < 40 μ
				99.5% < 45 μ
				100% < 50 μ

[a]Compiled from Ref. 46.

TABLE 3

Classification of Granular Formulations and Typical Number
of Particles per Square Inch

Tyler Mesh	Particle size range (μ)	"AA" RVM granules[a] (20 lb/acre) (particles/in.^{2})
8/15	2360–1080	0.4
15/30	1080–540	3.6
20/35	830–420	8.6
24/48	700–295	24.6
30/60	540–246	39.2

[a]Derived from Ref. 49.

size indicates that most of the material passed through a 15 mesh screen, and most of it was retained on a 30 mesh screen (Tyler Standard sieves). The adjacent column indicates the equivalent range of diameters in terms of microns. The last column indicates the approximate number of particles per square inch of flat surface area for a 20 lb/acre application of "AA" RVM attapulgite granule with a specific gravity of 2.45 (49).

Granular formulations are particularly suited for application of chemicals to soil or water surfaces, such as for mosquito larvae control. The large solid particles do not readily adhere to plant leaf surfaces and consequently will generally fall through a plant canopy and reach the surface area. One notable exception is the successful use of granules for control of European corn borer. In this case the corn plant forms a natural funnel which collects some of the granules in the whorl and leaf axils, exactly where the young borers feed before entering the stalk.

In regards to drift, granular formulations have the unique advantage that most of the small particles have been removed. It should be pointed out that the fine particles are not entirely eliminated since some may be carried through during the manufacturing process, and also later breakdown may occur because of handling as well as passage through the application equipment. Tests with an attapulgite RVM 30/60 granular (50) showed that the amount passing through a 60 mesh screen was 5.5% for un-metered material and ranged from 6.0 to 9.1% after it passed through various types of metering mechanisms. Thus, with proper application equipment at least 94% of the 30/60 mesh material would be larger than 250μ and would have a terminal velocity greater than 5 ft/sec. Greater drift control could be obtained with larger granules. For example, 15/30 mesh granules would have a terminal velocity in excess of 10 ft/sec.

The granular formulations offer a very high degree of drift control by use of carriers which are screened to provide a narrow band of large high density particles. The formulations have been successful for application of soil sterilants and pre-emergence herbicides, pre-emergence fungi-cides, and insecticides for control of soil-borne larvae, mosquito larvae, and European corn borer larvae. The use will probably continue to expand for the above applications, as will use with systemic or translocated pesti-cides which can be applied to the soil as granules and will move through the plant roots to control insects, weeds or fungi.

3. Spray Formulations

Sprays have emerged as the predominate formulation used for pesticide applications on plant foliage. In California the percentage of acres treated with spray (compared to the total treated by sprays and dusts) by commer-cial pest control operators increased from 59% in 1954 to 85% in 1960 and has remained at approximately that level up through 1967 (51). One of the major reasons for the shift from dust to spray can be attributed to a re-duction of drift hazards due to a significantly larger particle size spectrum. Unfortunately, most all agricultural spray nozzles produce a wide range of spray drop sizes. However, the pesticide applicator generally has a num-ber of factors he can select to control the mean drop diameter and to some extent the uniformity of the drop size. This section discusses some of the basic factors that affect the drop size spectrum, namely, fluid prop-erties, type of atomizer, and operating conditions. In addition, some of the atomization limitations are discussed and data are presented on typical drop size spectrums for various operating conditions with a jet, flat fan, hollow cone, rotary and twin fluid atomizers.

a. Fluid Properties. Selection of the fluid properties offers an im-portant means to control the drop size spectrum. The most important physical properties related to droplet size are the surface tension, vis-cosity, density, and vapor pressure.

(1) Surface tension. The surface tension represents a direct force that resists the formation of new surface area. The minimum energy

required for atomization is equal to the surface tension multiplied by the increased liquid surface area. Thus, it may represent a predominate force for certain types of atomization. Through dimensional analysis, the Weber number, $We = \rho_a \nu^2 d/\sigma$ (where σ is the surface tension) has been shown to be one of the useful dimensionless products that may be related to the drop size whenever surface tension forces are important. The Weber number represents a ratio of the inertial force to the surface tension force. The surface tension that is commonly encountered in sprays ranges from 73 dyn/cm for water to as low as 20 dyn/cm for some petroleum distillates (53). For most pure liquids the surface tension in contact with air decreases with an increase in temperature and is independent of the age of the surface. Since most pesticide sprays are mixtures of surfactants, carrier, and active ingredient it is important to note that the surface tension of a newly-formed surface is close to the value for the bulk of the liquid and with time reaches an equilibrium or static surface tension that is normally reported as the "surface tension". The term dynamic surface tension is the value obtained before equilibrium and is related to the age of the surface. In 1890 Rayleigh (54) reported that the dynamic surface tension of a newly formed water and soap solution was nearly that of pure water. In 1943 Addison (55) showed that the rate of the development of surface equilibrium varied with the nature of the solute. Later Addison (56) reported a useful method for measuring the dynamic surface tension by measurements of the vibrations of an elliptical jet. In respect to drop formation it has been observed (57) that the addition of a wetting agent to pure water gives no observable change in atomization, since the formation of drops occurs in a very short time after emission from the spray nozzle. Thus, the dynamic surface tension for the age of surface at the time of disintegration should be used for prediction of drop size. Kido and Stafford (58) reported the dynamic and static surface tension for a DDT emulsion with an anionic emulsifier and also values for additions of various metallic salts and cationic surface-active agents. The dynamic surface tension ranged from 41.3 to 72.1 dyn/cm, while the static values ranged from 25.2 to 31.7 dyn/cm. Although this chapter is primarily concerned with the drop size and drift aspects, it is interesting to note that the above laboratory studies (58) showed that in most cases an increase in the dynamic surface tension increased the amounts of spray deposit.

(2) Viscosity. The viscosity is one of the most important liquid properties that can affect the drop size spectrum. An increase in viscosity physically dampens the natural wave formations which generally delays disintegration and increases the droplet size. One set of possible dimensionless products related to drop size are the Weber number and the Reynolds number (52). The Reynolds number represents a ratio of the inertia force of the fluid to the viscous force and is defined as $Re = \rho d\nu/\mu$.

Products of the above numbers can be used to relate the atomization for various fluids and different operating conditions.

The viscosity of most spray solutions is relatively low and may range from 1.0 cp for water to 10 cp for some weed oils. Table 4 lists the values of viscosity for some common liquids. The viscosity of simple (or Newtonian) liquids is independent of the shear rate. The viscosity of liquids generally decreases with an increase in temperature, however, the absolute change varies with the type of liquid and temperature. For instance, the viscosity of oils changes much more with change of temperature than does the viscosity of water. It is important to note that for a complex (or non-Newtonian) fluid the viscosity is also a function of the shear rate.

In regard to drift control, one of the interesting developments in recent years has been the introduction of several new adjuvants to modify the viscosity of agricultural sprays. Several of the formulations have been introduced specifically to reduce the drift hazards. Most of the formulations produce non-Newtonian solutions and consequently it is necessary to specify the shear rate along with the apparent viscosity. In addition, it is very important to follow the recommended mixing procedures. For instance, some materials may be adversely affected by the pH or salt content of the water, additions of certain pesticides, temperature, order of mixing, type of mixing equipment, and rate of addition of materials to tank, as well as mixing time. The following materials have been commercially introduced (by companies noted) as drift control agents: invert emulsions, (Shell, Dow, Amchem, Hercules) and emulsifiers which can be used to produce viscous water/oil emulsions with up to 85% water content; a polysaccharide-gum, Dacogen (Diamond Shamrock), which has pseudoplastic and thixotropic properties; an alginate derivative, Keltex (Kelco); a dry granular water swellable polymer, Norbak (Dow Chemical Co.), which produces a "particulated solution"; and a hydroxyethyl cellulose, Vistik (Hercules), with viscoelastic properties. Van Valkenburg (59) reported an extensive study of the viscosities and mixing characteristics of invert emulsion formulations. Agricultural engineers at the University of California (60, 61) have reported the apparent viscosities of the above materials over a wide range of shear rates. A summary of the viscosities are shown in Table 4. The dramatic reduction in apparent viscosity with an increase in shear rate points out not only the desirable effect of lower pressure losses through the lines and nozzles but also the reduced effect of the non-Newtonian fluids on drop formation if the nozzles produce a high shear rate on the liquid film formed by the nozzle. Once the drops are formed very little secondary breakup will occur due to the increase in apparent viscosity at the lower shear rate. Thus, the type of nozzle will critically affect the drop size distribution produced with non-Newtonian fluids.

TABLE 4

Viscosity of Various Liquids at 68° F

	Approx. viscosity (cP)
Newtonian liquids	
Water	1.0
Fuel oil	10.0
S.A.E. 10 oil	100.0
Glycerin	800.0

Non-Newtonian liquids	"Apparent Viscosity"[a] at two shear rates, (cP)	
	50 sec^{-1}	4000 sec^{-1}
Dacagin, 6.0 lb/100 gal water, 2.5% Tordon	140	13
Invert emulsion, 5% emulsifier, 10% diesel fuel, 85% water	700	16
Keltex, 8.5 lb/100 gal water, 2.5% Tordon	260	39
Norbak, 6.2 lb/100 gal, 2.5% Tordon	680	45
Vistik, 6.5 lb/100 gal, 2.5% Tordon	450	16

[a]From Ref. 61, p. 3.

(3) Density. The spray density has little effect on the atomization due to the small range that is normally encountered in commercial spray formulations. The density generally ranges from a low of 0.8 g/ml for oil carriers to 1.2 g/ml for some technical materials while the gulk of the applications utilize a high percentage of water with a density of 1 g/ml. It should be remembered that the density may have an indirect effect if the

spray pressure is constant since the velocity is approximately equal to: $v = k\sqrt{p/\rho}$, where k is a constant, p is pressure, and ρ is the density. Again this would have a very small effect since the velocity varies inversely with the square root of the density.

(4) Vapor pressure. For most agricultural spray systems the vapor pressure has no effect upon the initial droplet size spectrum. However, it should be noted that some atomization systems may utilize a liquid carrier with a high vapor pressure such that upon release into the atmosphere the liquid jet may be shattered by the rapid growth of bubbles. This may occur when a superheated liquid passes into a lower pressure zone which requires some of the liquid to shift into the vapor state in order to maintain equilibrium. The process is called flashing and is dependent on a critical Weber number (62). Pressurized household atomizers for insecticides, hair sprays, and other materials use this principle to produce a fine spray or aerosol.

The vapor pressure gradient has a direct effect upon the rate of evaporation which consequently determines the size of a given drop with respect to time. Since a complete treatment of the theory of evaporation would be very lengthy, the following discussion attempts to point out the important variables that may influence the evaporation and consequently the drift of spray particles. Ranz and Marshall (63) reported an extensive study of the evaporation of water drops in still and moving air. Seymour (64) regrouped the constants to conveniently express the rate of change of diameter of a single drop in a large volume of surrounding air as follows:

$$\frac{\partial d}{\partial t} = -\frac{KC_v \, \Delta p}{2\pi \, d \, P}$$

where Δp = vapor pressure gradient between the surrounding air and the drop surface; P = partial pressure of air; K = diffusivity of water vapor in air at the ambient temperature; and C_v = effective transfer coefficient at instantaneous v. The value of the diffusivity times the transfer coefficient for water in (m^2/sec) is given by the following equation:

$$KC_v = (2.16)10^{-3} + (0.454)10^{-3}\sqrt{v}$$

It should be noted that KC_v must be evaluated for other liquids and is a function of diffusivity of the vapor in air, mean molecular weight of gas-vapor mixture in boundary layer, air viscosity, and drop diameter. The constants required for other liquids are discussed in detail in an excellent monograph by Marshall (65). The above equation points out that the driving force of evaporation can be expressed as the difference in vapor pressure at the surface and the vapor pressure in the surrounding air. The equation also shows that the rate of change in diameter is inversely proportional to the drop diameter at zero relative velocity. Thus, as would be expected, the evaporation would change the diameter of a small drop at a faster rate than a larger drop. Table 5 illustrates the vapor pressures for some common pesticides and carriers. The values for water illustrate the importance of temperature on the evaporation rate. It is interesting that the evaporation rate can also be expressed in terms of a driving force of the difference in temperature of the liquid surface and the surrounding air temperature multiplied by appropriate heat transfer coefficients (66). Again for a single drop it has been shown that the surface temperature quickly reaches the wet bulb temperature. Thus the evaporation rate can be expressed as a function of the wet bulb depression.

The evaporation of a mixture is more complex and is discussed by Marshall (65). In the initial stage the evaporation rate for suspensions of inert solids and some dissolved components can be treated as a pure liquid drop. This stage is followed by a period at which the rate decreased rapidly and the surface temperature increases continuously until it reaches the ambient temperature and an equilibrium condition. If all the water is lost from a wettable powder suspension, the dry particle may not adhere to the foliage even if it reaches the crop. Cunningham (66) investigated the evaporation from an airblast sprayer and found that more than 40% of the original volume of spray may be evaporated by the time the spray travels 36 feet from the sprayer. Seymour (64) presented curves of the drop diameter as a function of time for water drops falling in air with different humidities. The evaporation was based on the particles falling at a terminal velocity which varied with evaporation. The instantaneous velocity was based on Stokes Law, thus results are limited to particles less than 100 μ in diameter. Table 6 illustrates the time and vertical distance various size particles would fall while the drop reduced to 10% of the original volume. This represents a spray that contains water with 10% wettable powder which would consequently be dry at the indicated final diameter. The table represents a minimum time and distance since the data is based on an evaporation rate for a single pure water drop in a large atmosphere. The evaporation from the emission of a large number of drops in a spray would increase the partial vapor pressure in the surrounding air and increase the drying period. The effect of an increase in relative humidity from 30 to 70% is shown in Table 6. Duffie and Marshall (67) presented a technique to extend the evaporation calculations to drops falling in the intermediate range with a Reynolds number from 2 to 500, (100 to 1100 μ water drop).

TABLE 5

Vapor Pressure of Some Pesticides and Carriers

Material	Temperature ($^\circ$C)	Vapor pressure (mm Hg)
DDT	20	1.9×10^{-7}
Malathion	30	4.0×10^{-5}
Parathion	20	3.78×10^{-5}
2, 4-D (acid)	25	10.5×10^{-3}
Water	0	4.6
Water	10	9.2
Water	20	17.5
Water	30	31.8
Water	40	55.3

TABLE 6

Time and Vertical Fall Distance for Pure Water to Evaporate
from D_O to D_f at 78° F, 29.92 in. Hg

Initial diameter D_O (μ)	Final diameter equiv. to 10% of initial volume D_f (μ)	30% relative humidity $\Delta p = 0.68$ in. Hg t (sec)[a]	30% relative humidity Vert.[a] dist. (ft)	70% relative humidity $\Delta p = 0.29$ in. Hg t (sec)[a]	70% relative humidity Vert.[a] dist. (ft)
100	46	4.2	2.5	9.2	5.3
80	37	2.8	0.8	6.3	2.2
60	28	1.7	<.5	3.8	0.75
40	19	0.8	<.5	1.8	<.5

[a] Based on Ref. 64.

Mixtures of soluble pesticides with extremely low vapor pressures may significantly reduce the saturated vapor pressure of the carrier and reduce the evaporation rates. Also, it is apparent that introduction of certain adjuvants may significantly reduce the vapor pressure of the carrier and likewise reduce the evaporation. An example is the introduction of an additive containing an amine stearate mixture (68), Lovo, Fisons Pest Control, Ltd., which forms a film of stearic acid on the drop surface, reducing the evaporation of water. Laboratory tests (69) have shown that this material is particularly effective for use with spray suspensions of wettable powders. Tests with controlled temperature and humidity in a 20 ft spray tower showed that a spray of wettable powder reached the bottom of the tower as a nonadhesive dust whereas a formulation containing the amine stearate material showed that drops as small as 80 μ had no detectable evaporation. Field tests (70, 71) with Sevin and Bacillus thuringiensis have demonstrated that the amine stearate formulations again reduced the evaporation and increased the initial retention. The material is not effective if used with emulsifiers or other surfactants. The use of low volatile oil carriers of invert emulsion formulations of water in oil are additional examples where evaporation can be significantly reduced compared to a normal emulsion with water as the continuous phase. However, the cost and availability of nonphytotoxic oils may restrict the use of these formulations for certain post emergence applications on agricultural crops. Available information on evaporation of different o/w emulsions was summarized by Stig Johansson (72). Generally, it was found that the evaporation rates of o/w emulsions of insecticides were about the same as pure water drops.

(5) Other properties. There are a number of other liquid properties that may affect the particle size distribution. For instance, the elastic characteristics of the liquid may cause long thin ligaments to retract into a single drop. Another property is the particulate or granular type of liquid formulation with very little free water. The formulation of Norbark discussed above is very viscous but probably the most important feature is that the granular-like particles may reduce the number of fine drops.

b. Type of Atomizer. One of the most important means for the pesticide applicator to control the particle size is through selection of the type or design of the atomizer. Considerable research has been conducted on the theory and performance of atomization equipment in several professional areas including: physics, mathematics, mechanical engineering, chemical engineering, and agricultural engineering. A wide variety of atomizers are available and for agricultural spraying may be classified into three basic types: pressure nozzles, where the fluid is under pressure and is broken up by instability in the atmosphere or by impact on a plate or another jet, in which case the energy required for disintegration is imparted by the fluid; the two-fluid nozzle, where the liquid jet is broken up by a high velocity gas stream with the contact of the fluid and gas possibly taking place entirely outside of the nozzle or within a chamber, in which case the high velocity gas stream imparts the major portion of energy for atomization; and the rotary nozzle, where the liquid is introduced under low pressure to the center of a rotating cup, disk, or cage, and the centrifugal force breaks the liquid into drops as it leaves the periphery, in which case the rotating mechanism imparts the energy for atomization. The characteristics of the different types of nozzles are thoroughly discussed in excellent articles by Marshall (65) and Fraser (57).

c. Operating Conditions. In addition to selection of the proper nozzle it is just as important that the desired operating conditions be maintained. The equipment must provide a system that will produce and maintain a constant liquid pressure at the nozzle, and a constant air pressure or a constant speed of the rotary atomizer. In the case of some high shear nozzles and viscous liquids, it is also necessary to maintain a constant fluid temperature. Also, it should be emphasized that worn or damaged nozzle parts can substantially distort the spray distribution as well as alter the drop size characteristics.

d. Drop Size Spectrum. All spray nozzles used for agricultural applications produce a wide range of drop sizes. Major efforts to reduce drift hazards by reducing the number of fine drops for a given application have utilized one of the following approaches: (1) Production of a more uniform droplet size - attempts have been made to improve the uniformity by altering the liquid properties as well as by change in nozzle design and operating conditions. This would allow the mean diameter to be maintained and although a perfectly uniform spray may not be desired, a reduction in a large number of fine drops would reduce the drift and probably improve pesticide efficacy, resulting in a lower application rate. (2) Removal of the fine droplets - this approach utilizes present types of atomization equipment and liquids but attempts to remove the fine particles by coalescence or by physical forces. (3) Increase the drop size spectrum - use of larger drop size spectrums generally increases the entire spectrum with a significant reduction in the number of fine drops that may drift. In this case the total applied volume may need to be increased to maintain satisfactory coverage or distribution on the plant.

There has been considerable work on the development of useful measures of the average drop size. Concepts of mean diameter, volume mean diameter, Sauter mean diameter, Smd, number median diameter, nmd, and volume median diameter, vmd, (41) are frequently used to express the drop size. However, this information doesn't identify the most important factor related to drift -- the distribution or number of fine drops. Various types of

distribution functions have been introduced to mathematically identify the
drop size distributions. Some of the commonly used functions are the Rosin-
Rammler distribution, the Nukiyama-Tanasawa distribution, the log normal
distribution, and the square root normal distribution. Mugele and Evans
(73) reported an excellent review of the appropriateness of the various distribution functions. Many studies have shown that the drop size distribution
is approximately log normal. It should also be recognized that although
some studies have used the square root function, over a large range the
square root function is approximately the same shape as the log normal.
If the distribution is log normal, then the geometric standard deviation is a
measure of the uniformity of the spray. The geometric standard deviation
can be readily calculated from the slope of the data plotted on log-probabil-
ity paper (41). It should also be noted that the geometric standard deviation
is independent of the mean diameter and likewise is the same regardless of
whether the data is based on the number, area, or mass of the particles.
Liljedahl, Lovely, and Buchele (74) showed that the statistical coefficient of
variation, $100 \, \sigma/\bar{d}$, is mathematically related to the geometric standard deviation of a log normal distribution. Tate and Marshall (75) expressed the
uniformity of a square-root-normal distribution as

$$s' = \sqrt{d_{v \, 84.13}} - \sqrt{d_{v \, 50}}$$

where $\sqrt{d_{v \, 84.13}}$ refers to the value taken corresponding to 84.13% of volume less than this size. This is a modification of the standard deviation
and is based on the slope of the plot of the square-root of the drop size versus the cumulative probability. Fraser and Eisenklam reported the uniformity of a spray by a dispersion coefficient defined as: $q = (d_{95v} - d_{5v})/d_s$,
where d_{5v}, d_{95v} represents the fraction of drop size which represents 5 and
95 percent of the cumulative volume and d_s is the surface mean diameter.

The following section discusses the drop size distribution for each type
of nozzle with various liquids and different operating conditions.

(1) Jet breakup. Jet breakup is of considerable interest in drift control
since it offers the possibility of increased uniformity as well as producing
a maximum particle size with a minimum number of fine drops. Low speed
jet breakup is often referred to as Raleigh breakup after the pioneering work
of Lord Raleigh (76). He formulated a mathematical analysis of the breakup
of a nonviscous low velocity jet falling in air and considered surface tension
and fluid density as the major forces controlling breakup. For these conditions Raleigh deduced that the optimum wavelength for a water jet was 4.51
times the jet diameter. Several investigators have reported experimental
measurements that agree quite well with Raleigh's theoretical value. For
example, Tyler (77), Duffie and Marshall (67), and McDonald (78) reported
the approximate ratio of wavelength to jet diameter as 4.69, 4.6, and 4.6,
respectively. If each wavelength forms a single sphere, the diameter of the

drop for a wavelength of 4.51 would be 1.89 times the jet diameter. Merrington and Richardson (79) indicated that for jets at low velocity and smaller than 1000 μ the average drop diameter was 2.1 times as large as the jet diameter. Harmon (80) theoretically demonstrated that a laminar jet would produce a jet diameter equal to $\sqrt{3/2}$ times the nozzle diameter. Thus Harmon predicted a drop diameter (for a 4.51 d wavelength) equal to 1.63 times the nozzle diameter. Duffie and Marshall (67) conducted a detailed study of jet breakup at low velocity with a range of fluid viscosity from 0.84 to 36 cP; specific gravity from 1.0 to 1.3; nozzle diameter, d_n, from 145 to 300 μ; and a flow rate from 3.1 to 18.9 cc/min. They reported the geometric mean diameter of the drop, $\bar{d}_g$, was correlated by the following empirical relationship:

$$\bar{d}_g = 36 \, d_n^{0.56} \, R_e^{-0.10}$$

Miesse (81) measured jet disintegration characteristics and made critical comparisons with theories and experimental data of Raleigh (76), Weber (82), Tomotika (83), Tyler (77), Holroyd (84), Baron (85), Littaye (86). Although Merrington and Richardson reported that drops approached a uniform size for relative velocities below a critical velocity, Duffie and Marshall (67) reported the geometric standard deviation varied from 1.123 to 1.304. Later Schneider and Hendricks (87) demonstrated that extremely uniform size drops can be produced from a low velocity jet if a regular capillary wave is mechanically introduced that is close to the unstable wavelength predicted by Raleigh. By changing jet diameter, fluid velocity and frequency of disturbance, they were able to produce uniform size drops ranging from 50 to 2000 μ. Atkinson and Miller (88) illustrated a similar technique and produced uniform drops at a constant rate and ejection velocity. While the prospects appear exciting, the unit has only been used for laboratory tests and has several limitations for practical field applications of agricultural chemicals. For instance, the unit would require extremely small jets and a low relative velocity. Consequently, a large number would be required and clogging of the orifices may be very troublesome.

Low velocity jets have been introduced as a means of producing a minimum number of fine drops. One approach to obtain satisfactory distribution with jets on a tractor mounted boom sprayer incorporated jets directed backward, spaced at 2.5-in. intervals, operated at 2-4 psi, and vibrated laterally at approximately 540 cycles/minute (89). Following the same approach but instead of vibrating the boom, a multijet nozzle was designed with an electrical drive unit for each nozzle which produced a rotary oscillation of 25° at 4000 cycles/min., to be operated with a 6 psi spray pressure (90). Recently a multi-capillary jet unit was introduced for reducing drift from helicopter applications (91). This unit contains

3120 capillary tubes on a 26 ft boom. Presently two sizes are available, 0.013 and 0.028 in inside diameter, and are operated at 2 psi or less. When used on a helicopter, small drops, and thus drift, are reduced when the nozzles are properly aligned at 180° to the flight line, and operated at speeds less than 60 mph.

For fixed wing aircraft applications, the relative velocity between the jet and the air is generally high enough to substantially affect the jet disintegration process as well as produce secondary breakup of large droplets. Merrington and Richardson (79) reported that for relative velocities greater than the critical value, approximately 9 mph for water with a 3.8 mm jet, and with jet diameters from 1000 to 18,000 μ, the vmd, d, size was independent of nozzle size and a function of the relative velocity, $\underline{v}$, and liquid viscosity, μ, as given by the following empirical relationship:

$$d = \frac{500}{v} \left(\frac{\mu}{g}\right)^{1/5}$$

where $\underline{v}$ is the relative velocity between jet and surrounding air. The State of California (92) requires jet nozzles directed back or with the slipstream for applications of injurious herbicides in specified hazardous areas. For a 1/8-in. diameter jet at 30 psi and directed with a 100 mph air velocity (relative velocity equal to 55 mph) the volume median diameter is 900μ (61) and agrees quite well with the above empirical relationship. Fig. 1 illustrates the drop size distribution for this jet which emphasizes the reduction of potential drift since only 1.5% of the total volume of the spray is below 200μ.

Secondary breakup may occur due to aerodynamic forces on the drops and is important in determining the final drop size distribution. For instance, a mechanism may initially produce a spectrum with few fine drops; however, if some of the large drops are immediately broken by the aerodynamic forces, a large number of fine drops may be produced. Several investigators have studied the phenomenon and stability of drops subjected to steady and transient air velocities (93-97). Lane (93) pointed out that the mechanism was quite different for drops subjected to high transient conditions compared with steady air velocities. Hinze (94) related the stability of the drops to a Weber number, We = $\rho_g V^2 d/\sigma$, a viscosity number, Vi = $\mu_g/\sqrt{\rho_\ell \sigma d}$, and the type of air flow. For an inviscid liquid subjected to a shock exposure the critical Weber number is approximately equal to 13 while for the nearly steady state or falling drops the critical Weber number is approximately 22. The critical or maximum drop size for inviscid liquids, Vi < 0.01, exposed to different relative velocities and types of flow are shown in Table 7. It is interesting to note that the maximum size drop from the jet illustrated in Fig. 1 was 1650μ at a

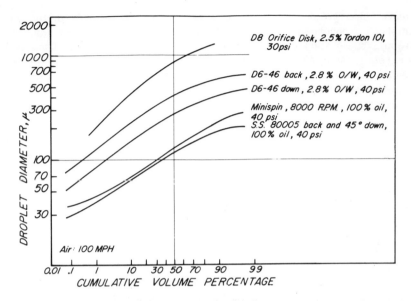

FIG. 1. Drop size distribution from various types of nozzles used on agricultural aircraft. From Refs. 61, 102, 129.

TABLE 7

Maximum Drop Size for Water Subjected to Various Air Velocities

Relative velocity V (mph)	Shock exposure critical $W_e \simeq 13$ d Max (μ)	Steady velocity or falling drop critical $W_e \simeq 22$ d Max (μ)
35	3150	5340
50	1550	2620
65	915	1550
85	535	905
100	386	654
150	172	291
200	97	163

relative velocity of 54.5 mph, which was between the value predicted for true shock exposure and a steady velocity predicted for a Weber number of 13 and 22, or 1300 and 2200μ maximum size, respectively. The explicit relationship of the critical Weber number and Viscosity number is given by Hinze (94). This relationship indicates a rapid increase of the maximum drop diameter with an increase of Viscosity number and that further disintegration will cease when the Viscosity number is greater than 2.0.

The atomization process for high speed jets, at Reynolds numbers from 2600 to 46,900, is considerably different and has been analyzed by numerous investigators. Harmon (98) presented an excellent review of the literatire and derived the following relationship to predict the Sauter mean drop size from an energy balance by use of dimensionless ratios and data from other investigators:

$$\overline{d}_{\text{sauter}} \propto \frac{\mu_g^{0.78} \, d_n^{0.3} \, \mu_\ell^{0.07}}{V^{0.55} \, \rho_g^{0.052} \, \rho_\ell^{0.648} \, \sigma^{0.15}}$$

The relationship predicts an increase in drop diameter with an increase in nozzle diameter and liquid viscosity, and a decrease in drop diameter with an increase in velocity, density of the liquid, and surface tension.

(2) Flat fan. The flat fan nozzle produces a spray in a nearly two dimensional plane and has found extensive use for application of agricultural sprays. Dorman (99) was one of the first to evaluate the drop size characteristics of a flat fan nozzle. He confined his work to kerosene and water and assumed that the small change in viscosity could be neglected. He analyzed the data through a simplified dimensional analysis and developed the following relationship:

$$d_s = 4.4 \, (Q/\theta)^{1/3} \, \sigma^{1/3} \, \rho^{1/6} \, p^{-1/2}$$

where Q is flowrate in cc/sec; θ is the spray angle in radians; σ is the surface tension in dyn/cm; p is pressure in Kg/cm^2; and ρ is the specific gravity. The vmd was found to be 2.7 times as great as the Smd. Yeo also used dimensional analysis to develop the following equation that is identical in form to the above relationship but extended it to include the ratio of velocity of emission, V_e, to the velocity of liquid relative to stationary air, V_o:

$$\frac{d_s V_e}{(Q\sigma/\rho_\ell)^{1/3}} = \chi \left(\frac{V_e}{V_o} \right)$$

By use of Yeo's data to obtain the function χ, the above relationship is useful to predict the drop size for low viscosity liquids applied by aircraft. Fraser and Eisenklam (101) presented the following empirical relationship explicitly for water in terms of readily measured parameters:

$$\log \overline{d}_s = 1.823 + \frac{4.42}{p} + 0.203\left(\frac{Q}{p}\right)$$

where pressure, p, ranges from 25 to 100 psi; flow rate, Q, is in imperial gal/min; (Q/p) ranges from 0.35 to 2.2; and the Smd is greater than 87 μ. The dispersion coefficient was closely grouped around 1.58. Thus, the uniformity of the spray was not materially affected by a change in pressure or size of nozzle.

Figure 1 illustrates the drop size distribution from a typical flat fan nozzle (102), Spraying Systems 80005, used for low volume application with agricultural aircraft. This nozzle produced a vmd of approximately 120 μ with the orifice directed back and $45°$ downward in respect to a horizontal 100 mph airstream. Note that the slope of the curve is nearly the same for various types of nozzles which indicates that the standard geometric deviation or uniformity of the spray are similar and selection of the nozzle type primarily affects the geometric mean. Likewise it is evident that an increase in the median diameter reduces the number of fine droplets. Additional data on drop size distributions from various types of flat fan nozzles used for application of agricultural chemicals by aircraft are given by Isler and Carlton (103), and Skoog and Cowan (104), and for ground equipment are given by Tate and Janssen (105) and Hedden (106).

There is considerable interest in the use of various thickened sprays for reducing drift. The flat fan nozzle has a particular advantage because the discharge coefficient remains nearly constant over a wide range of viscosities. Although some change in discharge coefficient may occur (0.40 to 0.54) for a wide range of viscosities from certain non-Newtonian liquids (105), a small change in the viscosity due to temperature or mixture variations in the field would probably not have a significant effect on the application rate. Information is available on the atomization characteristics produced by very viscous Newtonian (108) and non-Newtonian liquids (61, 64, 107). Ford and Furmidge (108) indicated that the characteristics of the change in mean drop diameter of a water/oil emulsion can be described in a manner similar to Newtonian liquids with the viscosity of the water/oil emulsion based upon the limiting viscosity under conditions of infinite shear. The effect of the viscosity of an invert emulsion on drop size from a flat fan nozzle at 20 psi revealed that the Smd stayed about the same for viscosities up to about 7 cP, decreased for a short intermediate region, then continued to increase as the viscosity increased. The maximum droplet size was reached with a formulation containing 14 parts water and one part oil. This produced a viscosity of approximately 200 cP and a droplet size approximately 2. 5 times as large as the Smd for a low viscosity liquid. The most discouraging information in regard to drift control was that the drop size distribution revealed a larger portion of small drops with the invert emulsion spray compared to a Newtonian liquid with nearly the same vmd but a higher surface tension, 35 and 85 dyn/cm, respectively. The higher number of fine particles was attributed to the low surface tension associated with the continuous oil phase of the invert emulsion.

Liljedahl and Lovely (52) postulated that the mean drop size and uniformity of the drop size can be related to the liquid properties and operational factors in terms of dimensionless products of the Weber number and Reynolds number. Preliminary results indicate that the coefficient of variation of the spray from a given flat fan nozzle decreases with a decrease in Reynolds number and at low Reynolds number the coefficient of variation also decreases as the Weber number decreases. Byrd and Seymour (64, 109) reported the relative number of fine drops produced with sprays containing various thickening agents. Results showed the particulate formulation substantially reduced the number of drops in the range of 90 to 150µ.

(3) Hollow cone nozzle. The hollow cone or centrifugal nozzle is commonly used in agricultural spraying by both ground and aircraft equipment. Several empirical and dimensionless relationships have been developed to describe the atomization characteristics (41, 57, 65, 75, 101, 110). Tate and Marshall (75) evaluated the effect of the tangential and vertical velocity components on the atomization characteristics and developed the following empirical relationship to predict the Smd:

$$\bar{d}_s = 286 \ (d_o + 0.17) \ \exp \ (13/V_v - 0.0094 \ V_t)$$

where d_o is the orifice diameter (0. 013 - 0. 04 in), V_v is the average axial velocity (40 - 150 ft/sec), V_t is the tangential velocity component at the inlet to the swirl chamber (0 - 50 ft/sec). The drop size distribution closely followed a square-root probability plot, and the slopes indicated that the uniformity of the drop size was not significantly affected by the axial velocity component. However, the modified standard deviation was found to decrease as the tangential velocity increased. A limited amount of work with the liquid viscosity ranging from 7 to 1 centistokes showed that the standard deviation increased with an increase in viscosity.

Straus (111) reported the following empirical expression for Smd in terms of easily measured parameters for atomization of water with a grooved core type hollow cone nozzle:

$$\overline{d}_s = 1.808 + \frac{6.94}{p} + 0.138\left(\frac{Q}{p}\right)$$

where p is pressure, 25 – 200 psi; Q is flow rate in imperial gal/min; and Q/p ranging from 0.05 to 2.0. Fraser and Eisenklam (101) noted that for small nozzles the flat fan and hollow cone nozzles produced nearly the same average particle size, but for larger nozzles (Q/p = 2.0) the hollow cone nozzles gave a smaller mean drop size, particularly at higher pressure. The average dispersion coefficient for the above type of hollow cone nozzle was 1.42. Thus, the spray was somewhat more uniform in drop size than from a similar size flat fan. It was also observed that the dispersion coefficient decreased with an increase in pressure.

Data on drop size distributions are available for hollow cone nozzles used with ground equipment (105, 106, 112) and aircraft equipment (61, 102, 103). Figure 1 illustrates some typical drop size distributions for aerial applications with hollow cone nozzles directed perpendicularly and in the same direction as the air flow. Again it can be observed that the slope or standard geometric deviation of the sprays from the various type nozzles were quite similar, and the number of fine drops can be reduced by use of a spray with a larger vmd.

Butler, Akesson, and Yates (61) reported the drop size distributions from nozzles in high velocity air streams and with several sprays containing commercially available adjuvants recommended for drift reduction. The results indicate that with the thickening agents the drop size may be excessively large, as great as $8000\,\mu$, with the use of jets directed back from agricultural aircraft. Figure 2 illustrates the drop size distribution for various thickened sprays atomized with a hollow cone nozzle and a similar vmd spray produced by a jet nozzle without an adjuvant. The curves indicate that the particulate type of formulation produced a more uniform spray with the least number of fine particles and not excessively large particles. Drop size distribution from hollow cone nozzles with a 2,4-D spray formulation containing a pseudo-plastic spray gel agent (113) and similar sprays without the thickening agent revealed a significant reduction of fine drops with the formulation containing the thickening agent, less than 0.1% below 100 μ with the adjuvant and approximately 8% below 100 μ for a 2,4-D spray. The vmd of the thickened sprays ranged from 950 to 1750 μ compared to about 225 μ for the spray without the adjuvant.

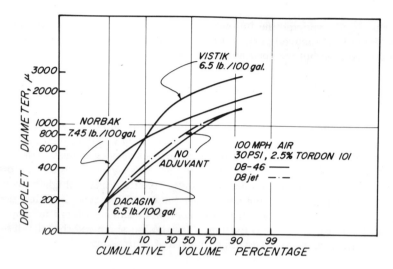

FIG. 2. Drop spectrums with various adjuvants compared with a jet containing no adjuvant. From Ref. <u>61</u>.

(4) Two fluid nozzle. A two fluid or pneumatic type nozzle employs a gaseous medium to supply additional energy for atomization. This type of nozzle has been the most satisfactory type for producing very fine sprays, vmd of 20 μ or less, at low liquid rates. The performance of a flat fan and hollow cone nozzle mounted on an aircraft could be considered a form of a two fluid atomizer but is not generally identified in this way. A typical two fluid nozzle provides an intimate contact of the liquid with a high velocity air stream. The design may employ either internal or external mixing and may utilize an air velocity of over 700 ft/sec. Nukiyama and Tanasawa (<u>114</u>) conducted a detailed study of the performance of small twin fluid atomizers and developed the following empirical relationship between the Smd and a wide range of liquid properties and operating conditions:

$$\overline{d}_s = \frac{1410 \sqrt{\sigma}}{V_a \sqrt{\rho_\ell}} + 191 \left(\frac{\mu}{\sqrt{\sigma \rho_\ell}} \right)^{0.45} \left(\frac{1000\, Q_\ell}{Q_a} \right)^{1.5}$$

where $\overline{d}_s$ is the Sauter mean diameter; σ is the surface tension (19 to 73 dyn/cm); μ is the liquid viscosity (0.3 to 50 cP); V_a is the relative velocity between air and liquid (up to sonic ft/sec); ρ_ℓ is liquid density (43.7 to 75 lb/cu ft); Q_ℓ is the volumetric flow rate of the liquid; and Q_a is the volumetric flow rate of the air. Atomization from venturi type nozzles

with a liquid jet directed concurrently with the airstream was evaluated by
Wetzel (115). His study was confined to the use of molten wax (ρ = 0.83
g/ml, μ = 9 cP, σ = 29.5 dyn/cm), and he expressed the vmd by the following equation:

$$d = 4.2 \times 10^6 \, (V_a - V_\ell)^{-1.68} \, d_o^{\,0.35}$$

where d is in microns, V is velocity in ft/sec, and d_o is jet diameter in
inches. Gretzinger and Marshall (116) presented a review of the literature on pneumatic (two-fluid) atomization and developed drop size correlations with operating conditions for production of sprays with a vmd
between 5 and 30 μ.

Although Nukiyama and Tanasawa (114) reported that the uniformity of
the drop size from a given type nozzle was nearly constant for a wide
range of conditions, Lewis et al. (117) indicated that the uniformity was
quite sensitive to nozzle design, type, and size. Wetzel and Marshall
found that the geometric standard deviation of drop spectrums from venturi type nozzles was proportional to the vmd raised to the 0.18
power, i.e., increased with an increase in drop size. Likewise,
Gretzinger and Marshall (116) reported that the geometric standard
deviation increased with the vmd; for converging nozzles the value
was proportional to the vmd raised to the 0.14 power and for pneumatic impingement nozzles it was proportional to the vmd raised to
the 0.16 power.

Two fluid nozzles have been used for some low volume applications of
agricultural chemicals to orchards (118) and field crops (119). However,
because of the extremely fine particle size, the use of this type of equipment is primarily confined to applications of relatively nontoxic materials
with very high tolerance levels or certain applications that may require
very low dosage rates.

(5) Rotary atomizers. Rotary atomizers represent another type of
atomizer that is used in some agricultural sprayers, spray drying equipment, and fuel burners. Atomization is achieved by supplying the liquid
at a low pressure to the center of a rotating disk, cup, or cage. High
speed photographs (57, 120) of spinning disk atomizers have clearly shown
that disintegration may occur in three different forms. At extremely low
flow rates, direct drop formation occurs at the periphery; at intermediate
flow rates, ligaments are formed around entire periphery; and at higher
flow rates, a film or sheet is formed beyond the periphery. The spinning
disk has been used in laboratory studies at a very low flow rate to produce
a band of uniform size drops. It should be noted that even with direct

drop formation, small satellite drops are formed by a small trailing liga-
ment produced at the time the major drop separates from the disk. However, under laboratory conditions the satellite and major drops are readily
separated. The horizontal movement of the satellite droplets are considerably shorter because of their lower inertia force and consequently the
predominate major drops all fall in a separate, very narrow umbrella-like
band around the disk. However, most commercial units are operated with
a flow rate that generally produces a thin film and results in a wide spectrum of drop sizes. The drop size characteristics have been investigated
by many scientists and summarized by Fraser ($\underline{57}$) and Marshall ($\underline{65}$).
Friedman, Gluckert, and Marshall ($\underline{121}$) conducted one of the most inclu-
sive studies and developed the following correlation of the ratio of the
Smd to the disk radius with several dimensionless groups:

$$\frac{\bar{d}_s}{r} = 0.4 \left(\frac{\Gamma}{\rho N r^2} \right)^{0.6} \left(\frac{\mu}{\Gamma} \right)^{0.2} \left(\frac{\sigma \rho L}{\Gamma^2} \right)^{0.1}$$

where Γ is the mass velocity (1.5 to 90 lb/min ft), L is wetted periphery
(0.36 to 19.2 in), μ is viscosity (1 to 9000 cP), ρ is liquid density (62.4 to
88 lb/ft^3), σ is surface tension (74 to 100 dyn/cm), N is disk speed (680 to
18,000 rpm), w is feed rate (0.55 to 67.5 lb/min), and r is disk radius (1
to 4 in).

Because it is used under controlled laboratory conditions, it is commonly thought that rotary atomizers produce a more uniform drop size
than other atomizers. Under low flow rates this may occur ($\underline{140}$), however, with comparable flow rates as used in commercial agricultural ap-
plications, the sprays have very similar drop size distributions ($\underline{57}$, $\underline{65}$,
$\underline{122}$, $\underline{123}$). Figure 1 illustrates a comparison ($\underline{122}$) of a small flat fan and
a rotary cage unit selected to give nearly the same median drop size. As
shown, the slope of the curve or uniformity of the sprays are nearly the
same. The rotary sprayer offers some attractive features for use in
agricultural applications. For instance, the unit is capable of producing
a very fine spray ($\underline{123}$), 10 to 40μ vmd, and generally it would have less
clogging problems than associated with very fine orifices. However, as a
rotary device it is subject to wear and breakdown not found with hydraulic
nozzles.

B. Electrostatic Force

The use of electrostatic force offers a unique and additional means
that the applicator may utilize to reduce drift by increasing the deposition of the small particles. Extensive research has been conducted on

the development and evaluation of electrostatic precipitation of pesticidal dusts. In 1959 Brazee and Buchele (124) reported a summary of 31 research publications related to electrostatic precipitation of dusts. Recently, Bowen and Splinter (125) presented results of field tests with improved electrostatic dusting equipment operated over a wide range of climatic conditions. In summary, deposits of dusts on cotton leaves from applications with charged dusts averaged from 2 to 3 times as much as similar applications with uncharged dusts. Although increased deposition would reduce the drift, the collection efficiency for very small charged dust particles may be significantly lower than for larger spray particles. Consequently, the drift from charged dust applications would probably remain significantly higher than most spray treatments. Recently there has been considerable interest in electrostatic charging of sprays which offers a technique for improving the collection efficiency of the fine portion of the droplet spectrum and consequently a reduction of drift from spray applications. Roth and Porterfield (126) constructed a multijet sprayer with a charged tubular element around each jet to deflect and capture small satellite drops. Results with the combination of a low velocity jet and electrostatic separation was found very effective in the laboratory. Although complete separation was not achieved in field operation, the unit significantly reduced the drift. Recently Law and Bowen (127) and Splinter (128) have introduced techniques for charging very fine sprays from hollow cone and pneumatic nozzles. In certain areas of a cotton plant the charged spray was found to enhance deposit over uncharged sprays by an average of 2.7 to 3.8. Carlton and Isler (129) developed an experimental device that charged a spray released from an airplane. Results however were disappointing.

The electrostatic force generally has a small effect on the large particles. Also the electrostatic charge does not affect the basic trajectory from the application equipment to the target. Thus, meteorological forces may deflect the trajectory such that it may miss the target. However, if the charged particle reaches the plant or target area the charge may significantly improve the collection efficiency of a fine particle that does not have sufficient inertia to cause impingement.

C. Application Equipment

Obviously the forces generated by the application equipment produce a major effect on the trajectory of the pesticide particles. Two basic forces that are utilized to achieve desired distribution and deposit are the inertia and aerodynamic drag forces. The inertia force can be very effective for large particles, and a desired distribution may be obtained with an emission velocity of sufficient magnitude and in the desired directions. However, to improve coverage and reduce volume rates the use of small particles may be desired. Since the inertia and gravitational forces may be very low for fine particles, movement of the air may be used to transport and impact the particles. As an airstream with spray particles approaches a plant or other obstruction some of the particles that are approaching the projected area of the object may be deflected and carried around the object while drops with sufficient inertia will impinge. The collection efficiency is defined as the fraction of the drops of a given size which strike the object in the total volume of air swept out by the object. The collection efficiency increases with an increase of drop size and air velocity. The collection efficiency also varies inversely with the size of the object and with its shape. Langmuir and Blodgett (131) developed a mathematical basis for predicting impingement. Several references contain summaries of typical results that may be related to agricultural applications (131-133). Brooks extended some of the results to relate the impaction of particles on 1/8- and 1/2-in. cylinders (132). Table 8 illustrates some of the above characteristics and indicates that a 100μ drop has a high collection efficiency for a wide range of conditions, but a drop size of 40μ or less may have a very low collection efficiency for certain conditions. Thus, even though the smaller particles may initially reach a given target, some of the fine particles may not impinge and may drift out of the area.

TABLE 8

Effect of Drop Size, Air Velocity, and Cylinder Diameter
on Collection Efficiency[a]

Drop size (μ)	1/8-in. cylinder (10 mph)	1/8-in. cylinder (20 mph)	1/2-in. cylinder (10 mph)
10	32%	48%	–
20	66%	77%	–
40	87%	92%	63%
100	98%	99%	90%

[a]From Ref. 132.

The air currents produced by an aircraft have a major effect on the trajectory of fine particles released from it. Although reference is frequently made to the downwash from an aircraft, it is perhaps better to consider the entire wake behind an aircraft. Basically, any aircraft, rotary or fixed wing, with a forward motion produces two trailing "wing tip" vortices with the centers located behind each wingtip and counter rotating to produce a downwash between the centers and an upwash beyond the wingtips. For a simplified rectangular spanwise loading the strength of the circulation is directly proportional to the weight per unit length of wing and inversely proportional to the forward velocity (134). In addition, a fixed wing aircraft has a strong propellor wake with some rotational component. A study of the circulation from low flying aircraft was made by Agricultural Engineers (122, 141) at the University of California. Movie cameras were used to record the trajectory of small, 4-in., gravitationally balanced balloons that were released from cages along a boom on the aircraft. Figures 3 and 4 illustrate the initial release point of the aircraft and balloon positions, and the subsequent paths of the free balloons. The figures show the general magnitude and direction of the air currents produced by a fixed and rotary wing aircraft. The results indicate somewhat similar circulation, and it is evident that some small particles may be carried 15 to 20 ft in the air even though the boom height may be less than 10 ft. Other tests with a helicopter at 55 mph (122) confirmed that a higher forward speed significantly reduced the strength of the vortices. Reed (135) reported a very comprehensive study of the trajectories of sprays released from agricultural aircraft. Results showed a looping trajectory with drops as large as 210μ that were released at 0.75 of the semispan from the centerline. A reduction in looping and lateral dispersion occurred with a lower flight elevation, a release closer to the centerline of the aircraft, and operation at a lower lift coefficient. Thus, it is apparent that to minimize drift, careful attention must be given to height of flight, speed and weight of aircraft, and release point relative to the aircraft.

Although spray applications by ground equipment are subject to drift, proper ground application equipment can significantly reduce the drift compared to aerial applications. Less drift can be achieved by use of large droplet spectrums with a high volume as well as confining the spray closer to the target area. Hoods or shields have been introduced to further reduce the drift for certain specific applications. Courshee (112) proposed a simple deflector to confine the trajectories and Edwards and Ripper (136) reported the successful use of an inflatable rubber boom cover. Tests with the rubber boom cover showed that drift was not eliminated but under strong winds of 9 to 17 mph the drift was reduced by 53 to 89% of the drift without a shield. Yates (137) developed a low drift system utilizing a combination of a hood and a very low pressure nozzle system. This system has been

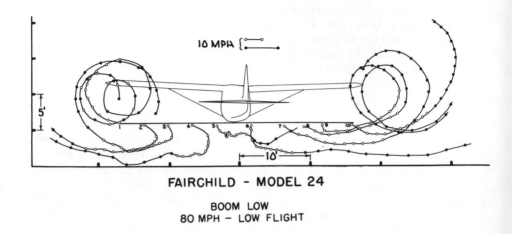

FAIRCHILD - MODEL 24

BOOM LOW
80 MPH — LOW FLIGHT

FIG. 3. Circulation from a high wing monoplane.

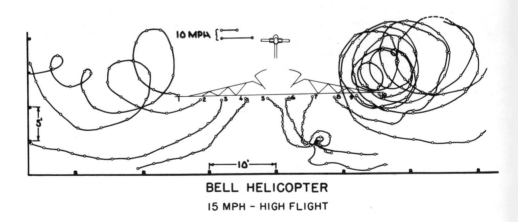

BELL HELICOPTER
15 MPH - HIGH FLIGHT

FIG. 4. Circulation from a helicopter.

used successfully in vineyards for the application of 2, 4-D sprays for control of morning glory. Also shielded sprayers have been used for application of 2, 4-D sprays for weed control in other sensitive crops such as lima
beans and tomatoes. Recently, experimental work has been conducted on a
new concept of applying herbicides in a foam (139). A foam generator with
water and surfactants was utilized to apply post emergence band applications
of herbicides to the base of cotton plants. Drift tests indicate the foam
agent itself will not reduce drift, but that coarse spray nozzles, as used
to make foam, will reduce drift.

D. Meteorological Conditions

The drift of agricultural chemicals is a direct result of the transport of
the particles by atmospheric movement. The diffusion transport and deposition characteristics of the wide range of particle sizes present in the
drift of agricultural chemicals are very complex, and the fundamental relationships for predicting drift concentrations are not fully established.
Some of the major meteorological parameters that affect drift are: wind
direction, wind velocity, air temperature, humidity, radiation, rain, and
several micrometeorological factors related to the stability of the atmosphere.

Wind direction is one of the most important and easily recognized parameters that can be used to prevent drift onto adjoining areas. Drift onto
a particular field can be avoided by timing the application during a period
the wind is coming from that particular field. In California smoke from
burning rubber tires is frequently used at the application site to provide a
visual indication of wind direction and enable the pilot to immediately cease
operation if the wind direction changes. Naturally in a diversified farming
area susceptible crops are usually located at some distance downwind, and
thus it is important to understand the parameters related to deposition at
various distances.

Wind velocity is of importance in determining transport distances and
may provide an estimate of movement under stable conditions. Table 9
illustrates the theoretical horizontal transport for nonturbulent conditions
with various size particles falling at terminal velocity for a vertical distance of 20 ft and displaced horizontally by an average wind velocity of 5
mph. The table serves only as a guide to illustrate the effect of particle
size and points out the dramatic increase in drift distance for particles below 100μ. It should be pointed out that the table was based on no evaporation and no turbulence as well as on a uniform wind velocity whereas in air
movement near the boundary layer the velocity decreases with a decrease
in height until it reaches zero at a height referred to as z_0, a value based

TABLE 9

Theoretical Drift Distance; No Evaporation

Drop diameter (μ)	Horizontal distance particle would be carried in a 5 mph wind while falling 20 ft, Sp. G. =1.0, no turbulence (theoretical)
1000	11.0 ft
500	21.6 ft
100	172.0 ft
50	587.0 ft
10	2.8 miles
5	253.0 miles

upon the roughness length. The wind velocity profile varies with surface roughness and atmospheric stability. Thus, it is important to measure the wind velocity profile to provide specific information to calculate the transport of particles near the ground surface. In addition, the wind velocity gradient has a direct effect on the turbulence and will be discussed under that section.

Air temperature and humidity may affect the evaporation rate of the particles. The evaporation rate can be calculated by techniques described in the previous section on evaporation. The evaporation rate of water base drops depends on both temperature and humidity of the atmosphere. For nonaqueous solvents or active pesticides, the humidity does not affect the evaporation and the air temperature is the major meteorological factor affecting evaporation rates.

Radiation from the sun does not have a very significant effect on evaporation but may introduce an important effect on certain chemicals. For very fine particles that remain suspended in air for long periods, the sun's radiation may aid in degradation of the pesticide to analogs that are less toxic.

Although chemical applications may not be conducted during a rainstorm, rain is one of the important mechanisms that removes some of the fine agricultural pesticide aerosols that are carried into the atmosphere.

It is well known that raindrops will impinge and remove small aerosol particles in the air. A recent meteorological manual published by the Atomic Energy Commission includes a summary of published articles that describe methods of calculating the washout coefficients for rain (143).

. One of the major meteorological parameters affecting drift characteristics is the turbulence. Turbulence is a rather complex phenomenon consisting of horizontal and vertical eddies which can mix and consequently dilute the concentration of fine particles released in the atmosphere. It should be remembered that chemical applications produce a wide spectrum of particle sizes which are transported by a combination of forces produced by gravity, an average wind velocity, and turbulence. For deposits at close ranges, gravity and mean wind velocity produce the predominate forces on the large drops. However, at a location some distance from the application, 1/2 mile or greater, an increase in turbulence will generally increase the vertical and horizontal size of the "drift" cloud, decrease the concentration per unit volume of air, and decrease the deposit on the ground.

Turbulence is related to the roughness of the ground surface, temperature gradient with height, and the wind velocity gradient with height. Turbulence near the ground is partially induced by the surface roughness characterized by a roughness length, z_0 (142), which is dependent on the size and distance between the protruding elements. Vertical and horizontal eddies are mechanically produced as the air stream flows over and around the protruding elements. In addition, mechanical turbulence is induced by the gradient of wind velocity as it produces wind shear. The velocity gradient is generally greater near the ground, increases with wind speed (for a given height in the boundary layer), and is also affected by the surface roughness. The temperature gradient is important since it represents the energy available for producing or depressing eddies by buoyancy forces. Figure 5 illustrates the dry adiabatic temperature gradient of $5.4°$ F per 1000 ft, which represents a neutral buoyancy condition. For instance, if a parcel of dry air was moved adiabatically from 1000 ft to 2000 ft, the parcel of air would be cooled $5.4°$ F as it expands to the corresponding standard atmospheric pressure at 2000 ft. Any temperature lapse greater than this value is called superadiabatic and produces unstable conditions with eddies formed by convection currents produced as the less dense air parcels near the ground surface accelerate upwards to a position of equilibrium. Air temperature profiles that have less lapse than the adiabatic condition are commonly referred to as inversions which dampen vertical displacements and produce stable conditions. The temperature profiles near the ground change diurnally. In the middle of the day a superadiabatic condition may exist near the ground because of high solar radiation. During the night, early morning, or late afternoon, a strong inversion may

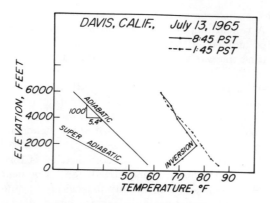

FIG. 5.   Temperature profiles.

exist.   Figure 5 shows a typical summertime condition over an alfalfa field
in the Sacramento Valley with a strong inversion present up to 2000 ft in
the morning and a nearly normal adiabatic condition present in the afternoon.   During the morning heating period a mixed layer may exist near the
ground with an inversion layer persisting aloft.   Sometimes an elevated inversion layer may persist throughout the day, particularly if one is present
at a higher elevation such as 4,000 to 5,000 ft.

The stability of the atmosphere is frequently characterized by the
Richardson number, Ri, given by the following relationship:

$$Ri = \frac{\dfrac{g}{T}\left(\dfrac{\partial T}{\partial z} + \Gamma\right)}{\dfrac{\partial \overline{u}}{\partial z}}$$

where g is the gravitational acceleration, T is the absolute temperature,
Γ is the adiabatic lapse rate, z is height, and $\overline{u}$ is the average wind velocity.   The Richardson number is a dimensionless parameter that relates
the consumption of turbulent energy by buoyancy forces to the rate of production of turbulent energy by wind shear.   A large negative value indicates
that convection predominates and is associated with strong vertical motion
which would diffuse fine particles both laterally and vertically.   As mechanical turbulence increases, the Richardson number approaches zero, and
intermediate mixing occurs.   Large positive values represent conditions
where the vertical motions are dampened and only horizontal eddies remain
with a minimum of diffusion.

Unfortunately, quantitative calculations of Richardson number require sophisticated instrumentation to accurately measure wind velocity and temperature gradients. The Stability Ratio (144), S.R., is a somewhat simplified index that has been satisfactorily correlated with drift deposit characteristics (145) and is given as follows:

$$S.R. = \frac{(T_{z_2} - T_{z_1})}{\bar{u}^2} \, 10^5$$

where T is in °C, $\bar{u}$ in cm/sec and measured at a height equidistant from z_1, and z_2 on a logarithmic scale, where $z_1 < z_2$. The Stability Ratio is not affected as much by changes in surface roughness as the Richardson number. Also the average wind velocity can be measured easier and more accurately than a velocity gradient.

Another approach to characterize the turbulence is to measure the three dimensional variations of velocity with fast response directional anemometers such as hot wire anemometers, acoustic anemometers, or directional bivanes with sensitive propellor. The intensity of turbulence can be defined as follows:

$$i_u = \frac{\sigma_u}{\bar{u}} \; ; \; i_v = \frac{\sigma_v}{\bar{u}} \; ; \text{ and } i_w = \frac{\sigma_w}{\bar{u}}$$

where σ is the standard deviation of the velocity variations of the component along the three axes, u is the component in the direction of the average horizontal velocity, v the crosswind component, and w the vertical component. Thus, the above approach requires a rather sophisticated system to measure three variable signals simultaneously. Then either an analog or digital scheme can be utilized to calculate the turbulence intensity. A somewhat simpler approach that has been used in several diffusion studies (143) was to measure the standard deviation of the angular variation of the two outputs from a bivane unit. There are portable analog systems available that will measure, calculate, and continuously record the standard deviations of the angular variations of a bivane unit. It can be shown that the angular fluctuations are approximately related to turbulence intensity as follows (143):

$$\sigma_\theta \approx \frac{\sigma_v}{\bar{u}} = i_v, \text{ and } \sigma_\phi \approx \frac{\sigma_w}{\bar{u}} = i_w$$

Thus, the latter provides a record and also an immediate indication of the intensity of turbulence from a single bivane system. It is important to recognize that the value of turbulence intensity is dependent on averaging time and that it requires experience to select the appropriate averaging period for the type and scale of diffusion under consideration. The standard deviation of the vertical wind direction is probably most important to the diffusion characteristics of a line source applied from agricultural aircraft. The standard deviation varies with height and wind speed. Neutral to moderately unstable conditions produce values between 5 and 10 degrees; extremely unstable conditions with light winds may approach 15 degrees; stable conditions generally result in a standard deviation of 2 to 5 degrees. Typical values for conditions over agricultural fields are given by Christensen et al. (146).

III. DRIFT RESIDUE CHARACTERISTICS FROM APPLICATIONS OF AGRICULTURAL CHEMICALS

Although research on pesticide drift problems was begun shortly after 2, 4-D became widely used (154), until recent years very little quantitative information was available on the drift and deposit characteristics of these applications. A few reports (64, 112, 135) have introduced techniques to calculate the trajectories of drops transported by gravity at an average wind velocity. The trajectories were for short distances of less than 50 ft, where the Stokes Law or ballistic fallout regime dominates. These did not introduce the important effect of diffusion and air transport which would be necessary for movement to greater distances.

The meteorology manual (143) has an excellent summary of the many different diffusion theories and results of extensive tests, some of model studies, on the diffusion of very fine particles in the air. However, due to the complexities of evaporation and distribution, and the difficulty in estimating the numbers of fine particles from a given application, an appropriate mathematical model has not been developed which will predict the distribution and fate of the total particle spectrum. Work has begun on such models, however, and hopefully these will prove to be of considerable benefit to all pesticide application programs, such as for mosquito control, as well as in agriculture (164). Certain spray drift residue information is available for short distances downwind of around 250 ft (112, 126, 147), and reports on residues to 1500 ft are to be found in respect to the drift spray programs related to tsetse fly and locust control (21, 123, 149).

Monitoring of downwind drift on the basis of a known source strength, and development of weather, formulation, particle size, and distribution parameters was begun as a research technique in 1960 by the Agricultural Engineers at the University of California at Davis. This was prompted by a continuing problem of 2, 4-D drift symptoms on grapes and cotton, but even more concern was being expressed about chlorinated hydrocarbon residues in dairy products as a result of feeding contaminated alfalfa and other forage to dairy cows (153). The overall fate or total budget account of the known dispersion of chemicals was stated as a basic research objective (154) of the project. However, this turned out to be an elusive factor, and the early work was largely directed toward determination of the amounts and the distances of drift in reference to residues on adjacent crops. For these tests we used the Stearman biplane aircraft with a 450 hp engine and typical application rates (7 to 10 gal/acre) and with the usual spray particle size (300 to 450 vmd) customary at that time.

The field monitoring program as developed at Davis can be broken down into the following: (a) the target area, which includes the actual aircraft swath and about 200 feet downwind where deposit takes place primarily by ballistic fallout; (b) the drift fallout zone, from about 200 feet to a mile or so downwind, which may receive some fallout, but by 1000 ft most of the material is airborne (aerosol size particles under 50 μ) (102), and the meteorological factors dominate the deposit of residues; finally (c) the general environmental area continuing from a mile or so onward, which becomes the sink for material transported in the atmosphere by gaseous diffusion and also as very fine aerosol particles of less than 10 to 15 μ. These may be deposited by settling and impingement, but may also be carried aloft and not returned to earth except by washout from precipitation of some form.

The experimental techniques and results of the rather elaborate aircraft trial runs have been reported in several publications (153-157). Each of the trials consisted of making 6 to 8 replicated passes over a single course, perpendicular to the wind direction. Figure 6 shows a typical test area as was used in 1968. Spray was emitted along a line 4750 ft long with sampling stations down to 10, 090 ft. Samples were collected in the target area on growing crops and in one gallon tin cans (6 in. diameter by 7 in. high). In the fallout zone and on into the environmental zone residues were collected on Mylar (plastic) sheets (6 by 18 in.). Air samplers were used, including the Unico and Casella cascade impactors (for drop size data), as well as the Staplex high volume glass fiber filter type devices for total sample. Plant crop samples were also taken for residue sampling in the downwind area and continuing into the environmental zone.

All deposits were corrected to a common basis of $\mu g/ft^2$ for an equivalent application of 1 lb active chemical per acre, per pass of the airplane. For drift considerations it may be desirable to express the data in terms of the emission per length of travel. The above corrections are equivalent to an emission of 1 lb active chemical per 1/4 mile pass or 343, 636 μg/ft

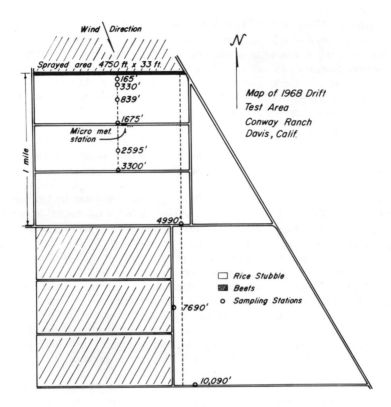

FIG. 6. Map of drift test area.

of travel. The 1968 meteorological measurements were taken in the rice
field at a point 1675 ft downwind from the applications.

A. Target Area

Deposited residues and application swath widths on the target area are
primarily related to the material particle size, cross–component of the
wind velocity, and the flight height. Figure 7 illustrates a typical average deposit pattern showing a sharp cutoff on the upwind side, a high and
somewhat irregular pattern in the propeller wake, and a rather rapid decrease to a substantially lower level within 100 ft downwind. Table 10
lists an estimate of the percent recovery of the total spray. It is significant to note that 77.9% was recovered within 160 ft. The recoveries for
the distant stations are included to point out the relative magnitude of the

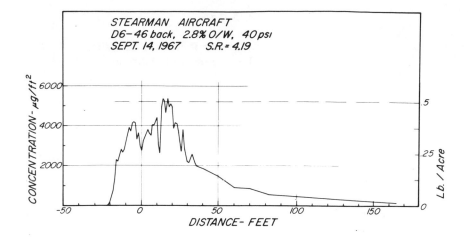

FIG. 7. Deposit in target area.

drift residues. As noted, the recovery from 160 to 2640 ft represented 6.71% of the spray. Since the application was under very stable weather conditions, this drift represents about the highest values to be expected for an application using a relatively coarse 420µ VMD spray. The total recovery was only 84.61% which appears to be due to poor collection efficiency of sampling units and loss to the atmosphere. Cans were used under the aircraft to prevent loss due to bounce of large drops, but the efficiency of this unit may be poor for small particles. The Mylar sheets may likewise result in some loss due to poor collection of fine particles or loss to droplet bounce. Although the collection efficiency may be somewhat lower than on the forage crop, previous data (155, 157) with Mylar sheets located from 82 to 2640 ft downwind have shown that deposits were proportional to residues collected on adjacent forage. Thus, results may not represent accurate absolute levels, but indicate relative drift characteristics and can be related to forage residues. Recent tests conducted by USDA personnel (158) with undiluted and water diluted insecticide applications with aircraft revealed a range in total recoveries of 18% to 62% with undiluted ULV applications and 46% to 96% with aqueous emulsion sprays. Tests run by the University of Arizona comparing aircraft and ground rig applications indicate that the aircraft tended to give higher drift levels when compared to a usual hydraulic type ground applicator. However, when compared to a very fine atomizing type air carrier type ground rig the aircraft showed less downwind drift. Presumably the very fine spray in this case caused the highest drift (160, 161).

TABLE 10

Spray Recoveries in Various Downwind Areas for
Two Different Meteorological Conditions

	Test date	
	9/14/67	11/4/67
Stability ratio	4.19 (very stable)	−0.14 (unstable)
Wind velocity (mph at 8 ft)	5.1	9.6
Temp. gradient (32 ft to 8 ft), °F	7.7	−0.6
Distance from centerline (ft)	Recovery percent of total emission	
−23 − 82	70.0	---
82 − 160	7.9	---
160 − 660	4.46	3.91
660 − 1320	1.01	0.43
1320 − 1980	0.66	0.16
1980 − 2640	0.58	0.09
Total	84.61%	

B.   Drift Fallout Zone

Our work has been primarily in the zone where we found the greatest concern for drift damage and high residues on adjacent crops. The following factors influence the drift residues in this zone: meteorological conditions, physical properties of pesticide formulations, spray atomization or particle size, and height of flight.

1.   Meteorological Conditions

The stability ratio was calculated for each test using the average wind velocity at 16 ft and the average temperature gradient between 8 ft and 32 ft. The statistical procedures described in previous reports (154, 155)

were used to calculate the regression curves relating the experimental
drift residue data to the downwind distance. Briefly, the regression curves
were calculated with an IBM 7044 computer using a stepwise regression
program to calculate the best least squares fit of data to the following type
of second degree polynomial equation: $\log y = b_0 + b_1 (\text{lod } d) + b_2 (\log d)^2$,
where y is the deposit on Mylar sheets, d is distance downwind, and b_0,
b_1, b_2 are regression coefficients. For convenience in discussions of drift
residue patterns over typical agricultural areas the atmospheric conditions
may be classified into the following categories: very stable, S.R. (stability ratio) greater than 1.2; stable, S.R. 0.1 to 1.2; nearly neutral, S.R.
-0.1 to 0.1; and unstable, -0.1 or less.

Drift residue patterns revealed that as the stability ratio increased, the
residues at the distance locations also increased. Figure 8 illustrates composite results of three tests during very stable conditions with nearly the
same stability ratio and five tests during nearly neutral conditions. Table
11 lists the basic meteorological data for the selected drift experiments.
Comparison of the curves in Fig. 8 shows very little difference in residues
as far as 200 ft downwind. However, at 1/2 mile downwind the regression
values are approximately 20 times higher under very stable conditions than
under neutral conditions.

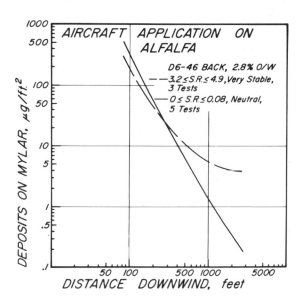

FIG. 8.   Effect on stability ratio on drift deposits.

TABLE 11

Meteorological Data for Tests Conducted Under Nearly Neutral and
Very Stable Conditions

Condition	Stability ratio	Velocity, 8 ft (mph)	Velocity, 16 ft (mph)	(Temp. at 32 ft- temp. at 8 ft) ($^\circ$ F)
Nearly neutral	0.06	13.9	16.1	0.5
	0.01	12.5	15.3	0.1
	0.08	12.7	15.4	0.6
	0.05	8.4	10.0	0.2
	0.00	13.3	14.9	0.0
Very stable	4.2	5.1	7.6	7.7
	4.9	3.1	4.2	3.1
	3.2	4.2	5.8	3.9

Table 10 lists the percentage of the total emission (from downwind portions as noted) that was recovered from two tests. The first was made during very stable conditions (S.R. 4.19) and the second during an unstable case (S.R. -0.14). The deposit recovery or fallout in the area from 160 ft to 2640 ft downwind represents a total of 6.71% for the very stable condition and 4.59% for the unstable case. It is also important to note that the recoveries were both nearly the same close to the swath, while at 1980 to 2640 ft the unstable condition produced 0.09% recovery compared to 0.58% during stable conditions. Thus, the stability ratio is a convenient index for estimating the drift residue pattern.

2. Physical Properties of Pesticide Formulations

The potential hazard of drift residues from application of dust formulations has been recognized for many years. In 1947 Brooks (132) reported information on drift hazards from dusts, and data from Kruse, Hess, and Metcalf (161) indicated as much as 72% of the dust may drift out of the treatment area. Some recent tests by MacCollom (162) confirmed increased downwind residues during inversion conditions as compared to applications during periods of normal lapse. One test in hilly terrain during a temperature inversion produced residues on forage of 0.2, 0.8, and 0.7 ppm at a distance of two miles downwind and over a lateral width of approximately one mile. This points out the possibility of dusts producing high residues at considerable distances downwind under certain topographical and temperature inversion conditions. Figure 9 illustrates regressions of drift residues measured from dust and spray tests conducted by the Agricultural Engineering Department, University of California at Davis. The dust application contained a calcium carbonate dust carrier impregnated with 4% Moricide insecticide. The application was during a very stable condition, S.R. of 4.9, and the residues were measured in parts per million on alfalfa samples. Results were corrected to an equivalent of $\mu g/ft^2$ based on previous correlations of deposits on Mylar and alfalfa. Thus, the results can be compared with the previous regression from spray tests under similar conditions, S.R. of 3.2 to 4.9. The regression curves indicate the dust application produced over 8 times as much residue as the 420μ VMD spray in the area from 350 ft to 800 ft downwind, and residues approached similar levels at 80 ft and 1 mile downwind.

As previously mentioned, the drift potential may be reduced by use of thickening agents to reduce the number of fine particles. Figure 10 illustrates the regression data of the drift residues for a drift test with a particulating agent during a Stability Ratio of 0.3. For comparison, the previous

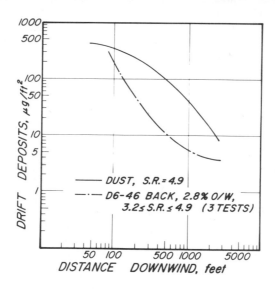

FIG. 9. Drift of dust and sprays under very stable conditions.

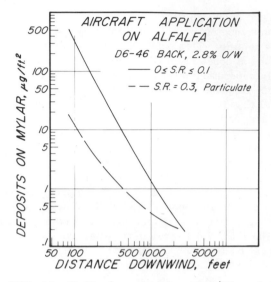

FIG. 10.   Drift of particulate and O/W emulsion.

regression of residues from 5 drift tests under nearly neutral conditions, S.R. 0.0 - 0.1, is included. All applications were with D6-46 hollow cone nozzles operated at 40 psi. The stability ratio was nearly the same, and the total drift residue was dramatically reduced by the thickened spray; the residue at 100 ft downwind was approximately 30 fold lower than the usual formulation. However, it is important to note that the particulating agent didn't eliminate drift, and apparently some very fine particles were produced as the residue at 2640 ft approached the level from conventional oil in water emulsions.

3.   Nozzle Type, Size, and Orientation

The selection and operation of the spray nozzles is an important means of controlling the drift residue pattern. Because of the variable atmospheric conditions, the tests were not under identical stability conditions; however, the tests do serve to illustrate the general trends and important differences in the drift residues from various atomization systems. Figure 11 illustrates the regression curves from the following nozzle systems under the respective stability conditions: 80015, flat fan nozzles, down, S.R. -0.7; D6-46 hollow cone nozzles, back, S.R. 0.0 - 0.1; D6-45 hollow cone nozzles, down, S.R. 0.2 - 0.7; and 3/32 in. orifices directed back, S.R. 0.4. The residues at the distances downwind follow the order

AIRCRAFT APPLICATION					
CURVE ID	NOZZLE TYPE	S.R.	% NON VOLATILE	GROUND COVER	NO. OF TESTS
– – –	80015 down	-0.7	100	Stubble	1
———	D6-45 down	0.2 to 0.7	2.8	Alfalfa	2
—·—	D6-46 back	0.0 to 0.1	2.8	Alfalfa	5
- - - -	D6 jet back	0.4	2.8	Stubble	1

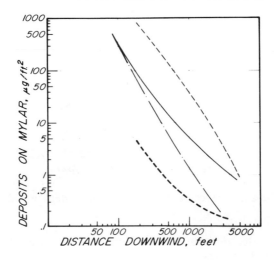

FIG. 11. Effect of nozzle types on drift residues.

expected, based on the drop size spectrums (Fig. 1). As a means of comparison, the residues for the three tests above, measured at 1000 ft downwind, were divided by the residue at the same distance as found from the
fourth test, or the D6-46 nozzles directed back. The respective values of
the ratios for each are: 80014, down (36.3); D6-45 down, (4.5); and 3/32
in. jet, back (0.3). From this it can be seen that a choice of nozzle system can give a wide range of drop size and consequent effect on downwind
drift.

4. Height of Flight

The release height is an important factor that must be considered in
confining spray to this close fallout area. Although an increase in
height is sometimes used to increase the swath width by allowing the wind
to carry the material downwind, conversely the elevation should be mini-
mized to reduce the drift residue hazard. Applications of injurious herbicides in California must be released at less than a 10 ft elevation. Figure
12 indicates results from recent tests that show some increases in residues,

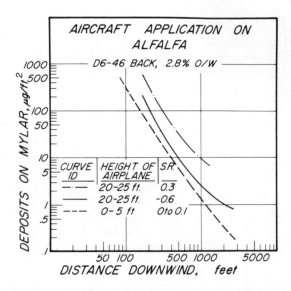

FIG. 12.   Effect of flight elevation on drift residues.

to at least 1/2 mile, for applications at 20-25 ft compared to low level re-
leases at 0-5 ft elevation. All applications were with D6-46 nozzle directed
back. The stability ratios were not the same for the runs; the low level run
falls between the two at the greater height. The residue from the 20-25 ft
height with unstable conditions (solid line graph) was about 2 times as much
as the 0-5 ft height at 1000 ft. However, a further comparison of the effect
of stability showed that the residues 1000 ft downwind from the release at
a 20 to 25 ft elevation under the more stable versus unstable conditions
(S.R. 0.3 versus -0.6) increased the residue eight fold.

5.   Estimated Cumulative Residues

The above results are useful in comparing relative drift hazards and
can further be used to estimate the maximum cumulative residue downwind
from a progressive multiple swath application on a given crop. Previous
reports (154, 155) explain the statistical and mathematical procedures utilized in estimating the maximum residues. Briefly the upper 99% confidence
limit for the regression line was used to estimate the maximum probable
residue on Mylar from each pass. The next step involved converting the
residue on Mylar to the equivalent residue on forage in ppm. Again, the
upper 99% confidence limit for the regression relating Mylar deposits and
crop residues was utilized. The total cumulative residue was calculated
by summing the residues contributed by each pass for typical treated areas.

Figure 13 illustrates the maximum residues that would likely occur from a 20 swath application spaced at a 33 ft swath width or an equivalent of 20 acres with a 1/4 mile length of field. The residues are based on one pound active pesticide applied per acre. Thus, the graph summarizes typical results in terms of maximum estimated residues for typical applications under different conditions of weather, nozzle types, formulations, and height of flight. Examining the curves of Fig. 13, it can be seen, for example, at the top of the drawing, that very high downwind drift residues (to a distance of 1/2 mile) are obtained with the very fine atomizing 80015 nozzles using a 100% low volatility oil. The drop size remains largely fixed with such a formulation, which is similar to a dust (fixed size) and is shown in the next curve down. Note, however, that the S.R. is much higher for the dust run which increases drift. Thus, lower drift values might have been shown, perhaps below the fine spray curve, if the S.R. had been the same. The next curve is for the D6-46 nozzle, back, and shows the significant drop of over 5 times at 1000 ft in the drift residues between this and the dust at comparable S.R. The next curve (solid line) shows the significance of the high S.R. or stable weather. A temperature inversion and low wind velocity is probably the most hazardous weather condition in which to put on an application, if nearby drift residues are to be avoided. A higher wind velocity can be shown to spread out the airborne materials more rapidly than low winds. If in addition to the low wind there is a temperature inversion there will be no vertical mixing, and concentrations are highest at all points downwind (157).

The next curve is for the low drift particulate type spray material and indicates lowered residues out to about 1/2 mile. Beyond this the small particles in the spray are almost as frequent as with the finer spray of the D6-46 curve and so drift residues are about the same. The last curve is for the jet nozzles which are recommended for plant hormone type sprays such as 2,4-D and others. The very coarse drops composing such a spray are not suitable for any except translocated or systemic type materials as the coverage would be too low for most contact type chemicals (164).

C. Environmental Contamination

The rapid expansion of pesticide use coupled with increasing research evidence of pesticide contamination in the environment has given rise to a wave of public concern, and in many cases condemnation of pesticide chemicals. The depository of pesticides is of course found to be the oceans, lakes, and inland waterways as well as the soil itself. Pesticide chemicals become hazardous contaminants based on their tendency to vaporize, their degree of solubility in water or other solvents, and their resistance to the normal degradation process (165). While it was generally thought that

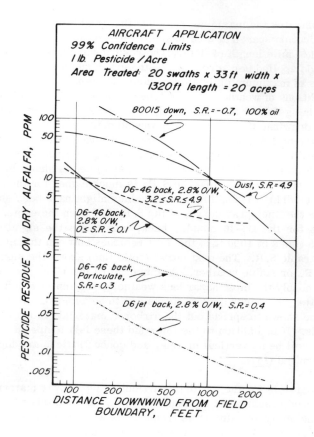

FIG. 13. Predicted cumulative drift residues on alfalfa.

most pesticide chemicals degraded rather rapidly, it has of course been
clearly proven that such materials as DDT and its metabolites can remain
in the environment for many years as persistent toxicants. Even the
rapidly degradable materials such as parathion have been found to persist for a much longer period than earlier studies led us to believe.

Our work has been largely concerned with the aerial transport of pesticides from the area of treatment to relatively nearby (usually within one
mile or so) food and feed crops, rangeland, and pastures. Our concern
was primarily with herbicide damage to growing crops, and insecticide
residues on the foods and feeds. However, in gaining information on the
mode and levels of aerial transport that moves hazardous chemicals onto

nearby crops we have also found correlation of our work with the more typical air pollution research methods (150).

Some materials such as the phenoxy acid herbicides have been known for years to be transported typically by prevailing winds within an air shed and have caused symptoms, and in some cases damage to sensitive crops, many miles from the application site. Bamesberger and Adams conducted monitoring studies on movement of 2,4-D in the air at sampling sites in the middle of the Palouse grain area of Eastern Washington (165). The results of monitoring for 100 days during the spray season shows characteristic high and low levels of air burden dependent on place and amount of 2,4-D applied, and the basic weather pattern of the air shed. For example, at one of the stations, 24-hour averaged levels in the air were as high as $2 \ \mu g/M^3$ in early May during the height of the spraying season. However, even as late as the end of July peaks of $0.2 - 0.3 \ \mu g/M^3$ continued to appear. During our air monitoring studies conducted in 1969 with the herbicide propanil the aerial transport of this relatively nonvolatile material appeared as distinctive yellow spot symptoms on sensitive prune trees which were set out as test plants at various distances to over 20 miles from the sprayed rice area of about 9500 acres (166). Data taken on both high volume and cascade impactor air samplers indicated that the largest portion of this material was being transported as very small (10-50 μ diam) nonevaporative, liquid drops. Levels in the air at about 5 miles downwind (prevailing diurnal sea breeze) rose to a peak of $0.066 \ \mu g/M^3$ (8-hr sample av) for a period of 3 days during the peak of the spray work, but dropped back to $0.0016 \ \mu g/M^3$ and remained there for about another week. The U.S. Public Health Service, Pesticides Monitoring Project has also reported very low, but detectable levels in the air, particularly in agricultural communities (152).

During many of our fixed line source tests we have had air samples located at downwind points to as much as 5 miles. Figure 14 shows total air sample collection (for the time of spraying) of fluorescent dye as the dashed line graph in the upper right hand side of the figure. This application was made with a Stearman type aircraft with coarse spray (approx. 420 μ vmd), a 2.8% oil in water emulsion, and under nearly neutral (neither stable nor turbulent) weather conditions. The ground fallout is shown on the heavy line and displays the normally expected decrease with distance of nearly 100-fold between 100 ft and 3000 ft downwind. However, the point of interest is the levels of total airborne chemical leaving the area which remain at 30-40 $\mu g/ft^2$ of the high volume air sampler with a glass fiber filter face. This filter is capable of removing about 98% of all particles above 0.1 μ diameter. This data is normalized to one pass of the aircraft across the field and a rate of application of 1 lb/acre. While it was pointed

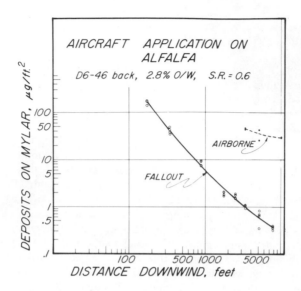

FIG. 14. Comparison of drift fallout and airborne concentrations.

out earlier that a test run with coarse spray such as this would expect to lose only a few percent of the applied spray from the treated field (varying from 5-15%), the airborne levels, although maintained for only a few minutes from each pass of the aircraft, show significant amounts of 50 or more times higher than what falls out at the 1 and 2 mile stations. Of course, this airborne material is carried and spread out over several miles of diluting air and land surface. But when large acreages are treated, as in the case of the Washington studies on 2, 4-D and our work on propanil, then the entire air shed becomes contaminated with low, but significant levels of pesticide chemicals.

In summary, it would appear that unless better control of the aerial transport of drift is not made in the very near future, the increasing evidence of pollution will force greater restriction on the use of many valuable but potentially hazardous chemicals. Present research points toward the hope that drift can be controlled through more specific confinement of smaller effective dosages to the treated field by (1) reducing the numbers of small drops in the applied spray, both through vaporization and evaporation loss control and through mechanical cutoff of the fine drop spectrum, and (2) utilizing the best weather conditions where turbulent ventilating weather can be safely used to aid in removing the airborne portion of the atomized pesticide in the direction least likely to cause downwind pollution problems. It must be recognized, however, that to use pesticide chemicals means to lose small amounts of these to the environment. For even though we restricted 100% of the applied material at the time of application to the treated field, the residues on the crops and in soils will be moved in part with the harvested crop and with a remainder left in the soil to be exposed to potential dust, surface washing, or ground water transport from the fields to the general environment. It should be obvious that rapidly de-gradable chemicals, and use of nonhazardous biological means such as microbial or virus sprays, and other nonchemical pest control techniques, must ultimately become of greater importance in the total system of plant protection.

REFERENCES

1. Anon., "Pesticides in the Environment," Cleaning Our Environment: The Chemical Basis for Action, Am. Chem. Soc., Washington, D. C., 1969, pp. 193-234.

2. W. J. Hayes, Jr., Scientific Aspects of Pest Control, Nat. Acad. Sci., Nat. Res. Council, 1966, pp. 314-342.

3. E. G. Hunt, Scientific Aspects of Pest Control, Nat. Acad. Sci., 1966, pp. 251-262.

4. R. F. Smith and R. van den Bosch, Pest Control: Biological Physical and Selected Chemical Methods, Academic, New York, 1967, pp. 295-340.

5. Anon., "Bioenvironmental Control of Pests," Restoring the Quality of Our Environment, Rpt. President's Science Advisory Comm. The White House, 1965, pp. 227-291.

6. Anon., "Uses and Benefits of Pesticides," Report of the Secretary's Commission on Pesticides and their Relationship to Environmental Health, U. S. Dept. of Health, Education and Welfare (Mrak Report), U. S. Govt. Printing Office, Washington, D. C., Dec. 1969, pp. 21, 46, 180.

7. V. M. Stern, R. F. Smith, R. van den Bosch, and K. S. Hagen, "The Integrated Control Concept," Hilgardia, 29, 81-101 (1959).

8. R. F. Smith, Agr. Science Rev., USDA, 7, 1-6, (1969).

9. E. P. Savage, "Ecology of Pesticides," Training Handbook, HEW, Public Health Service, Div. of Community Studies, 1969.

10. H. H. Shephard, J. N. Manhan, and D. L. Fowler, The Pesticide Review 1966, USDA, SSCS, Washington, D.C., 1966, pp. 1-33.

11. R. E. Duggan, "Compliance Activities for Pesticide Residues in Foods," Pesticides and Public Health Short Course, HEW, PHS, Consumer Protection and Environmental Health Service, Atlanta, Georgia, 1969.

12. R. E. Duggan and K. Dawson, Pesticides, "A Report on Residues in Food," FDA Papers, U.S. Govt. Printing Office, Washington, D.C., June 1967.

13. G. M. Woodwell, The Scientific American, 216, 24-31 (March 1967).

14. J. O. Keith and E. G. Hunt, Trans. 31st. N. American Wildlife and Natural Resources Conf., Wildlife Management Inst., Washington, D.C., March 1966.

15. T. E. Bailey and J. R. Hannum, Proc. Am. Soc. Civil Engrs., SA5, Sanitary Engr. Div., October 1967.

16. E. D. Lichtenstein, Scientific Aspects of Pest Control, Nat. Acad. Sci., Washington, D.C., 1966.

17. H. P. Nicholson, Science, 158, 871-876 (Nov. 1967).

18. R. W. Risebrough, R. J. Huggett, J. J. Griffen, and E. D. Goldberg, Science, 159, 1233-1235 (March 1968).

19. W. E. Westlake and F. A. Gunther, Organic Pesticides in the Environment, Am. Chem. Soc., Advances in Chemistry Series 60, Washington, D.C., 1966, pp. 110-121.

20. D. L. Gunn, 5th Commonwealth Ent. Conf., London, 1948, pp. 54-59.

21. R. J. Courshee, Bull. Entomol. Res., 50, 355-369 (1959).

22. K. S. Hocking, D. Yeo, and D. G. Anstey, Bull. Entomol. Res., 45, 585-603 (1954).

23. A. W. A. Brown and L. G. Putnam, J. Econ. Entomol., 40, 606 (1947).

24.   K. Messenger, Agr. Chem., 30 (February, 1953).

25.   F. E. Skoog and F. T. Cowan, "Flight Height, Droplet Size, and
       Moisture Influence on Grasshopper Control Achieved with Malathion
       Applied Aerially at U. L. V.," Jour. Econ. Ento., 61, 1000-1003
       (August 1968).

26.   J. S. Yuill and D. A. Isler, J. Forestry, 57, 263-266 (1959).

27.   C. M. Himel and A. D. Moore, Science, 156, 3779 (1967).

28.   W. N. Sullivan, L. D. Goodhue, and J. H. Fales, "Insecticide Dis-
       persing; a New Method of Dispersing Pyrethrum and Rotenone in the
       Air," Soap and Sanitary Chemicals, 16, 121-125 (1940).

29.   W. E. Burgoyne, K. G. Whitesell, and N. B. Akesson, Down to
       Earth, 24, 13-15 (1968).

30.   F. A. Brooks, Agr. Eng., 28, 233-239 (1947).

31.   B. E. Day, E. Johnson, and J. L. Dewlen, "Volatility of Herbicides
       Under Field Conditions," Hildgardia, Univ. of Calif., 28 (February
       1959).

32.   A. N. Kasimatis, R. J. Weaver, and R. M. Pool, J. Enology and
       Viticulture, 19, 194-204 (1968).

33.   W. A. Norris, Stanford Law Review, 6, 69-90 (1953).

34.   G. Douglas Barbe and John C. Hillis, "Effects of the Herbicide Pro-
       panil on Stone Fruit Trees in California," California Dept. of Agr.
       Special Publication, Sacramento, Calif., June 1969, 69-1.

35.   R. S. Jensen, Hastings Law J., 19, 476-493 (Jan. 1968).

36.   S. W. Turner, Proc. Western Regional Pesticide Chemical Applica-
       tion Short Course, University of California, Berkeley, 1967, pp. IV,
       1-15.

37.   C. E. Lapple and Shepherd, Ind. Eng. Chem., 32, 605-617 (1940).

38.   J. H. Perry (ed.), Chemical Engineer's Handbook, McGraw-Hill,
       New York, 1950, pp. 1017-1021.

39. C. E. Lapple, Fluid and Particle Mechanics, University of Delaware, 1954.

40. H. L. Green and W. R. Lane, Particulate Clouds: Dusts, Smokes, and Mists, E. and F. N. Spon Ltd., London, 1964, pp. 62-69.

41. C. Orr, Particulate Technology, MacMillan Co., New York, 1966.

42. E. K. Edgerton and J. R. Killian, Flash Seeing the Unseen by Ultra-high-Speed Flash, Hale Publishing Co., 1939.

43. W. R. Lane, Ind. Eng. Chem., 43, 1312-1317 (1951).

44. J. L. Buzzard and R. M. Nedderman, Chem. Eng. Sci., 22, 1577-1586 (1967).

45. R. Gunn and G. D. Kinzer, J. Meteorology, 6, 243-248, 1949.

46. T. C. Watkins, L. B. Norton, D. E. Weidhaas, and J. L. Brann, Jr., Handbook of Insecticide Dust Diluents and Carriers, Dorland Books, Caldwell, N. J., 1955.

47. S. F. Potts, Bur. of Ent. and Plant Quarantine, E-508, USDA, 20 pages, Washington, D.C., 1940.

48. C. W. Kruse, A. D. Hesse, and R. L. Metcalf, Nat. Malaria Soc. J., 3, 197-209 (Sept. 1944).

49. Minerals and Chemical Corp. of America, Tech. Information No. 153, Menlo Park, New Jersey, 1959.

50. H. A. Meyers and W. G. Lovely, Agr. Eng., 38, 298-301, 316-319 (1957).

51. State of California, Announcement PC-183, Dept. of Agriculture, Field Crops and Agricultural Chemicals, March 1965.

52. L. A. Liljedahl and W. G. Lovely, Paper No. 68-137, American Society of Agricultural Engineering, June 1968.

53. A. S. Crafts and H. G. Reiber, Hilgardia, 18, 77-156 (1948).

54. Lord Rayleigh, Proc. Royal Soc., XLVII, 281 (1890).

55. C. C. Addison, J. Chem. Soc., 535 (1943).

56. C. C. Addison, Philosophical Magazine, 36, 73-100 (1945).

57. R. P. Fraser, Advances in Pest Control Research, Interscience, New York, 1958, pp. 1-106.

58. H. Kido and E. M. Stafford, J. Econ. Entomol., 59, 454-460 (1966).

59. W. Van Valkenburg, Proc. Northeastern Weed Control Conference, 17, 363-372 (1963).

60. C. R. Kaupke and W. E. Yates, Transactions of ASAE, 9, 797-799, 802 (1966).

61. B. J. Butler, N. B. Akesson, and W. E. Yates, Transactions of ASAE, 12, (1969).

62. R. Brown and J. L. York, A.I.Ch. E. J., 8, 149-153 (1962).

63. W. E. Ranz and W. R. Marshall, Chem. Eng. Progr., 48, 141-146, 173-180 (1952).

64. K. G. Seymour, Pesticidal Formulations Research Advances in Chemistry Series, A.C.S., 86, 135-154 (1969).

65. W. R. Marshall, Jr., Chem. Eng. Progr. Monograph Series, No. 2, 50, 1-122 (1954).

66. R. T. Cunningham, J. L. Brann, Jr., and G. A. Fleming, J. Econ. Entomol., 55, 192-199 (1962).

67. J. A. Duffie and W. R. Marshall, Jr., Chem. Eng. Progr., 49, 417-423 (1953).

68. Fisons Pest Control Ltd. (England), Lovo Spray Formulants, 1961.

69. R. C. Amsden, Agr. Aviation, 4, 88-93 (1962).

70. G. A. Geering and J. H. Lloyd, Overdruk uit de Mededelingen van de Landbourokogeschool en de opzolkingsstations van de staat, Deel XXVI, 1961, pp. 1471-1483.

71. G. A. Geering and J. H. Lloyd, J. Econ. Entomol., 55, 786-790 (1962).

72. S. Johansson, National Institute for Plant Protection Contributions, Stockholm, 12, 88 (1962).

73. R. A. Mugele and H. D. Evans, Ind. Engr. Chem., 43, 1317-1324 (1951).

74. L. A. Liljedahl, W. G. Lovely, and W. F. Buchele, American Society of Agr. Engineering, Paper No. 67-661, 1967.

75. R. W. Tate and W. R. Marshall, Jr., Chem. Eng. Progr., 49, 169-174, 226-234 (1953).

76. Lord Rayleigh, Proc. London Math. Soc., 10, 4-13 (1878).

77. E. Tyler, Philosophical Magazine, 16, 504 (1933).

78. McDonald, Private Communication with W. R. Marshall, Chemical Engineering Monograph Series, No. 2, 50, 7 (1954).

79. A. C. Merrington and E. G. Richardson, Proc. Physical Society, 59, 1-13 (1947).

80. D. B. Harmon, Jr., J. Franklin Institute, 259, 519-522 (1955).

81. C. C. Miesse, Paper 88-53, American Rocket Society, 1953.

82. C. Weber, Z. Angew. Math u. Mech., 11, 136 (1931).

83. S. Tomotika, Proc. Royal Soc. (London), A150, 322; A153, 302 (1935).

84. H. B. Holroyd, J. Franklin Institute, 215, 93-97 (1933).

85. T. Baron, University of Illinois Technical Report 4, 1949.

86. G. Littaye, Compt. rend. (France), 217, October 11, 1943; No. 4, 99, 340, July 6, 1943.

87. J. M. Schneider and C. D. Hendricks, Rev. Sci. Instr., 35, 1349-1350 (1964).

88. W. R. Atkinson and A. H. Miller, Rev. Sci. Instr., 36, 846, 847 (1965).

89. C. A. Reimer, Down to Earth, 20, 3-6 (1964).

90. Plant Protection Limited, Instructions for Use of Experimental "Vibrajet," Imperial Chemical Industries Ltd., Fernhurst, Haslemere, Surrey, England.

91. J. Kirch, J. Waldrum, and P. Bishop, Proc. Northeastern Weed Control Conference, 1969.

92. California Department of Agriculture, Field Crops and Agricultural Chemicals, California Administrative Code, Section 2454.

93. W. R. Lane, Ind. Eng. Chem., 43, 1312–1317 (1951).

94. J. O. Hinze, A.I.Ch.E. J., 1, 289–295 (1955).

95. G. D. Gordon, J. Appl. Phys., 30, 1759–1761 (1959).

96. G. Morrell, NASA, Tech. Note D-677, 1961.

97. J. D. Wilcox and R. K. June, J. Franklin Institute, 271, 169–183 (1961).

98. D. B. Harmon, University of California Publ. in Engr., 5, 145–158 (1955).

99. R. G. Dorman, Brit. J. Appl. Phys., 3, 189–192 (1952).

100. D. Yeo, J. Ag. Engr. Res., 4, 93–99 (1959).

101. R. P. Fraser and P. Eisenklam, Trans. Inst. Chem. Engrs., 34, 294–319 (1956).

102. H. Coutts and W. Yates, Transactions of ASAE, 11, 25–27 (1968).

103. D. A. Isler and J. B. Carlton, Transactions of ASAE, 8, 590, 591, 593 (1965).

104. F. Skoog and F. Cowan, J. Econ. Entomol., 61, 1000–1003 (1968).

105. R. Tate and L. Janssen, Transactions of ASAE, 9, 303–305, 308 (1966).

106. O. Hedden, Transactions of ASAE, 4, (Special Power and Mach. Ed.), 158, 159, 163 (1961).

107. N. Dombrowki, P. Eisenklam, and R. Fraser, J. Institute Fuel, 30 (July 1957).

108. R. Ford and G. Furmidge, Brit. J. Appl. Phys., 18, 335–348, 491–501 (1967).

109. B. Byrd and K. Seymour, American Society of Agr. Engr., 1964, Paper No. 64-609C and No. 64-609D.

110. P. Nelson and W. Stevens, A.I.Ch.E. J., 7, 80-86 (1961).

111. R. Straus, Ph.D. Thesis, University of London, 1949.

112. R. Courshee, J. Agr. Engr. Res., 4, 144-152, 229-242 (1959).

113. Diamond Shamrock Co., Technical Bulletin TC-36-6, Cleveland, Ohio.

114. Nukiyama and Y. Tanasawa, Trans. Soc. Mech. Engr. (Japan), 4, No. 14, 86, 15, 138 (1938); 5, No. 18, 63, 68 (1939); 6, No. 22, II-7 and No. 23, II-8 (1940).

115. R. Wetzel, Ph.D. Thesis, University of Wisconsin, 1952.

116. J. Gretzinger and W. Marshall, A.I.Ch.E. J., 7, 312-318 (1961).

117. H. Lewis, D. Edwards, H. Goglia, R. Rice, and L. Smith, Ind. Eng. Chem., 40, 67 (1948).

118. J. Byass, G. Weaving, G. Charlton, J. Agr. Engr. Res., 5, 94-108 (1960).

119. L. Bode, M. Gebhardt, and C. Day, Transactions of ASAE, 11, 754-756 (1968).

120. W. Yates, M.S. Thesis, University of California, 1951.

121. S. Friedman, F. Guckert, and W. Marshall, Jr., Chem. Eng. Progr., 48, 181 (1952).

122. W. Yates and N. Akesson, Third International Agricultural Aviation Congress, Int. Ag. Aviation Center, The Hague, 1966, pp. 129-141, 251-277.

123. C. Lee, H. H. Coutts, and J. D. Parker, Agr. Aviation, 11, 12-17 (1969).

124. R. Brazee and W. Buchele, USDA, ARS-42-29, July 1959.

125. H. Bowen and W. Splinter, American Society of Agr. Engr., 1968, Paper No. 68-150

126. L. Roth and J. Porterfield, Transactions of ASAE, 9, 553-555 (1966).

127. S. Law and H. Bowen, Transactions of ASAE, 9, 501-506 (1966).

128. W. E. Splinter, Transactions of ASAE, 11, 487-490 (1968).

129. J. Carlton and D. Isler, Agr. Aviation, 8, 44-51 (19 6).

130. I. Langmuir and K. Blodgett, Report RL 225, General Electric Research Lab., 1944-1945.

131. R. Magill, F. Holden, and C. Ackley, Air Pollution Handbook, McGraw Hill, New York, 1956, pp. 13-35.

132. F. Brooks, Agr. Eng., 28, 233-239, 244 (1947).

133. N. B. Akesson, W. E. Yates, and R. W. Brazelton, Proc. the Application of Pesticides in Concentrated Form, ASAE Council on Pesticides Application, Oct. 1966, pp. 33-54.

134. Millikan, Aerodynamics of the Airplane, Wiley, New York, 1941.

135. W. Reed, III, NACA Tech- Note 3032, 1953.

136. C. Edwards and W. Ripper, Proc. Brit. Weed Control Conf., 1953, pp. 348-367.

137. W. Yates, Down to Earth, 16, 15-19 (Fall 1960).

138. E. Stevenson, Proc. 6th Annual California Weed Conf., 1954, pp. 33-36.

139. C. Ball, Farm Journal, November 1968, p. 26.

140. H. J. Sayer, Agr. Aviation, Int. Agr. Av. Center, The Hague, 11, 78-85 (July 1969).

141. Norman B. Akesson and Wesley E. Yates, J. Royal Aero. Soc., 67, 760-767 (December 1963).

142. O. Sutton, Micrometeorology, McGraw Hill, New York, 1953.

143. D. Slade, Meteorology and Atomic Energy, U.S. Atomic Energy Commission, 1968.

144. R. Munn, Descriptive Micrometeorology, Academic, New York, 1966.

145. W. Yates, N. Akesson, and P. Coutts, Transactions of ASAE, 9, 389-393 (1969).

146. P. Christensen, W. Yates, and N. Akesson, Proc. 4th Int. Agr. Avia. Conf., Int. Agr. Aviation Centre, The Hague, 1969.

147. B. Byrd and C. Reimer, Proc. Northeastern Weed Control Conf., 1966, pp. 422-436.

148. J. Byass and G. Charlton, J. Ag. Engr. Res., 9, 48-59 (1964).

149. D. Yeo, Proc. First Int. Agr. Avia. Conf., Int. Agr. Aviation Centre, The Hague, 1959.

150. Elbert C. Tabor, Air Pollution Control Assoc. J., 15, 415-418 (Sept., 1965).

151. W. H. Yule and A. F. W. Cole, Proc. 4th Int. Agr. Aviation Congress, August 1969, Int. Aviation Centre, The Hague.

152. C. A. Powell, Pesticides in the Air, Pesticides and Public Health Short Course, U.S. Dept. H.E.W., Atlanta, Georgia, May 1969.

153. N. Akesson and W. Yates, Annual Rev. Entomology, 9, 285-318 (1964).

154. W. Yates and N. Akesson, Transactions of ASAE, 6, 104-107, 114 (1963).

155. W. Yates, N. Akesson, and H. Coutts, Transactions of ASAE, 10, 628-632 (1967).

156. W. Yates and N. Akesson, Proc. Third Int. Agr. Avia. Congress, 1966, Int. Agr. Aviation Centre, The Hague, pp. 129-141.

157. W. Yates, N. Akesson, and K. Cheng, ASAE, Transcript No. 67-155, 1967.

158. R. Argauer, H. Mason, C. Corley, A. Higgins, J. Sauls, and L. Liljedahl, J. Econ. Entomol., 61, 1015-1020 (1968).

159. G. W. Ware, et al., J. Econ. Entomol., 62, 840-846 (August 1969).

160. K. R. Frost and G. W. Ware, ASAE, Paper No. 69-615, St. Joseph, Michigan, Dec. 1969.

161. C. Kruse, A. Hess, and R. Metcalf, J. Nat. Malaria Soc., 3, 197-209 (1944).

162. G. MacCollom, Agrichemical West, 12, 10, 12, 14 (May 1969).

163. J. Maybank and K. Yoshida, Transactions ASAE, St. Joseph, Mich., 12, 759-762 (November 1969).

164. N. B. Akesson, W. E. Yates, and S. E. Wilce, Proc. 4th Int. Avia. Congress, August 1969, Int. Agr. Aviation Centre, The Hague.

165. W. L. Bamesberger and D. F. Adams, "Organic Pesticides in the Environment," ACS Advances in Chemistry Series, No. 60, 1966, pp. 219-227.

166. N. B. Akesson, D. E. Bayer, C. L. Elmore, and W. Winterlin, Unpublished reports on the Propanil Research Project, University of California, College of Agriculture, Davis, California, 1970.

Chapter 8

SPREADING AND RETENTION OF AGRICULTURAL
SPRAYS ON FOLIAGE

D. R. Johnstone*

Ministry of Overseas Development
Tropical Pesticides Research Unit
Porton Down, Salisbury, Wiltshire, England

*Present affiliation: Overseas Development Administration, Centre for Overseas Pest Research, Division of Chemical Control.

343

I. INTRODUCTION

The biological activity of pesticidal sprays on foliage is a function of
many factors relating to the physico-chemical interaction of toxicant formulation and plant material. One important aspect of this interaction which
largely determines the initial character and subsequent availability of the
deposit is the behavior of liquid droplets as they impact and accumulate on
the foliage surface, before they subsequently penetrate or evaporate to
leave solid residue.

Since the behavior that requires examination differs according to the
amount of spray applied, it is useful to classify the character of spray
cover according to high, medium and low volume. It should be noted that
these terms have acquired somewhat varied meanings when used to describe methods of application, particularly with regard to aerial and
ground spray treatments. Byass (1) has given an appraisal of their scope
when referred to current application practice. For the high volume type of
cover the target is sprayed until surface saturation occurs and the excess
spray liquid runs off, whereas in low volume application the spray is deposited thinly as discrete droplets. At medium volume the surface density
of deposit will be increasing through partial coalescence occurring where
droplets spread and overlap to form larger globules. The limiting volumes
for these types of cover will relate to the amount of leaf area per unit area
of ground. Run-off conditions may be initiated at as little as 100 liters of
spray per hectare in a low sparsely leaved crop, as opposed to volumes of
over 1000 liters per hectare in more densely leaved bush or tree planta-
tions.

It is generally held that when applying active materials of nonsystemic
nature the degree of spray cover which will prove effective will be determined by the disposition and movement of the pest, so that the more static
the pest the more complete the spray cover required at the sites of attack.
On the other hand, redistribution of the initial deposit by drainage and splash
may sometimes assist the active material, especially fungicides, to reach
inaccessible sites [E. C. Hislop, (53)]. Conversely, systemic materials
may only require to be brought into contact with those areas of the plant
through which they can be absorbed and translocated through the vascular
system.

The wider aspects of properties of interfaces have received detailed
treatment elsewhere [Adam (2), Bikerman (3), Davies and Rideal (4)],
but adopting a similar fundamental approach has led to considerable progress by a number of specialist workers in resolving the physical processes
which concern the interaction of sprays on plant surfaces, since an earlier
review by David (5) in 1959. At high volume application, as previously noted,

individual droplets will coalesce on the surface of leaves to form larger globules, until these commence to run off. The degree of cover is then determined by the volume of liquid which may be retained and the manner in which the surface drains and possibly rewets. This process has been considered in fundamental physical terms by Furmidge (6, 7) and a complementary statistical approach to droplet coalescence has been given by Johnstone (8). The physical properties to be taken into account are those specific to the spray liquid, i.e., its surface tension, viscosity, and density, and those describing the composition and texture of the surface presented to the droplets, its porosity and fine structure. Challen (9) has described a method of preparing carnauba wax replicas of leaf surfaces to illustrate how macroscopic surface roughness relates to wetting. Juniper (10) has described a carbon replica technique which permits resolution by the transmission electron microscope of the leaf ultra-structure, which has possibly a more important influence on the mutual property of contact angle. Holloway (11) has subsequently developed a scanning electron microscope technique for examination of leaf ultra-structure in greater detail. On an inclined collector both the difference between the advancing and receding contact angles and their mean value are important in relating the gravitational stability of droplets to the slope of the target, as discussed by Furmidge (6) and Kawasaki (12). Contact angle has been shown to vary, not only with species and position on the leaf [Furmidge (7)], but also, for a given species, with age [by Fogg (13, 14)] and environment [by Linskens (15)]. Challen (16), Bikerman (17) and Furmidge (7) have investigated the effects of hairiness and surface roughness on wetting behavior.

Impact velocity requires consideration in relation to droplet spread, bounce and shatter, although the latter may be excluded for droplets smaller than 1000 μm in diameter under normal application conditions. Experimental studies have been carried out by Brunskill (18) and Ford and Furmidge (19), the latter rationalizing their work in fundamental terms. The influence of airflow over the target on the collection efficiency of small droplets has been investigated in some detail by Langmuir and Blodgett (20) and Gregory (21), and reviewed by Davies (22). Studies of its effect on maximum initial retention have been made by Suzuki and Uesugi (23).

The extent of surface cover in relation to density of deposit is an important criterion of distribution, and the current trend to reduce volume of insecticidal application down to ultimate limits through the use of highly concentrated or possibly undiluted technical materials requires special examination in this respect. At low volume, target cover will be determined by volume application rate, spray droplet size, and in addition, those physical factors already mentioned which control the spread on foliage.

The heterogeneity of leaf surfaces and the complexity of their wetting behavior has hindered the elucidation of physical principles, which can be more readily studied at an initial stage on simpler and more reproducible solid surfaces which simulate in part the behavior of a leaf. Clean glass and coated glass slides, as well as cellulose acetate sheet and photographic bromide paper, are among the variety of artificial surfaces which have been used by a number of workers to study the effects of change in physical properties on impaction, spread and spray retention. Relevant information has accrued during the study of techniques for sampling spray droplets for size classification [May (24, 25), Jarman and King (26) and Johnstone (27)].

II. FACTORS AFFECTING THE SPREAD OF DROPLETS ON IMPINGEMENT WITH SOLID SURFACES

A. Preliminary Observations on Spread

In sampling spray droplets on absorbent filter paper for droplet size assessment, Jarman (28) found that the spread factor (F), i.e., the ratio of the diameter (D) of the area wetted to the diameter (d) of the original droplet, was not constant, but was determined by the relation:

$$d = aD^b$$

The constants a, b appeared to be properties of the liquids and surfaces used and were unaffected by impact velocity, in agreement with the observations of Merrington and Richardson (29). Johnstone (27) subsequently found that a similar relation held for the relatively impervious fixed glazed photographic bromide paper used in droplet sampling, but that F was a function of droplet momentum and liquid surface tension as shown in Fig. 1.

B. Spread Factor

If the impacted droplet is assumed to have the form of a spherical cap (a close approximation for small droplets with contact angle θ at the liquid/solid/air boundary < about $110°$), then the spread factor F may be expressed by the equation

$$F = D/d = \left[\frac{4 \sin^3\theta}{(1 - \cos \theta)^2 (2 + \cos \theta)} \right]^{1/3} \tag{1}$$

Values of F for a number of values of θ are given in Table 1.

The appropriate value of θ in determining F in Eq. (1) depends on the conditions under which the droplet impacts and is considered in detail in the ensuing sections C., D. and E. The "spreading power" of a liquid

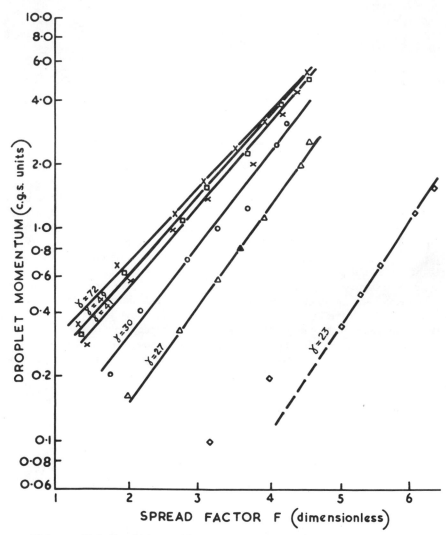

FIG. 1. Relation between droplet momentum, spread factor and surface tension (γ) for acetone/water mixtures on glazed photographic bromide paper. (Droplet size 1.9-2.6 mm.)

on a surface, or its ability to form a liquid/solid interface, is determined by the maximum angle of contact or advancing angle (θ_A). Spreading power is distinct from the "wetting" characteristic, or ability to form persistent liquid/solid interface when excess liquid has drained, which is determined by the minimum or receding angle of contact (θ_R). A high degree of wetting is only possible as θ_R tends to zero.

TABLE 1

Variation of Spread Factor with Angle of Contact

$\theta°$	5	10	20	35	60	75	90	110	130	150
F	3.90	3.12	2.48	2.07	1.61	1.43	1.26	1.04	0.79	0.51

C. The Impaction Process

A droplet impinging on a solid impervious surface may behave in several ways. When a droplet is placed on a collecting surface with little or no momentum, the boundary of the droplet in contact with the collector and air is not forced beyond the maximum advancing angle of contact (θ_A) and equilibrium is established with the spread factor at a minimum given by Eq. (1), where $\theta = \theta_A$. If θ_A is zero, as may occur with certain surfaces if the air/liquid surface tension (γ) is low, the droplet wets the surface and spreads irregularly to form a thin film. For $\theta_A > 0$, an increase in momentum causes the droplet to flatten on impact, but the transient increase in spread is limited by surface tensional forces. Provided impact momentum is insufficient to shatter the droplet, elastic recovery follows, and at this stage the recoil may in certain instances cause detachment from the surface and "bounce." More often a series of rapid oscillations takes place, damped by viscous forces, and at equilibrium the resulting angle of contact of the lens-shaped droplet will be less than the advancing angle noted for zero impact momentum, with consequent increase in spread.

D. Energy Considerations

Ford and Furmidge (19), considered the energy transitions involved in these interactions and distinguished three stages in the impaction of a droplet as shown in Fig. 2.

1. Initial Spreading [Fig. 2(a)].

The initial flattening on impact, in which the incident energy is partly dissipated by work against viscous forces and partly converted to surface energy by extending the droplet surfaces in contact with air and solid.

2. Retraction [Fig. 2(b)].

The recoil stage, which may pass through the equilibrium spherical cap position (stage 4) to a roughly conical form whose top may in certain

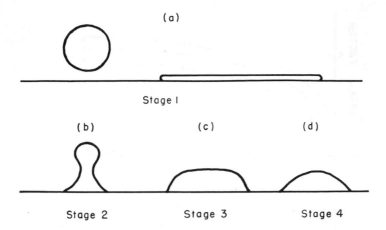

FIG. 2. Stages in the impaction of a droplet. Reprinted from Ford and Furmidge (19), p. 421. By courtesy of the Society of Chemical Industry, London.

circumstances detach. The liquid will retract across the target surface to cover a smaller area than that wetted at the end of stage 1 if the contact angle reaches the receding angle θ_R.

3. Secondary Spreading [Fig. 2(c)].

Recoil from stage 2 towards and possibly through the equilibrium position, further spread ensuing if θ_A is reached. Stages 2 and 3 may be repeated, with reduced amplitude, until a final equilibrium is reached at stage 4.

At stage 1, considering impaction on a rigid surface, the energy balance is

$$E_K + E_P + E_S = E_P{}' + E_S{}' + E_D$$

where E_K, E_P and E_S are respectively kinetic, potential and surface energies prior to impact, and $E_P{}'$ and $E_S{}'$ the same in the spread condition; E_D is the energy expended in deformation of the droplet.

Reduction in θ_A (or reduction in γ which has the same effect) causes only slight increase in spread at stage 1. Figure 3 illustrates stage 1 spread factor plotted against impact energy. It was concluded that the kinetic energy of the impacting droplet was the most important factor governing stage 1.

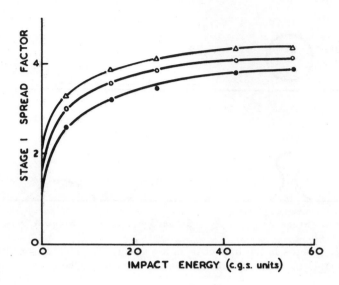

FIG. 3. Variation in spread factor in stage 1 with impact energy. Re-printed from Ford and Furmidge (19), p. 424. Δ water on glass $\theta_A = 27°$. o water on cellulose acetate $\theta_A = 62°$. ● water on beeswax $\theta_A = 111°$. By courtesy of the Society of Chemical Industry, London.

The retraction in stage 2 was shown to depend on the energy available for retraction (E_R), and assuming a minimum surface energy condition to correspond with a spherical cap and $\theta = \theta_R$,

$$E_R = \frac{\pi}{4} d^2 \gamma \left[F_1{}^2 (1 - \cos \theta_A) + \frac{8}{3F_1} - \frac{5.04 (1 - \cos \theta_R) - 2.52 \cos \theta_R \sin^2 \theta_R}{(1 - \cos \theta_R)^2 (2 + \cos \theta_R)^{2/3}} \right]$$

where $F_1 = D_1/d$ is the spread factor for stage 1. Because of the large dif-ference or hysteresis which may exist between θ_A and θ_R the area of solid/liquid contact may not be reduced in the retraction process, and in general the energy available in stage 2 is very much less than in stage 1.

Stage 3 and beyond have proved difficult to treat theoretically because of the problem of assessing the excess energy at the end of stage 2. The shape of the retracted droplet may vary from a slightly deformed spheri-cal cap to the shape shown in Fig. 2(b). Bounce may occasionally be ob-served provided the excess energy is high, and this normally requires that the advancing angle of contact exceeds about 140° [Brunskill (18)] with small hysteresis. The area of solid/liquid contact only alters in stage 3 if the retraction in stage 2 reduces the area to a value less than that given by the relevant conditions in Eq. (1), which is only likely to occur if impact energy and θ_A are high and $\theta_A - \theta_R$ is again small. In all other cases small internal oscillations absorb the excess energy owing to the passage through the equilibrium position in stage 3.

E. Some Practical Effects

The impact energies of water droplets travelling at their terminal velocities in air, plotted against droplet size, are shown in Fig. 4, curve B. Figure 4 (curve A) indicates that increase in wind speed to 5 m/sec increases impact energy considerably and droplets > 250 μm may enter stage 2 although droplets < 100 μm are unlikely to spread beyond the minimum conditions of Eq. (1) unless assisted by very considerable air blast.

For any given droplet size, when θ_R is finite, more cover of the surface will be obtained when impact energy is just sufficient to cause the droplet to spread in stage 1 to the extent given by the appropriate value of θ_R. When $\theta_R = 0$, stage 2 cannot occur and cover depends solely on θ_A and the initial impact energy.

Because the impaction process is very rapid, being completed in < 0.05 secs (< 0.01 sec for stage 1), wetting agents may not have the anticipated influence in extending initial spread, although subsequent adsorption to the air/water interface [Burcik (30), Posner and Alexander (31)] with reduction in γ and contact angle [Ford, Furmidge and Montagne (32)] should cause a delayed increase in spread, the extent of which may be inferred from Table 1.

Ford and Furmidge (19) suggest that droplets below 500 μm are unlikely to reach the limit of spread in stage 1 unless $\theta_A - \theta_R$ is very small, but that when this occurs at high θ bounce is likely at the termination of stage 2. Brunskill's experiments on pea leaves (18) showed a dramatic increase in percent retention of impinging droplets of aqueous mixtures of acetic acid and methanol as surface tension was reduced, undoubtedly due to the accompanying lowering of θ_A and increase in $\theta_A - \theta_R$. At constant γ (< 45 dyn/cm) high retentions were recorded for droplets < 150 μm impacting at terminal velocity, E_R being insufficient to produce bounce.

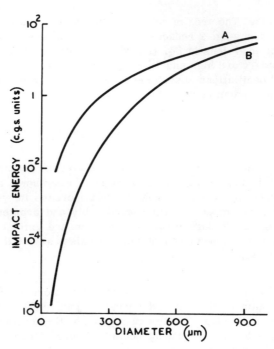

FIG. 4. Impact energies of water droplets. Reprinted from Ford and Furmidge (19), p. 428. Curve A: droplets impacting at 5 m/sec. Curve B: droplets impacting at terminal velocity in air. By courtesy of the Society of Chemical Industry, London.

III. FACTORS AFFECTING LOW AND INTERMEDIATE VOLUME SPRAY COVER

A. Influence of Droplet Size and Contact Angle on Low Volume Spray Cover

Equation (1) may be used to examine the variation in fractional cover with changing contact angle and droplet size of the spray provided coalescence can be excluded, i.e., for fractional cover less than about 0.1.

If a surface is sprayed with homogeneously sized droplets diameter d and 100% deposition is achieved with uniform cover at an application rate V, (volume/unit area) then the droplet density or number/unit area (n) is given by $6V/\pi d^3$ and if "a" is the circular area, diameter D, covered by an individual droplet on spreading, the total wetted area/unit area of the surface is given by

$$na = n\pi d^2/4 = n\pi F^2 d^2/4$$

where, as before $F = D/d$. In terms of volume application rate, substituting for n, the wetted area/unit area, or fractional cover, c, is given by

$$c = 3VF^2/2d \qquad (2)$$

Figure 5 indicates the fraction of the surface area covered by liquid as droplet size and contact angle vary, with no overlap, when droplets are uniformly applied to the surface at 10 liters/hectare. Fractional cover in excess of about 0.08 has been excluded.

As an example, the figure indicates that under the low volume condition of 10 liters/hectare application, 250 μm droplets with an effective contact angle of 90° provide a fractional cover, c = 0.0095. At the same droplet size, reduction in contact angle to 35° (which may be obtained with aqueous formulation through reduction in surface tension brought about by the addition of a wetting agent) gives c = 0.0250, indicating a two and a half fold increase in cover, while halving the droplet size in addition to reducing contact angle gives c = 0.0500, or over fivefold increase. The distribution of deposit in terms of number of droplets/unit area is a more easily visualized criterion of efficiency than area cover for very low volume application. Ultra-low volume technique has been tried at rates down to about 0.5 liters/hectare, and at this figure "ideal" application provides just under 10 droplets/cm^2 at 100 μm droplet size, and c, taken to correspond with the high spread factor of F = 3 for an oily solution, is 0.00675. Halving droplet diameter to 50 μm gives 76 droplets/cm^2 and c = 0.0135. The extent to which ultra-low volume application necessitates exceptionally fine atomization is immediately evident.

These examples indicate the degree of control over distribution and extent of cover which may be exercised by suitable variation of the two factors γ and d.

B. Measurements of Effect of Droplet Size on Spray Cover With Increasing Volume Application

In practice agricultural sprays are not monodisperse but are composed of a more or less wide range of droplet size, frequently characterized by parameters derived from the mass or volume size distribution. Thus, the mass (or volume) median diameter (m.m.d. or v.m.d.), is the droplet diameter dividing the mass (or volume) of the spray into two equal halves; the 16 and 84%, or 25 and 75% diameter may be quoted as a measure of the spectrum or range.

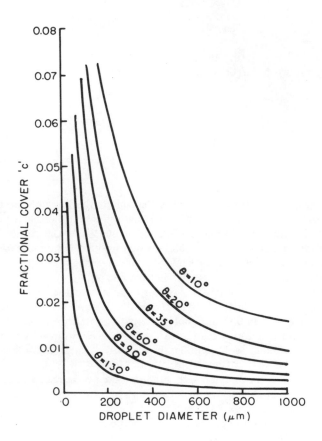

FIG. 5. Fraction of surface area covered by liquid as droplet size and contact angle vary, for an "ideal" application at 10 L/hectare.

Figure 6 illustrates the change in cover measured using a dyed aqueous spray, collected on cellulose acetate targets, for five hydraulic fan-pattern nozzles providing volume median diameters between 120 and 360 μm. Fractional cover was determined from the size distribution of the stains for a series of samples sprayed to different levels between 10 and 5000 L/hectare (0.1-50 mg/cm^2), and the overall droplet density was determined colorimetrically.

Below about 400 L/hectare reduction in volume median droplet size from 260 to 120 μm effectively doubles surface cover as predicted by the simplified treatment deriving Eq. (2). Median diameter sprays of 165 and 180 μm give intermediate cover. The somewhat anomalous higher fractional

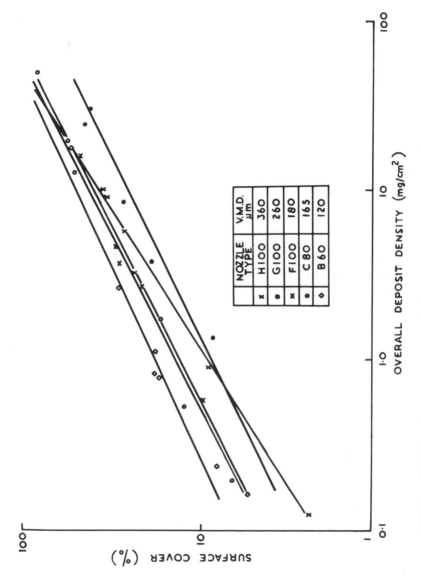

FIG. 6. Variation in spray cover with changing deposit density and droplet spectrum.

cover evident for the coarsest spray, v.m.d. 360 μm, above 500 L/hectare is attributed to increased impaction effects from droplets depositing at speeds in excess of terminal velocity.

C. Droplet Coalescence as a Prelude to Run-off

As deposition proceeds beyond the low volume level droplets begin to overlap and coalesce. With zero receding contact angle and low advancing angle the surface may be expected to wet out rapidly and be covered with a thin drainage film. At higher angles of contact, recession of liquid occurs as droplets fuse, so maintaining the permitted receding contact angle. This allows much larger globules of spray to collect over the surface (illustrated for water on cellulose acetate targets in Fig. 7), until one such globule becomes gravitationally unstable and on an inclined collector rolls down the surface to discharge from the lower edge, marking the onset of run-off.

Further spraying causes additional coalesced globules to reach critical size and commence to roll, while if the surface is not left wet, accumulation recommences in the cleared tracks (Fig. 8). It appears that the moment the first globule reaches critical size - the point of incipient run-off - the volume of spray retained on the surface is a maximum. Since maximum initial retention is governed by the movement of liquid globules, this phenomena must be considered in some detail.

IV. FACTORS AFFECTING RETENTION AT HIGH VOLUME SPRAY COVER

A. The Movement of Coalesced Globules on Inclined Surfaces

The effects of target inclination and interfacial surface properties on the gravitational stability of isolated deposited globules has been investigated by a number of workers in different fields and their results have been discussed by Furmidge (7). The basic equation for incipient motion takes the form:

$$mg \sin \alpha / W = \text{constant} = C \gamma \tag{3}$$

where m is the mass of the globule, α is the angle of tilt, W is the width of the globule, γ is the air/liquid surface tension, and C is a constant.

When a globule is placed on a horizontal surface which is then tilted, it begins to deform and the leading edge advances down the slope. Deformation increases with tilt until, at a critical angle, the trailing edge also

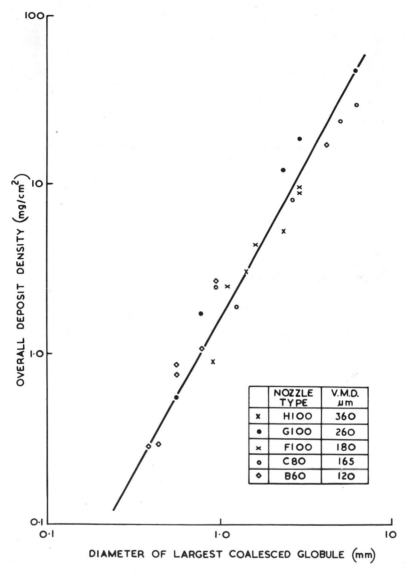

FIG. 7. Relation between size of largest coalesced globule and overall deposit density for water on cellulose acetate.

begins to move. A typical transition for an aqueous globule is illustrated in Fig. 9. Once clear of the original area of contact an increase in velocity is usually observed which may be checked by reduction in angle of tilt.

FIG. 8. Accumulation of spray up to and beyond run-off, water on coffee leaf (upper surface).

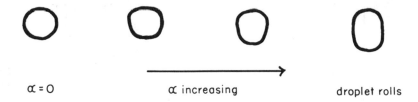

α = 0   α increasing   droplet rolls

FIG. 9. Transition in globule shape with increasing tilt (plan view).

The advancing and receding angles of contact and other dimensions of the globule may be determined by measurements taken from photographs of the globule profile and plan (Figs. 10 and 11).

Furmidge (7) investigated the physical significance of the constant C in Eq. (3) in terms of the work done in wetting and dewetting unit area of surface during the rolling process and deduced that:

$$mg \sin \alpha / W = \gamma (\cos \theta_R - \cos \theta_A) \qquad (4)$$

His derivation was based on the assumption of a rectangular plan, an approximation which may not be justified for all solid/liquid systems. Equations of similar form based on the alternative approximation of circular plan in which W has been replaced by the spread diameter D have been given by Rosano (33) and Buzágh and Wolfram (34). Close agreement between measured values of the constant determined from the parameters in Eq. (3) and calculated values determined from the right hand side of Eq. (4) was reported by Furmidge for aqueous solutions of surfactants on beeswax and

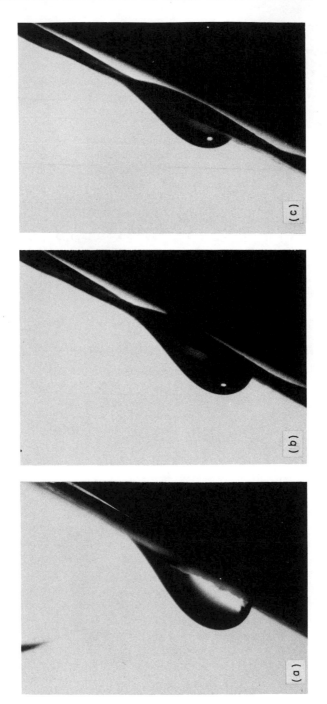

FIG. 10. Globules of water and aqueous suspensions of copper fungicide rolling down the upper surface of a coffee leaf, (all same scale). (a) water, (b) 2% suspension of fungicide, (c) 8% suspension of fungicide.

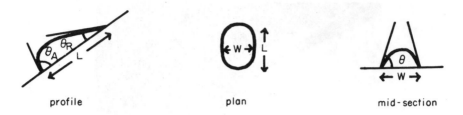

profile   plan   mid-section

FIG. 11. Size parameters of a rolling globule.

cellulose acetate. However, it was noted that the value may be expected to relate to the conditions of measurement, since θ_A and θ_R may be affected by globule velocity and the time of contact between the globule and the underlying surface [Ablett (35), Yarnold and Mason (36), Kawasaki (37)].

Measurements by Johnstone (38), attempting to relate the limiting volume ($V\alpha$) of an incipiently rolling globule to the surface slope (α), showed that a general empirical relation, which may be expressed as:

$$V_\alpha = V_{90°} (\sin \alpha)^{-q}$$

holds for a variety of liquids on artificial and plant surfaces over a wide range of slope (Fig. 12).

A similar relation was found for slow, steady rolling droplets where slightly lower values of $V_{90°}$ and q were indicated. Figure 13 illustrates the effect of increasing concentration of a proprietary copper fungicide (and hence of wetting agent incorporated) for globules on the upper surface of coffee leaves.

It was found that q ranged between the limits $1.0 < q < 2.1$, though for water on uniform artificial surfaces the range was narrower ($1.48 < q < 1.63$). These findings are less precise than those of Wolfram (39) whose work puts $q = 1.5$. If the shape of the globule is approximated to a spherical cap, base diameter D, then this value of q is also indicated by Eq. (4), since

$$V_\alpha = \pi (D/F)^3/6 = m/\rho \tag{5}$$

where ρ is the liquid density and

$$D \simeq W = \left[\frac{6 F^3 \gamma (\cos \theta_R - \cos \theta_A)}{\pi \rho g \sin \alpha} \right]^{1/2} \tag{6}$$

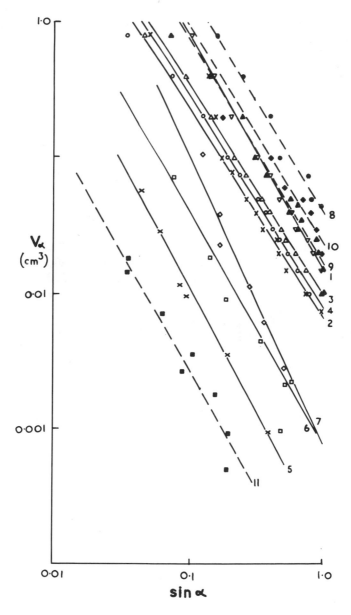

FIG. 12. Experimental relation between the limiting globule volume
(Vα) required to initiate rolling at surface slope α, and sin α, for various
liquids on artificial and leaf surfaces. (See Table 2.) Artificial sur-
faces:_____, Leaf surfaces: ----------.

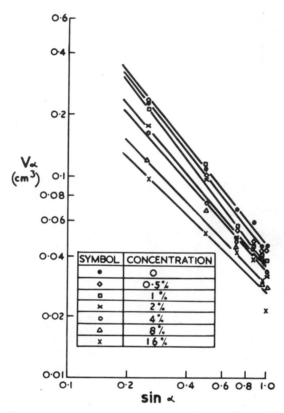

FIG. 13. Experimental relation between the limiting globule volume (Vα) required to initiate rolling at surface slope α, and sin α, for suspensions of copper fungicide on coffee leaves (upper surface).

so that

$$V\alpha = \left(\frac{6}{\pi}\right)^{1/2} \left[\frac{F \gamma (\cos \theta_R - \cos \theta_A)}{\rho \, g \, \sin \alpha}\right]^{3/2} \tag{7}$$

The extreme variations in q occurred in two different ways as indicated in Table 2. At high advancing and receding contact angles (mercury on artificial surfaces, water on nasturtium) high values of q were obtained. The theoretical treatment breaks down in this instance due to erroneous assumption regarding globule shape on highly unwettable surfaces. Low values of q were obtained with globules containing suspended copper, the sedimentation of which presumably interfered with the dewetting of the trailing edge of the globule, possibly a general feature in the behavior of suspensions.

TABLE 2

Some Properties of Liquids and Surfaces

(*)	Surface[a]	Liquid	γ dyn/ cm	ρ g/cm^3	θ_A	θ_R	θ_M	$(\theta_A -\theta_R)$	q
1.	C.A.	Water	74.0	1.000	74	13	44	61	1.63
2.	C.A. (p)	Water	74.0	1.000	64	8	36	56	1.57
3.	C.A. (s)	Water	74.0	1.000	109	57	83	52	1.48
4.	C.A.	1% Teepol	30.1	1.002	59	1	30	58	1.51
5.	C.A.	Mercury	48.5	13.65	146	129	138	17	1.8
6.	C.A. (p)	Mercury	48.5	13.65	140	128	134	12	1.7
7.	Glass	Mercury	48.5	13.65	131	125	125	6	2.0
8.	Lilac	Water	74.0	1.000	90	6	48	84	1.57
9.	Dock	Water	74.0	1.000	86	15	50	71	1.60
10.	Plantain	Water	74.0	1.000	84	20	52	64	1.51
11.	Nasturtium	Water	74.0	1.000	148	131	140	17	1.73
12.	Coffee	Water	74.0	1.000	96	29	63	67	1.24
13.	Coffee	1/2% Cu fungicide	71.9	1.002	96	18	57	78	1.26
14.	Coffee	1% Cu fungicide	68.6	1.005	95	15	55	80	1.3
15.	Coffee	2% Cu fungicide	63.5	1.012	102	13	57	89	1.13
16.	Coffee	4% Cu fungicide	57.7	1.025	98	14	56	84	1.11
17.	Coffee	8% Cu fungicide	55.4	1.050	92	13	53	79	1.02
18.	Coffee	16% Cu fungicide	51.5	1.103	96	12	54	84	1.00

[a] Artificial surfaces were as follows: C.A., untreated rolled cellulose acetate sheet; C.A. (p), the same, but with the surface treated by dipping in a solution of cellulose acetate (2% w/v) in 75/25 acetone/dimethylphthalate and drained dry; C.A. (s), as C.A., but polished with a commercial silicone polish. Surface tensions (mercury excepted) were measured by the ring detachment method of Du Nouy (40), in conjunction with the correction factors of Harkins and Jordan (41). The value for mercury was assumed [Staicopolus (42)].

(*) Identifies liquid/surface combinations used in Figs. 12 and 13.

The aqueous suspensions of a proprietary copper fungicide (50% copper, as cuprous oxide, with undisclosed wetter) were examined on coffee leaves in connection with field observations on the effect of concentration on retention during high and low volume ground application to coffee bushes in East Africa [Anon, (43)]. These indicated a reduction in average level of deposit of about 35% for a low volume treatment of 5 lb of fungicide in 20 gal/acre (2.5%), as compared with the same weight in 120 gal/acre (0.4%). Reference to the measured properties of the fungicide suspensions (Table 2) suggests that change in γ/ρ overrides relatively small changes in contact angles and accounts for the progressive measured reduction in V_α at given slope as concentration is increased. The reduction in V_α corresponding to an increase in concentration from 1/2 to 4%, is about 30% and tends to confirm that early localized run-off could have contributed to the decrease in retention observed with the experimental field application at reduced volume.

Correlation between V_α (experimental) and values calculated from Eq. (7) has been examined over a wide range of V_α on a log-log plot (Fig. 14). The points show considerable scatter about the linear regression V_α (experimental) = V_α (calculated), departing by as much as a factor of 4 on either side. In view of the uncertain numerical agreement in some cases, the validity of the substitution of the parameter D for W has been investigated. Table 3 compares values of W and D calculated from Eqs. (6) and (5), respectively. In general it appears that D overestimates W by an average of 5-10%, causing a corresponding increase in V_α calculated from Eq. (7). Larger variations do occur where surfaces give rise to very high contact angles with low hysteresis, e.g., mercury on artificial surfaces, water on nasturtium. In these cases small discrepancies in measured contact angles can lead to large changes in the difference $\cos \theta_R - \cos \theta_A$ and hence in the calculated value of W. The low values of the ratio W/D for the siliconed cellulose acetate surfaces may arise similarly, but the values for the prepared cellulose acetate surfaces which exhibit lower contact angles may be due to water adsorption modifying the surface properties [Kawasaki, (37)].

In applying Eq. (5) and (6) the spread factor F was taken as a function of the advancing contact angle (θ_A), rather than the mean contact angle (θ_M), to obtain a better approximation to globule width (W). Reference to Fig. 11 shows that the angle of contact at the mid-section, which determines W, will tend to approach θ_A, while a spread factor based on θ_M will enable a more realistic prediction of the length (L) and hence the area of the base of the globule to be made. Thus it appears appropriate to consider F as a function of θ_A in Eq. (5) when D is to be substituted for W in Eqs. (4) and (7), whereas in any relation including W^2 as a measure of the base area of the globule, the appropriate value of D to substitute for W should be based on F calculated as a function of θ_M.

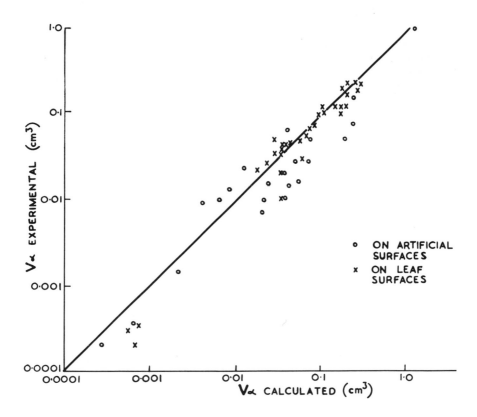

FIG. 14. Correlation between the values of limiting globule volume (V_α) determined by experiment and by calculation from Eq. (7).

B. Retention Theory

Furmidge (6, 7) has based a theory of retention on the assumption that the overall deposit density on a sloping surface sprayed to the point of run-off is proportional to the period of spraying and hence to the density of deposit given by the ratio of the volume of an incipiently rolling globule (V_α) to its base area $\pi W^2/4$; i.e., on slope α

$$R = k\, V_\alpha/W^2 \qquad (8)$$

where R is the maximum volume of spray which can be retained per unit area and k is a constant.

Consideration of the dynamics of rolling in terms of the work done by gravity against surface tensional forces, as outlined in the previous section

TABLE 3

Calculated Values of Globule Width (W) Given by Eq. (6), and the Diameter
(D) of the Wetted Circle of Contact for Droplets Approximated in Shape
to a Spherical Cap [Eq. (5)], for Various Liquid/Surface
Combinations and Inclinations (α)

Surface[a]	Liquid	$\alpha°$	W (cm)	D (cm)	W/D
		Artificial surfaces			
C.A.	Water	90	0.34	0.45	0.76
		60	0.30	0.46	0.65
		45	0.34	0.51	0.67
		30	0.47	0.64	0.74
		15	0.70	0.74	0.95
		5	1.57	1.74	0.90
C.A. (p)	Water	90	0.19	0.38	0.50
		50	0.18	0.41	0.44
		35	0.22	0.47	0.46
		15	0.29	0.69	0.42
C.A. (s)	Water	75	0.16	0.28	0.58
		50	0.15	0.30	0.50
		30	0.22	0.41	0.54
		15	0.27	0.55	0.50
C.A.	1% Teepol in water	60	0.42	0.43	0.98
		45	0.67	0.49	1.36
		30	0.58	0.52	1.11
		15	1.15	0.82	1.40
C.A.	Mercury	45	0.026	0.045	0.58
		8	0.24	0.15	1.60
C.A. (p)	Mercury	45	0.34	0.089	3.8
Glass	Mercury	40	0.44	0.103	4.3
		Leaf surfaces			
Nasturtium	Water	45	0.015	0.033	0.46
		30	0.030	0.046	0.65
		5	0.32	0.11	2.9

TABLE 3 (continued)

Surface[a]	Liquid	α°	W (cm)	D (cm)	W/D
Plantain	Water	90	0.32	0.44	0.73
		20	0.70	0.83	0.85
Dock	Water	90	0.16	0.35	0.46
		50	0.35	0.53	0.66
		30	0.88	0.60	1.46
		15	1.26	1.10	1.14
Lilac	Water	75	0.56	0.54	1.03
		60	0.53	0.55	0.97
		30	0.74	0.79	0.94
Coffee	Water	90	0.56	0.51	1.10
		30	0.71	0.69	1.03
		15	1.06	0.95	1.11
Coffee	1/2% Cu fungicide in water	90	0.61	0.54	1.12
		30	0.69	0.70	0.99
		15	1.03	1.00	1.03
Coffee	1% Cu fungicide	90	0.51	0.54	0.94
		30	0.65	0.65	1.00
		15	0.79	0.91	0.87
Coffee	2% Cu fungicide	90	0.47	0.47	1.00
		30	0.62	0.64	0.97
		15	0.62	0.81	0.77
Coffee	4% Cu fungicide	90	0.59	0.50	1.17
		30	0.68	0.67	1.01
		15	0.70	0.83	0.84
Coffee	8% Cu fungicide	90	0.54	0.67	1.15
		30	0.60	0.61	0.98
		15	0.53	0.75	0.71
Coffee	16% Cu fungicide	90	0.54	0.40	1.35
		30	0.62	0.61	1.02
		15	0.48	0.68	0.71

[a]Abbreviations as in Table 2.

enabled V_α and W to be replaced by expressions involving only the surface properties of advancing and receding contact angles θ_A and θ_R, the air/ liquid surface tension γ, liquid density ρ and slope α, to give

$$R = k \left[\frac{\pi \gamma (\cos \theta_R - \cos \theta_A)}{6 \, F^3 \, \rho \, g \, \sin \alpha} \right]^{1/2} \tag{9}$$

Furmidge (6) considered whether the spread factor should be taken as a function of θ_A, θ_R or the mean angle $(\theta_A + \theta_R)/2 = \theta_M$. He deduced that F_A (in terms of θ_A) could underestimate impaction effects, while the use of F_R (in terms of θ_R) overcompensates for such effects, especially on readily wetted surfaces. Thus for reasons similar to those given in the last paragraph of the previous section, it seems preferable to accept F_M (in terms of θ_M) as the best compromise. Since

$$(4/F^3)^{1/2} \simeq \theta_M$$

Furmidge used the simpler retention factor (f) given by

$$f = \theta_M (\gamma (\cos \theta_R - \cos \theta_A)/\rho)^{1/2} \tag{10}$$

to examine maximum initial retention on artificial surfaces (6) and subsequently on a variety of leaf surfaces (7).

Good linear correlation between measured maximum initial retention (R) and retention factor (f) was obtained for aqueous solutions on the artificial surfaces, beeswax and cellulose acetate (Fig. 15), when interfacial surface properties and hence the values of f were varied by change in the concentration of wetting agents. The ratio R/f was found to be the same for both surfaces.

In the case of foliar sprays the problem is complicated by the large variations in surface properties found not only between leaves of different species, but also within one species, resulting from maturation and conditions of growth. The leaf surface itself is normally a composite structure and the veins may prove to be more readily wetted than the intervening zones, while edge effects can cause uneven retention. Macro- as well as micro-roughness plays an important part in determining retention at and beyond run-off, especially the presence of leaf hairs.

Furmidge (7) made a detailed study of the retention at and beyond run-off on a number of leaf types of different roughness, including apple, black-currant, coffee, banana, and rubber, at constant spray droplet size and leaf slope. Liquid and interfacial surface properties were varied by the addition

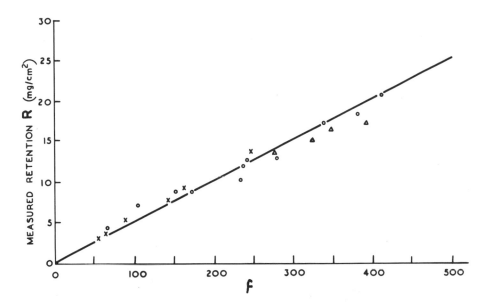

FIG. 15. Measured retention (R) plotted against retention factor (f) for aqueous solutions of wetting agents on artificial surfaces. Reprinted from Furmidge (6), p. 316. By courtesy of Journal of Colloid Science (copyright Academic Press).

of nonionic, anionic, and cationic surface active agents to the aqueous spray. Leaf roughness modified the simpler picture obtained when spraying artificial surfaces. With certain exceptions approximately linear plots of R against f were obtained (Fig. 16), but with lower gradient ratios R/f.

In explanation, it appears that surface irregularities tend to increase premature run-off at high f, while acting as a barrier to run-off at low f, flattening the R/f plot. Variation in retention was found to be closely associated with macroscopic surface irregularities and the highest measured retention at any f value was found on the lower surface of blackcurrant leaves whose prominent vein structure caused considerable quantities of liquid to be trapped.

Nonlinear plots were associated with very high advance and receding contact angles, e.g., water on banana, and again for very large differences between advancing and receding angles, e.g., water plus wetter on banana. Errors in the latter case may relate to the arbitrary use of θ_M in Eqs. (9) and (10), but despite the nonlinearity R still shows increase with increase in f.

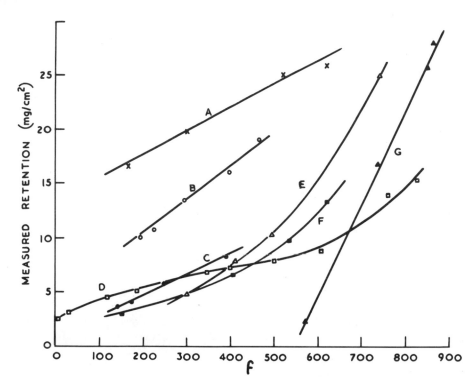

FIG. 16. Measured retention (R) plotted against retention factor (f) for solutions of the wetting agent CPC with safranine dye on various leaf surfaces. Reprinted from Furmidge (7), p. 133. Curve A: blackcurrant, lower surface; B: blackcurrant, upper surface; C: coffee, upper surface; D: banana, upper surface; E: rubber, upper surface; F: rubber, lower surface; and G: apple, lower surface. By courtesy of the Society of Chemical Industry, London.

While it was not found possible to use f to make direct comparison between species without reference to structure, f was shown to accurately predict the change in retention associated with increasing concentration of surface active agent. Since both γ and contact angles usually decrease with increasing concentration of surface active agent both f and R may be anticipated to do the same. This was found to be the case for all surfaces showing some hysteresis of contact angle with water, the only exceptions being such highly unwettable surfaces as the lower surfaces of rubber and banana leaves, also nasturtium and certain brassica leaves having small hysteresis and high contact angle. Such leaves may actually exhibit an increase in retention of spray with low concentration of wetters predicted by rise in f, before the subsequent decrease at higher wetter concentrations (Fig. 17).

Similar considerations rationalize the earlier experimental observations of Blackman, Bruce and Holly (44), who found that reduction in surface tension reduced retention for species which were easily wetted and affected an increase for those which were not.

The theory presented by Furmidge is concerned only with the effect of interfacial properties on maximum initial retention from solutions, or where emulsions or suspensions are applied, with retention of the continuous phase. With the onset of run-off there is usually a considerable fall in the volume of retained liquid and fluctuations about this reduced level occur as further buildup and drainage proceeds. With emulsion and wettable powder formulations there is a possibility that phase separation will occur and preferential retention of oil or solid phase may result, to an extent which depends on the properties of the formulation and leaf surface. Although the primary reason for incorporation of surface active agents is to increase spread and cover, they may frequently be included to enhance formulation stability, i.e., to facilitate dispersion of suspended solid material, or emulsification of oil phase. The presence of wetter then affects not only retention but also persistence, availability, and phytotoxicity of the deposit by modifying the form of the residue [Hartley, (45)]. The influence of wetting agents on preferential retention of the oil phase of oil-in-water emulsions sprayed at high volume has been investigated by Furmidge (46, 47), and Haydon (48) has provided a qualitative explanation of the former's results in terms of the competitive adsorption of wetter and emulsifier components and its effect on the signs of the electrostatic potentials of leaf and droplet surfaces. Furmidge found preferential deposition from oil phase during run-off occurred most markedly when wetter or emulsifier concentration was low and that in these circumstances the effect was greatest with cationic surfactants which adsorb strongly on many leaf surfaces.

Ashworth and Lloyd (49) have described laboratory and field tests for evaluating the efficiency of anionic and nonionic wetting agents used with agricultural sprays at high volume. They showed that the concentration of wetter required to wet a standard length of cotton tape was related to the degree of wetting of a cabbage leaf determined by visual assessment of a leaf dipping test. The test may be standardized for other leaves. Conibear and Furmidge (50) have examined the errors arising in visual assessment of leaf wetting and have shown that a technique using a standard wetting chart illustrating the limits of wetting ranges gives much improved accuracy. Furmidge (51) carried out a survey of methods of measuring the

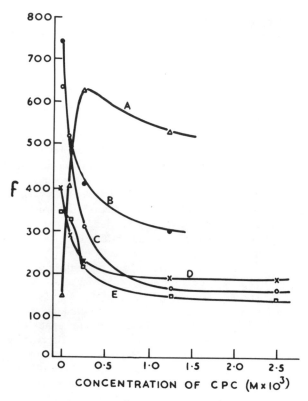

FIG. 17. Variation of the retention factor (f) on various leaf surfaces with concentration of the wetting agent CPC. Reprinted from Furmidge (7), p. 137. Curve A: rubber, lower surface; B: rubber, upper surface; C: blackcurrant, lower surface; D: blackcurrant, upper surface; and E: bean, upper surface. By courtesy of the Society of Chemical Industry, London.

wetting ability of high volume spray formulations and concluded that a satisfactory wetting test must involve the appropriate spray target and also take into account the effect of spray droplet impaction; these requirements are best satisfied by a test based on visual assessment of wetting under the practical spraying conditions.

C. Effect of Target Area, Shape, Inclination, and Impinging Droplet Size on Retention

In addition to the liquid/surface interactions which determine the spreading and wetting characteristics of the formulation, retention is also dependent on the orientation and physical dimensions of the target and size of the

spray droplets. Leaves vary enormously between species in form and disposition but in most crops in which spraying is practiced the leaves are of
fairly regular well-defined shape, though varying widely in size according
to the stage of growth, while in many instances there appears to be a marked correlation between leaf inclination and age, since the latter is closely
associated with position on the plant. To clarify the interpretation of field
trial data, particularly regarding the location of sampling sites, some
measurements have been made of changes in maximum initial retention
brought about by change in target area, shape and inclination, together
with the influence of the impinging droplet size distribution.

In order to exclude variations due to surface macro-roughness, measurements were made with aqueous sprays on the uniform artificial surface
cellulose acetate which simulates the behavior of a smooth partially wettable leaf surface. Targets were cut by guillotine from unblemished rolled
cellulose acetate sheet, approximately 1.5 mm in thickness. The surface
was untreated, apart from an initial detergent wash, followed by thorough
rinsing.

For "effect of area" measurements, squares of 64, 32, 16, 8, 4, 2, 1,
0.5 and 0.25 cm^2, were employed in replicates of 4, 5, or 6. For "effect
of shape" measurements, the target area was in all cases 4 cm^2. The following shapes were used in replicates of 4: circle, square, equilateral
triangle (base low), same (apex low), rectangle 1 x 4 cm (wide edge low),
same (narrow edge low), rectangle 1/2 x 8 cm (wide edge low), same (narrow edge low). Both the above factors were investigated at fixed slope
(45°) and droplet size (the diameters corresponding to 16, 50 and 84% by
volume of the spray being respectively 380, 590, and 860 μm).

For "effect of slope" and "droplet size," squares of 32 cm^2 were used,
again with 4 replicates. Slopes were 15, 30, 45, 60, and 75 degrees, using
the same spray, i.e., 590 μm v.m.d. Effect of change in droplet size was
measured at a fixed slope of 45°, for seven different droplet spectra derived from twin-fluid and hydraulic fan nozzles, and ranging in size from
60 to 590 μm v.m.d.

The spraying device consisted essentially of a nozzle placed over a
variable width slot in the base of a spray trough, with a rotating table,
gear-driven from an electric motor set beneath, and so positioned that a
band of spray was received from the slit over a width of 15 cm bordering
on the periphery of the table, excess spray being retained in the trough.
The cellulose acetate targets were supported at the appropriate inclination on the rotating table by aluminum mounts faced with absorbent paper.
(The smaller targets were held in position by capillarity so that their lower
forward edge was not in contact with the lip of the support and thus not subject to immediate drainage.) For the hydraulic nozzles the procedure for

obtaining maximum deposit density was to observe the number of passes through the spray required to reach the run-off condition, and then, having weighed the targets dry, to carry out repeated exposures (not less than six per target) varying the duration about this level by means of the moveable slide, and rapidly weighing the targets in the loaded condition using a "chainomatic" balance. The highest weight difference recorded was taken as the maximum load, and divided by the target area to give the overall deposit density. The lower output of the twin-fluid nozzle (adjusted for full fan pattern), together with the effective nondrip pneumatic cut-off, rendered the turntable unnecessary, and exposures were made over a static target on a time basis and by observation to the point of incipient run-off. The effect of varying in turn the four parameters, target area, shape, inclination and impinging droplet size brought about the following significant changes in maximum spray retention.

1. Effect of Change in Target Area (Fig. 18).

No significant difference was apparent between the mean maximum retention (R) for squares of 16, 32, and 64 cm^2, or between R for 8 and 16 cm^2, though R for 8 cm^2 was significantly greater than that for 32 and 64 cm^2. However, successive reductions in area below 8 cm^2 by a factor of 2 down to 0.25 cm^2 resulted in a continued significant increase in R.

Two factors contribute to an explanation of these observed changes. Firstly, assuming that the maximum retention occurs at the point of incipient run-off, R should rise as the target area becomes comparable with the base area (A_α) of an incipiently rolling coalesced globule on slope α. The limiting value of R would appear to be determined by the ratio of the volume (V_α) of the critically sized globule, to its area of contact (A_α). In the present case the increase in R does not become apparent until A is reduced below about 25 A_α, $A_\alpha = 0.34$ cm^2, but thereafter R continues to rise, even unexpectedly, when $A < A_\alpha$. It was observed that with the smaller target areas retention continued to increase beyond the point of incipient run-off. Thus a critically sized globule may move down the surface, but be restrained by the sharp edge at the lower boundary of the target, and continue to accumulate by direct collection and drainage until it finally topples. In the case of targets larger than 16 cm^2, any globule which moved down the target through more than 1-2 cm accumulated sufficient volume and momentum to immediately discharge at the lower edge, so that this did not take place. However the importance of target area at low A should be noted, especially in connection with the influence of target shape. Practically interpreted, these results suggest that small immature leaves may retain significantly higher deposits in high volume spraying, possibly providing enhanced protection to meet the demands of progressively increasing area due to growth.

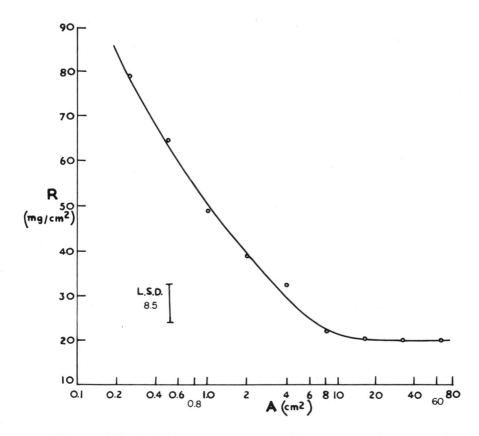

FIG. 18. Effect of change in target area (A) on measured retention (R), water on cellulose acetate.

2. The Effect of Change in Target Shape (Fig. 19).

These measurements were made with $A < 25 \, A_\alpha$ so that R may be anticipated to exceed the more or less constant values measured for squares with higher area. The order of measured R is indicated in the figure, but not all adjacent differences are significant (they might become so at somewhat lower values of A). Thus R for the square target is only significantly greater than that for the transversely disposed rectangles and for the apex-down triangle, while R for the latter is significantly less than that of all targets except that for the laterally disposed narrow rectangle. With all targets except two, maximum retention occurred beyond the point of incipient run-off, as earlier defined. The apex-down triangle was an obvious

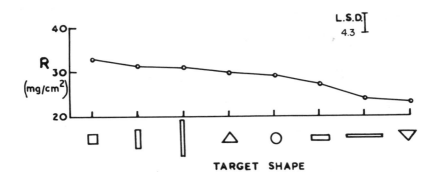

FIG. 19.   Effect of change in target shape on retention (R), water on cellulose acetate.

exception.   The reason for the lower value of R prevailing with the narrow transversely mounted rectangle must be associated with its restricted width, which could not accommodate a very large sized globule without exceeding the permitted maximum receding contact angle at its upper edge.   In relation to leaf structure it is apparent that the pointed apex characteristic of many leaves facilitates drainage, but that otherwise the shape is probably of less significance in determining maximum spray retention than surface irregularities, which may trap liquid and impede drainage.

3.   The Effect of Change in Slope (Fig. 20).

As anticipated, these measurements indicated significant increase in R as the inclination (α) is reduced.   Since the total volume retained must be a function of the size distribution of the coalesced globules the relation between R and the volume of the limiting sized globule (V_α) has been examined, and good correlation obtained on a semilog plot, indicating $R \propto \log V_\alpha$ (Fig. 21).   However, in Sec. IV.A. it has been shown that V_α and $\sin \alpha$ are linked by a power law, so that R should be proportional to $\log \sin \alpha$. Measured variation of R against $\sin \alpha$ on a semilog plot, shown in Fig. 20 confirms this, but the theoretically indicated relation $R \propto (\sin \alpha)^{-1/2}$ does not appear to hold over the wide range of α here examined.

In practice, the influence of slope on maximum retention will be decided partly by the manner of spraying and partly by the flexibility of the sprayed foliage.   Thus atomizers using air blast to transport the droplets to the target leaf may cause considerable movement of larger lightly supported leaves, preventing high levels of retention arising from shallow inclination.   Young compact leaves on the other hand may be little disturbed.

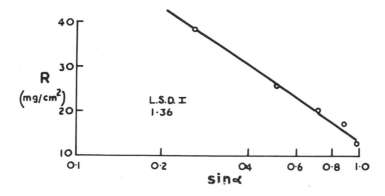

FIG. 20. Effect of change in target inclination (α), plotted as sin α, on retention (R), water on cellulose acetate.

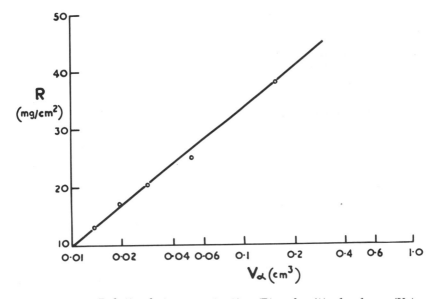

FIG. 21. Relation between retention (R) and critical volume (Vα).

Laboratory studies of run-off induced by air-blast spraying have been made by Suzuki and Uesugi (23) using artificial targets. Retention decreased on gelatine, paper and paraffined plane targets, in that order, for dyed solutions without wetter, but progressive addition of surfactant increased retention on the paraffined targets without change in retention on paper and gelatine. Wind velocities less than 3.1 m/sec did not affect retention.

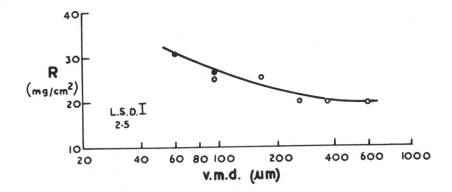

FIG. 22. Effect of change in impinging droplet size on retention (R),
water on cellulose acetate. ● pneumatic nozzle; ○ hydraulic fan nozzle.

4. The Effect of Change in Impinging Droplet Size (Fig. 22).

Furmidge (6) noted that reduction in impinging droplet size increased R
and this was confirmed in the present work. This increase is to be expected
since reduction in droplet diameter should favor a more even pattern of coalescence and an increase in the average volume of coalesced globules at
incipient run-off. In terms of volume median diameter the droplet size
range examined was approximately 60-600 μm. For sprays of median diameter 590, 360, and 260 μm negligible change in R was observed. Hydraulically atomized sprays of 165 and 95 μm gave a significant increase in
R (25%) over the coarser sprays, but between themselves were not significantly different. The 95 μm spray was produced from a 60° fan nozzle
whereas the 165 μm spray was derived from an 80° fan, so that some
changes in impact velocity may have restricted the increase in R anticipated
at the smaller droplet size.

The spray from the twin-fluid nozzle at 95 μm v.m.d. gave slightly
higher R than the equivalent hydraulically atomized spray, but at 60 μm
v.m.d. the increase was more marked and amounted to 25% over the spray
of 165 μm v.m.d.

In terms of high volume spraying, choice of droplet size is partly a
compromise, since very fine atomization is invariably obtained at the ex-
pense of reduced atomizer output, or alternatively much increased power
demand. However, a mean droplet size less than 150 μm appears to be
desirable as a means of preventing onset of premature run-off at unduly
low R.

V. A STATISTICAL MODEL FOR THE BUILD-UP OF DEPOSIT BY DROPLET COALESCENCE

While the relative levels of spray retention have been estimated using Eqs. (9) and (10) of section IV. B., prediction of the absolute level of R requires the interpretation of the constant k in Eq. (9). Empirically k appears to be a measure of the globule size distribution and surface cover at incipient run-off, and a function of impinging droplet size and velocity. Furmidge (6) observed that k increased from 0.25 to 0.31 for reduction in mean droplet size of the impacting spray from 520 to 120 μm for water on cellulose acetate.

If the entire deposit can be considered as made up of critically sized globules of volume V_α, the equivalent fractional cover (c) may be obtained by comparing the area wetted by a single globule and the area equivalent to the volume V_α, as given by the overall deposit density R. This indicates that

$$c = \frac{\pi W^2/4}{W^2/k} = \pi k/4 \tag{11}$$

The range of c corresponding to the above observed k values is then 0.195 – 0.245, a little lower than might be anticipated. The problem of rationalizing the influence of droplet size and target area on the maximum initial levels of retention may be approached by devising a statistical model based on the concept of random deposition.

A. Theory

Consider an inclined collecting surface, area A, slope α, subjected to random deposition from a spray of homogeneously-sized droplets, diameter d. At any instant prior to the onset of run-off the overall deposit density ϕ on a segment (a) of the total area A may be defined as

$$\phi = nv/a$$

where n is the number of impacting droplets of diameter d, volume v, which have fallen within a.

If incipient run-off, determining maximum initial retention, is defined as before as the condition of the surface when coalescence first produces a globule of limiting volume V_α, (accumulated by coalescence of n_α impacted droplets in diameter d, drawn from a portion of the surface, area A_α), it becomes necessary to determine the most probable average number of droplets n' which must impact on each portion of the surface, area A_α, so that on just one such site, n should exceed n_α.

The surface may be considered to possess A/A_α such possible sites for the accumulation of globules of volume V_α, and thus the onset of run-off is most likely when the probability of n' exceeding n_α is A_α/A, i.e., the event occurs most probably only once on A.

Since in general n_α will be large, the Poissonian distribution of n over the A/A_α sites will approximate closely to a normal distribution, and the standard deviation σ may be equated to the square root of the mean, i.e., $\sigma = \sqrt{n'}$. Therefore if t is the number of standard deviations which cut off one tail (fractional area A_α/A) of the normal curve, the most probable value of n' for incipient run-off is given by

$$n_\alpha = n' + t\sqrt{n'}$$

or $\qquad n' = n_\alpha + \dfrac{t^2}{2} - \dfrac{t}{2}\sqrt{(4n_\alpha + t^2)}$ \hfill (12)

(negative root only, since n' < n).

and for large n, neglecting t^2,

$$n' \simeq n_\alpha - t\sqrt{n_\alpha}$$ \hfill (13)

Equations (12) or (13) may be solved, provided A_α and n_α are known, and hence the maximum initial retention R may be obtained since

$$R = \phi \max = n'v/A_\alpha$$ \hfill (14)

n_α is given by

$$n_\alpha = \frac{V_\alpha}{V} = (d_\alpha/d)^3$$ \hfill (15)

where $d_\alpha = (6V_\alpha/\pi)^{1/3}$ is the equivalent spherical diameter of the globule and V_α is obtained theoretically from Eq. (7), where the spread factor F_A is taken as a function of the advancing contact angle, or by experimental measurement.

Normally the value of A_α will be somewhat greater than the area of the base of the limiting sized globule, since recession of the liquid/solid interface accompanies coalescence of droplets (unless the receding contact angle is very small). Thus the value of A_α used here has been derived using the spread factor F_R as a function of the receding contact angle, i.e.:

$$A_\alpha = \pi/4 \ (F_R d_\alpha)^2$$ \hfill (16)

The equivalent fractional spray cover (c) has been shown to approximate to $c = \pi k/4$ (Eq. 11). Using Eq. 8, if k is expressed as $k = R (F_M d_\alpha)^2/V_\alpha$, where W is deduced from d_α, using the spread factor F_M based on θ_M (the logic of this choice is discussed in Sec. IV.A.), then:

$$c = \pi R \ (F_M d_\alpha)^2/4V_\alpha$$

or

$$c = R/\phi_\alpha \qquad (17)$$

where $\phi_\alpha = V_\alpha/(\pi (F_M d_\alpha)^2/4)$ is the average deposit density under the limiting sized globule.

B. Practical Applications

The foregoing relations have been used to determine how R and c may vary with change in spray droplet size, target area, and surface inclination for water on an artificial surface (cellulose acetate) and the data are recorded in Tables 4 and 5.

TABLE 4

Predicted Variation in Maximum Initial Retention R (mg/cm^2) and Surface Cover c (Bracketed Figures), with Change in Spray Droplet Size d (μm) and Target Area A (cm^2) for Water on Cellulose Acetate Inclined at 45°

d＼A	2	8	32	128
50	32.4 (0.442)	32.4 (0.442)	32.3 (0.441)	32.3 (0.441)
100	32.3 (0.441)	32.0 (0.436)	31.9 (0.435)	31.8 (0.430)
200	32.1 (0.438)	31.7 (0.432)	31.4 (0.428)	31.2 (0.425)
500	31.7 (0.433)	30.3 (0.414)	29.1 (0.397)	28.2 (0.384)
1000	31.1 (0.424)	26.3 (0.359)	23.2 (0.322)	20.8 (0.284)
2000	32.1 (0.438)	21.7 (0.296)	16.2 (0.221)	13.6 (0.186)

TABLE 5

Predicted Variation in Maximum Initial Retention R (mg/cm^2) and
Surface Cover c (Bracketed Figures), with Change in Target
Slope and Droplet Diameter d (μm) for Water on
Cellulose Acetate Targets, Area 32 cm^2

d	α 15°	30°	45°	60°	75°	90°
500	42.2	31.3	29.3	28.3	27.2	27.9
	(0.351)	(0.388)	(0.400)	(0.425)	(0.450)	(0.480)
1000	39.5	27.4	23.4	20.8	19.2	18.9
	(0.332)	(0.386)	(0.320)	(0.312)	(0.316)	(0.325)

The predicted variation in R resulting from change in concentration of suspensions of a copper fungicide on the upper surface of coffee leaves at several droplet sizes is shown in Fig. 23. Physical properties for the liquid/solid interfaces were taken from the data recorded in Table 2 and Figs. 12 and 13.

Change in target area (A) has little effect on R at small droplet size (d), but as d increases a significant increase in R is predicted for reduction in A. Analogous changes are shown for surface cover. The effect of decreasing target inclination (α) is to increase R, but not necessarily c. Provided A is large compared with A$_\alpha$, R varies inversely with impinging droplet size in nearly linear fashion.

These predicted changes may be compared with some measured changes in R recorded in Sec. IV.C., for water on cellulose acetate sheet, obtained using polydisperse sprays from hydraulic and twin-fluid atomizers. (1) On an area of 32 cm^2 and at a slope of 45°, R was observed to increase from 20.1 to 31.1 mg/cm^2 as the v.m.d. of the spray was reduced from 590 to 60 μm. The predicted increase in R for reduction in impinging droplet size for monodisperse sprays of 1000 to 50 μm is 23.2 to 32.3 mg/cm^2. (2) The observed increase in R as A was reduced from 64 to 2 cm^2 at α = 45°, and with v.m.d = 590 μm was from 20.2 to 39.1 mg/cm^2. The predicted change in R for reduction in area from 128 to 2 cm^2 at 1000 μm droplet size is 20.8 to 31.1 mg/cm^2. The very large increase in measured R as A approaches A$_\alpha$ was attributed to post run-off accumulation at the lower edge of the small targets, behavior not covered by the present theoretical approach. (3) The increase in R measured as α was reduced from 75 to 15°, for A = 32 cm^2

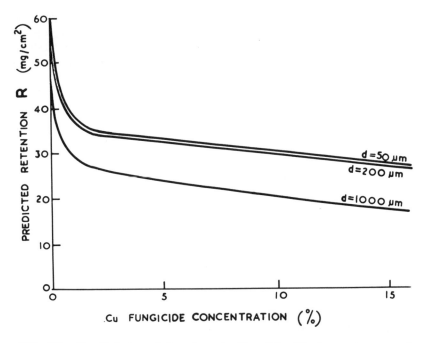

FIG. 23.   Predicted variation in retention (R) with change in concentration of a copper fungicide incorporating wetting agent for A = 32 cm^2, α = 90° at d = 50, 200 and 1000 μm. (Values of V_α used in calculating R have been taken from Eq. (7) using the measured physical properties recorded in Table 2).

and with v. m. d. of 590 μm was from 13.1 to 38.1 mg/cm^2. The predicted increase for a monodisperse spray of 1000 μm on the same target area is from 18.9 to 39.5 mg/cm^2. The theoretical treatment thus gives results in reasonably close numerical accord with those obtained by direct measurement.

No direct measurements of the variation of R for the suspensions of the proprietary copper fungicide on coffee are currently available, but the predicted changes shown in Fig. 23 are of interest in relation to the field observations commented on in Sec. IV.A. and also in regard to Somers's investigations (52) into factors influencing the run-off deposition of copper fungicides. The desirability of keeping d less than 200 μm in order to achieve highest levels of R is apparent. For increasing concentrations of fungicide and consequently concentration of the component wetting agent, the theory predicts a marked fall in R, the indicated reduction at d = 200 μm being about 40% for change in fungicide concentration from 0 to 2%. The corresponding reduction in surface cover c is about 25%. Somers reported an analogous large measured decrease in R (about 60%) over the anionic wetter concentration range 0.05 to 0.17% for Burgundy mixture and sodium di-octyl and di-nonyl sulphosuccinates on laurel leaves. As before, the practical interpretation is that where high volume spraying is required, the aim should be to keep droplet size and concentration of wetter down to the minimum permitted by other application requirements, if early localized run-off is to be avoided.

REFERENCES

1.   J. B. Byass, P.A.N.S., London (A), 14, 249 (1968).

2.   N. K. Adam, Physics and Chemistry of Surfaces, 3rd Ed., O.U.P., London, 1941.

3.   J. J. Bikerman, Surface Chemistry, Academic, New York, 1958.

4.   J. T. Davies and E. K. Rideal, Interfacial Phenomena, Academic, New York and London, 1961.

5.   W. A. L. David, Outlook Agr., 2, 127 (1959).

6.   C. G. L. Furmidge, J. Colloid Sci., 17, 309 (1962).

7.   C. G. L. Furmidge, J. Sci. Food Agr., 13, 127 (1962).

8.   D. R. Johnstone, T.P.R.U. Porton Report No. 251, 1963.

9.   S. B. Challen, J. Pharm. Pharmacol., 12, 307 (1960).

10.   B. E. Juniper, Endeavour, 18, 20 (1959).

11.   P. J. Holloway, Ph. D. Thesis, School of Pharmacy, London University, 1967.

12.   K. Kawasaki, J. Colloid Sci., 17, 228 (1962).

13.   G. E. Fogg, Proc. Roy. Soc. (B), 134, 503 (1947).

14.   G. E. Fogg, Discussions Faraday Soc., 3, 162 (1948).

15.   H. F. Linskens, Planta, 41, 40 (1952).

16.   S. B. Challen, J. Pharm. Pharmacol., 14, 207 (1962).

17.   J. J. Bikerman, J. Phys. Colloid Chem., 54, 653 (1950).

18.   R. T. Brunskill, Proc. 3rd British Weed Control Conference, 2, 593 (1956).

19.   R. E. Ford and C. G. L. Furmidge, Soc. Chem. Ind. Monograph No. 25, 417 (1966).

20.   I. Langmuir and K. B. Blodgett, General Electric Res. Lab. Report No. RL 225 (1949).

21.   P. H. Gregory, Nature, London, 166, 487 (1950).

22.   C. N. Davies, Aerosol Science, Academic, London, 1966, p. 422

23.   T. Suzuki and Y. Uesugi, Japan. J. Appl. Ent. Zool., 1, 219 (1957).

24.   K. R. May, J. Sci. Instr., 22, 187 (1945).

25.   K. R. May, J. Sci. Instr., 27, 128 (1950).

26.   R. T. Jarman and P. T. King, C.I.R.U. Porton Report No. 114 (1956).

27.   D. R. Johnstone, C.P.R.U. Porton Report No. 177 (1960).

28.   R. T. Jarman, Quart. J. Roy. Meteorol. Soc., 82, 352 (1956).

29.   A. C. Merrington and E. G. Richardson, Proc. Phys. Soc., 59, 1 (1947).

30.   E. J. Burcik, J. Colloid Sci., 5, 421 (1950).

31.   H. M. Posner and A. E. Alexander, J. Colloid Sci., 8, 585 (1953).

32.   R. E. Ford, C. G. L. Furmidge and J. Th. W. Montagne, Soc. Chem. Ind. Monograph No. 19, 214 (1965).

33.   H. L. Rosano, Mém. Services Chim. état (Paris), 36, 437 (1951).

34.   A. Buzágh and E. Wolfram, Kolloid-Z., 149, 125 (1958).

35.   R. Ablett, Phil. Mag., (6), 46, 224 (1923).

36.   G. D. Yarnold and B. J. Mason, Proc. Phys. Soc. (London), 62, 125 (1949).

37.   K. Kawasaki, J. Colloid Sci., 17, 169 (1963).

38.   D. R. Johnstone, T. P. R. U. Porton Report No. 249 (1963).

39.   E. Wolfram, J. Colloid Sci., 16, 195 (1961).

40.   Du Nouy, J. Gen. Physiol., 1, 521 (1919).

41.   W. D. Harkins and H. F. Jordan, J. Am. Chem. Soc., 52, 1756 (1930).

42.   D. N. Staicopolus, J. Colloid Sci., 17, 439 (1962).

43.   Anon., T. P. R. I. Arusha Progress Report No. 30, 1962, p. 21.

44.   G. E. Blackman, R. S. Bruce and K. Holly, J. Expt. Botany, 9, (26), 175 (1958).

45.   G. S. Hartley, Chem. and Ind., 575 (1967).

46.   C. G. L. Furmidge, J. Sci. Fd. Agric., 15, 542 (1964).

47.   C. G. L. Furmidge, Chem. and Ind., 1917 (1962).

48.   D. A. Haydon, Chem. and Ind., 1922 (1962).

49.   R. de B. Ashworth and G. A. Lloyd, J. Sci. Food Agr., 3, 234 (1961).

50.   D. I. Conibear and C. G. L. Furmidge, J. Sci. Food Agr., 3, 144 (1965).

51.   C. G. L. Furmidge, J. Sci. Food Agr., 16, 134 (1965).

52.   E. Somers, J. Sci. Food Agr., 8, 520 (1957).

53.   E. C. Hislop, Chem. Ind., 42, 1498 (1969).

Chapter 9

PENETRATION AND TRANSLOCATION OF HERBICIDES

D. E. Bayer and J. M. Lumb

Department of Botany
University of California
Davis, California

I. INTRODUCTION

Pesticides are formulated according to convenience and ease of handling by the user and to maximize biological performance. Studies conducted during recent years have emphasized the need to consider certain physico-chemical principles in relation to the biological specimen. Deposition and penetration into the target organism must be considered in relation to loss to the environment. Once inside the plant many of the pesticides must be moved to the site where the desired effect can be exerted.

The current approach to translocation of herbicides embraces: (1) usage of herbicides in weed control practices, (2) translocation studies involving radioactive labeled herbicides, (3) structural studies using the light and electron microscopes, and (4) translocation studies involving compounds other than herbicides.

The structure–function relationships of the phloem must be understood to visualize how water and solutes can move throughout the tissues of the entire plant. Translocation of herbicides follows the same pattern as with naturally occurring materials, except when the herbicide induces a toxic effect. The significance of this toxic effect will depend on the herbicide and its use. This understanding is invaluable in making proper use of modern tools of agriculture such as herbicides.

II. PESTICIDES AND FOLIAR ABSORPTION

The penetration and absorption of pesticides applied to the foliage of plants is influenced by a variety of factors. These factors may somewhat arbitrarily be placed into two groups: (1) properties of the pesticide and the medium in which it is presented to the plant surface and (2) properties of the plant. The interactions which occur between pesticide and the plant which regulate the absorption process and understanding these interactions is made more difficult by the introduction of environmental effects, many of which are uncontrollable. However, regardless of the complexities of the process, foliar application has proved extremely effective as a means of transferring pesticides from an external phase (usually liquid) to the site of biological activity within the plant. And study of specific systems operating under defined conditions has contributed greatly to the understanding of the absorption process.

It is well established that the biological activity of a compound is influenced by the manner of presentation to the leaf and by the particular derivative of the compound chosen, but the interactions between the plant and the pesticide which affect the quantity that enters the cell and the rate of entry are poorly defined. It is the foliar properties, as they influence the absorption characteristics of pesticide molecules, that will be discussed in the ensuing section.

A. Barriers to Penetration

To move from the external surface of a leaf to within the protoplast it is necessary for the pesticide molecule to permeate a series of chemically and structurally differing layers or strata (Fig. 1). These strata are referred to as the cuticular layers, the cell wall, and finally the membrane bounding the protoplast, the plasmalemma in order from the external surface. With the exception of the plasmalemma, there would appear to be

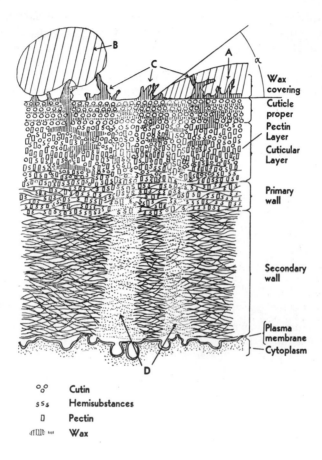

$\overset{\circ}{\underset{\circ}{\circ}}$	Cutin
$s\,s_s$	Hemisubstances
$\square$	Pectin
⨍⫿⫾⫾ ⁗	Wax

FIG. 1. Simplified scheme of the outer wall of an epidermal cell: A = water droplet with detergent, B = without detergent, C = wax rodlets, D = ectodesmata as nonplasmatic structures, and ∝ = contact angle. Reprinted by permission from Ref. 1, p. 284, courtesy of Annual Reviews, Inc.

a gradual transition between one layer and the next in many plant species (1, 2). However, this is by no means a general rule as in some species the pectinaceous layer adjacent to the cell wall is greatly reduced and is apparently absent in some xerophytic plants (3).

The terminology applied to the extra cellular layers is somewhat confusing. Crafts and Foy (4) have defined the cuticular layers as the several lamellae of the surface covering of cells that have become embedded with

wax and cutin, including the cellulose of the outer walls of the epidermal
cells when this layer has become impregnated with lipid substances or with
hemicelluloses (1). The cuticle refers to the outermost of the cuticular
layers and consists of cutin in which are inserted lamellae of wax. Although not stated, it would seem reasonable to include the epicuticular or
surface waxes as part of the cuticle. A somewhat different terminology
is frequently used by those investigating the properties of isolated cuticle
and cuticular layers in that all layers isolated are referred to, for convenience, as cuticle.

Cuticular layers cover all the external portions of the shoot (5), and
even the intercellular space system is thought to be lined with a noncellu-
losic cutin-like substance (6). The cuticle has also been shown to extend
through stomatal openings (7) and is probably continuous within the stomatal cavity appearing in isolated cuticular membranes as a stomatal
sac (8). Thus, molecules placed on the surface of the aboveground portions of plants must pass through a cuticle or cuticle-like barrier in order
to penetrate underlying cells.

B. Epicuticular Waxes

The waxy layer external to the cuticle is generally thought to be the
first barrier through which foreign molecules must pass. However, it
is not clear to what extent the surface waxes actually impede penetration
(9), as fissures and insect punctures occur in the surface and there
are areas where the thickness is reduced due to limited deposition or
sloughing off with age (4, 10, 11).

Surface wax is not present on all leaves. In a survey by Schieferstein
and Loomis (12), wax was absent in about half of the mature leaves of the
50 species investigated. They also noted that wax accumulation in the
older leaves generally occurred in the cuticle and cuticular layers rather
than on the surface. Mueller et al. (13) showed that weathering reduced
the concentration of surfactant required to completely wet the leaf surface.
In field grown plants, the contact angle (a measure of the hydrophobic nature of the leaf surface), although greater than plants grown in the green-
house, declined rapidly after reaching a maximum (14). Thus, the restric-
tions imposed upon penetration through the cuticle by the epicuticular waxes
would appear to be lessened with the advancing maturity of the leaf, yet it
is commonly observed that the absorption of foliarly applied pesticides is
reduced with increasing age of the leaf (15) despite the increased hydrophilicity of the surface.

Using enzymatically isolated pear leaf cuticular layers, Norris and
Bukovac (7) showed by polarized light microscopy that both upper and

lower layers exhibited areas of negative birefringence indicative of oriented wax molecules embedded in the cuticle. In the isolated upper cuticle there was an almost continuous layer of birefringent wax while the birefringent wax of the lower cuticle was not continuously distributed throughout the cuticular layers. After extraction with chloroform for periods of 1 minute and 24 hours, measurement of the extracted wax indicated that there was a greater quantity of epicuticular wax on the lower cuticle than on the upper, but little difference between the quantity of embedded waxes. As the birefringent waxes were not measurably reduced after a 10-minute extraction period it was assumed that the 1-minute extraction did in fact remove the epicuticular waxes and possibly some of the low melting point "soft wax" consisting of long chain alcohols, aldehydes, esters, free acids and hydrocarbons (16). However, as the authors point out, the data were obtained from isolated cuticle and extraction of wax components could have occurred from both the inner and outer cuticular surfaces. In a following paper, Norris and Bukovac (17) showed that the penetration of NAA was greater through a lower astomatous cuticle even though the lower cuticle had been found to be thicker and covered with a greater quantity of epicuticular wax than the upper cuticle (7).

Thus, the distribution and molecular orientation of the embedded waxes may be of particular importance in influencing the rate of cuticular penetration, especially in older leaves where the epicuticular waxes have been reduced in thickness and changed in chemical composition by weathering processes, or where the effectiveness of these waxes as a barrier to penetration has been reduced by fissures extending at least into the underlying cuticle.

Schieferstein and Loomis (12) suggest that deposition of surface waxes occurs at, or shortly after, the period of cell elongation and location of the deposited waxes is related to leaf development and solidification of the cuticle. Later formed waxes cannot reach the surface and constitute the embedded waxes. Although it is known that the surface waxes of young leaves generally constitute a more continuous layer it has not been established whether the distribution and orientation of the embedded waxes is similar to that found in older leaves.

C. Epicuticular Waxes—Physicochemical Characteristics

The chemistry and morphology of the cuticular surface waxes has received considerable attention and is of unquestioned importance in the retention and wetting characteristics of leaves. The morphology of the waxy surface of plants shows considerable variation between genera (18) and even between closely related species (13). DeBary (19), nearly a century ago, suggested that surface waxes could be divided into four morphological types: (1) heaped wax layers, (2) single granulated layers, (3) rodlet layers, and (4) membranous or crusty wax layers. However, more recently carbon replica techniques and electron microscopy have been used to describe the waxy configurations of plant surfaces in far greater detail than was possible with the light microscope, and so variable are the forms observed that a clarification based on these findings has not yet been attempted. The use of such morphological features in taxonomy is difficult, although some forms will be characteristic of a species, genus or even family. There are similarities between taxonomically distinct plants and surface ultra-structure that may vary according to the portion of the plant examined (20).

There is also considerable variation in the response to weathering of waxy surfaces and also in the ability to recover from such damage (21). If the surface of Chrysanthemum segetum is damaged by brushing, the contact angle falls to below 100°. Within a week the surface regenerates and the contact angle rises to above 140°. However, this only occurred when the surface damage took place before leaf expansion had ceased. In the majority of species with waxy surfaces regeneration did not take place at all (21).

The liquid-solid interactions that occur on leaf surfaces are in part dependent upon the properties of the formulated droplet and in part upon three properties of the surface as follows (15): (1) the degree of surface roughness, (2) the presence or absence of an air film between the pesti-cide droplet and the leaf surface, and (3) the nature of the chemical groups at the leaf surface. Thus, the chemical composition of the epicuticular waxes assumes a certain importance. It is apparent (18) that these waxes are complex mixtures of long chain alkanes, alcohols, ketones, aldehydes, acetals, esters and acids with variation in the positioning and number of functional groups, the degree of chain branching, and unsaturation. The chain length of the homologues may vary from C_{20} to C_{37} and both even and odd carbon numbers may occur (Table 1).

Generally the mono- and di-ketones and secondary alcohols have pre-dominantly odd carbon number chains while the carboxylic acids, the pri-mary alcohols, the aldehydes, the α and ω-dioles, the hydroxy acids and the dicarboxylic acids are more frequently present in even carbon number chains. Other long chain constituents have also been identified, but these do not commonly occur in the species so far investigated.

It has been established by x-ray diffraction that the aliphatic molecules of waxes crystallize in uni-molecular layers with carbon chains parallel and in an upright or tilted position (22). Thus the degree of substitution in the chain will influence the density of the exposed groups. Adam and

TABLE 1

Long-Chain Constituents of Leaf Waxes

In the natural, unsaponified wax the alcohol and carboxylic acid functions are often present as esters. Olefinic and branched constituents have been omitted. [a]

Dominant carbon number[b]	
Even	Odd
$CH_3(CH_2)_nCO_2H$	$CH_3(CH_2)_nCH_3$
$CH_3(CH_2)_nOH$	$CH_3(CH_2)_nCHOH(CH_2)_mCH_3$
$CH_3(CH_2)_nCHO$	$CH_3(CH_2)_nCO(CH_2)_mCH_3$
$HO(CH_2)_nOH$	$CH_3(CH_2)_nCX(CH_2)_4CX(CH_2)_mCH_3$
$HO(CH_2)_nCO_2H$	$CH_3(CH_2)_nCOCH_2CO(CH_2)_mCH_3$
$HO_2C(CH_2)_nCO_2H$	

[a]Reprinted from (18), p. 1325, by courtesy of American Association for the Advancement of Science.

[b]The n and m are appropriately odd or even, the chain length being in the range C_{20} to C_{37}, generally being from C_{20} or C_{30}. +CX is $> C = O$ or $> CHOH$.

Elliott (23) examined a number of compounds, investigating the relationship between the exposed chemical group and the contact angle between water and pure saturated hydrocarbons. They found that surfaces containing exposed methylene groups exhibited a 30% higher attraction between the surface and water than did a hydrocarbon surface in which methyl groups were exposed. Thus, the hydrophilicity of plant surfaces is a function of the chemical groups exposed and also the number of hydrophobic groups per unit of surface area.

The effect of chemical composition on the smooth surface contact angles was investigated by Holloway (24). After measuring the contact angles of 51 plant wax constituents belonging to 10 chemical classes he concluded that no class was very water repellent, since the maximum contact angle, shown by the alkanes, was about 108°, and this did not differ significantly with chain length within a class. He suggested that the variation in wettability, as reflected by the contact angle, results mainly from differences in the packing of exposed groups. Alkanes and esters with no substitution in the chain and with exposed methyl groups closely packed at the surface were less easily wetted than classes in which substitution prevented close packing or in which there was a terminal substitution. Although wetting of the surface is a function of exposed groups, it is important to note that the range of contact angles given by the chemical classes most commonly occurring in surface waxes was relatively small. Similar angles resulted from mixtures of waxes containing various proportions of alkanes, esters, ketones and secondary alcohols, e.g., Allium porrum with a wax contact angle of 105° contains 19.8% alkanes, 27.2% ester and 30.7% ketones with respective contact angles of 107.5°, 105° and 104.5°, while Brassica oleraceae wax shows a contact of 104.5° but contains 32.3% alkanes, 9.8% esters, 15.9% ketones and 13.9% secondary alcohols with contact angles of 108°, 104.5°, 104° and 104° respectively.

Smooth hydrocarbon surfaces generally result in contact angles of about 105°, and increasing the roughness of a surface of the same composition usually increases the contact angle (25). When leaf surfaces rather than supported thin films are used, contact angles of less than 80° have been obtained for leaves without surface waxes, while waxy leaves have shown contact angles for water to up to 140° (26, 27). Therefore, in terms of variation in the contact angle of aqueous solutions on waxy leaf surfaces, the chemical composition of the wax and orientation with regard to the exposed groups would appear less important than the morphological configurations of the waxy projections.

D. Leaf Surface Phenomena

The primary objective in transferring a biologically active compound from the external environment to within the plant cell is to lodge a sufficient quantity on the leaf surface such that, even though only a small proportion of that applied reaches the site of action, it is adequate to produce the desired response. In spite of the chemical and morphological variation known to occur in the leaf surfaces we have a better understanding of the events proceeding at the environmental-spray-droplet-leaf surface interfaces than of the subsequent events involving the penetration of the compound into and through the cuticle.

The effectiveness of deposition on the leaf surface of foliarly absorbed pesticides can be considered in terms of three criteria: the quantity of active material deposited on the target surface, the extent to which the

target surface is covered (28), and the persistence of the pesticide in a
readily absorbed form on the leaf surface.

The retention of spray droplets on a leaf surface is dependent upon
the number of droplets lost by reflection and/or run-off after aggregation
(the latter being more important in the case of high volume spraying). In
the theoretical situation where there are no adhesive forces between the
liquid and the solid and when the liquid has zero viscosity, the initial spherical drop will be flattened on impact with the surface. With increase in
surface area, the surface energy will also increase. This leads to an
elastic recoil towards the spherical form and then beyond it to an extended
spheroidal form during which there is a transformation of energy into up-
wardly directed kinetic energy.

In practice however, there will be some loss of energy when the droplet impacts on the leaf surface and distortion will be reduced due to viscous forces. This reduction in droplet deformation will also cause the
adhesive forces to decrease. The adhesive forces are in turn determined
by the area of contact between the drop and the surface. This surface contact area will increase until the advancing contact angle is reached. Therefore reflection will only occur if the sum of the forces operating results in
a recoil energy (29). The recoil energy is a function of droplet size, surface tension of the spray liquid and as previously mentioned, the advancing
contact angle (30).

Hartley and Brunskill (29) found that reflection only occurred from surfaces exhibiting "micro roughness," that is, roughness due to wax projections and small contaminants as opposed to the "macro roughness" of the
veins, etc. They considered the necessary condition for reflection was a
large advancing contact angle. No reflection was detected from smooth
surfaces and the reflecting property of leaves could be destroyed by removal of the waxy deposits.

Run-off or drainage from leaves may occur when pesticides are delivered in high volumes. Under these conditions spray retention is governed
by the advancing and receding contact angles, by the degree of contact
angle hysteresis (the contact angle variation between the advancing and
receding states), and by the surface tension of the spray medium. The
problem of evaluating the loss of spray chemical by drainage is complicated by the variable nature of the leaf surface, since leaves with prominent vein structures retain more spray than leaves in which the veins are
flush with the surface or only slightly raised. It must also be considered
whether some of the chemical remains on a leaf from which the drop has
drained. Hartley (9) suggests that if the receding as well as the advancing
contact angle is finite, no chemical remains. If the receding angle is zero

some chemical may be retained but this is not due to the zero contact angle
but to some other unexplained phenomena.

The extent to which the target surface is covered and the extent to which
it is wet are separate considerations. Evans and Martin (31) have defined
wetting as the ability of the liquid to form a persistent liquid-solid interface after the excess liquid has drained off, while spreading is the ability
of the liquid to form a liquid-solid interface over a surface by virtue of its
surface activity.

Wetting then may be considered in terms of the contact angle formed by
a drop standing on a leaf surface. The particular contact angle formed is
a result of the equilibrium established between the surface tensions of the
solid and the liquid and the solid-liquid interfacial tension (25). Thus the
surface tension of a liquid, one of three interfacial tensions contributing
to the contact angle, has been found unsatisfactory for comparison of the
wetting ability of various liquids (32).

One equation that has been useful as a measure of the attraction between
liquids and solid surfaces is Young's equation (33) where the work of ad-
hesion (W_a) is expressed in terms of the contact angle (θ) and the surface
tension of the liquid (γ_1).

$$W_a = \gamma_1 \, (1 + \cos \theta)$$

The contact angle generally exhibits hysteresis and because of this, the
equilibrium value (θ_e) (average of the advancing (θ_a) and receding (θ_r) contact angles) is sometimes used (34).

$$\cos \theta_e = (\cos \theta_a + \cos \theta_r)/2$$

The contact angle varies from one portion of a leaf surface to another
and surface contamination and roughness results in considerable variation
between the contact angles measured in the laboratory and those in the
field (25). The degree to which surface roughness influences the contact
angle has been investigated by Wenzel (35) and the following relationship
proposed.

$$\cos \theta_s = R \cos \theta_r$$

where θ_s and θ_r are the average contact angles on smooth and rough surfaces of the same composition and R is the ratio of real to apparent surface.

As Dimond (25) points out, consideration of this equation shows that if the contact angle on a smooth surface is less than 90°, the effect of surface roughness is to decrease the contact angle and cause the surface to become less water repellent. As the contact angle in most species bearing epicuticular wax is greater than 90°, any increase in surface roughness will increase water repellency compared with those species which do not appear to bear surface waxes and which may have a contact angle as low as 80° (14).

The effects of solvent and solvent volatility on absorption are ill-defined. It is generally considered that sufficient amount of the pesticide must penetrate before evaporation of the carrier solvent and crystallization of the compound if biological activity is to be adequately expressed (15, 36, 37). However, it has been demonstrated that during the process of solvent evaporation, concentration of the pesticide component will occur and this may result in an increased penetration for a period of time (17). Increased concentration will not always enhance penetration because, as is pointed out by Hartley (9), if an active acid is applied as a soluble salt an increase in concentration by evaporation may in fact retard the penetration of the undissociated molecules. This occurs if the concentration of the undissociated form is lowered by the presence of the salt forming alkali.

Evaporation occurs when the kinetic energy of the molecules exceeds the cohesive forces. Freed and Witt (38) have suggested that the Clausius-Clapeyron equation may be useful in calculating the latent heat of vaporization of pesticide solutions.

$$\ln P = (\Delta H/R)\,(1/T)$$

where $\ln P$ is the natural log of the vapor pressure of the solution at an absolute temperature T, ΔH is the latent heat of vaporization and R is the ideal gas constant.

By plotting the log of the vapor pressure against the reciprocal of the absolute temperature, the slope of the resultant plot represents $\Delta H/R$. Thus, to decrease the loss by vapor it is apparent that ΔH, the latent heat of vaporization must be increased. Such an increase may be obtained by choice of a particular derivative (Table 2).

The other alternative is to add to the solution a nonvolatile solute as according to Raoult's Law, the vapor pressure of a solution containing a solute is directly proportional to the concentration of the solvent. The

TABLE 2

Vapor Pressure and Latent Heat of Vaporization of Esters
of Growth Regulator Herbicides[a]

Compound	Mol. wt.	Vapor pressure 25° C(mm Hg) (x 10^{-3})	Heat of vaporization cal/mole
MCP-isopropyl	242.69	21.6	3,510
MCP-butoxyethyl	300.67	4.8	7,400
2,4-D-isopropyl	263.12	10.5	2,390
2,4-D-butoxyethyl	321.20	1.7	8,400
2,4-D-butyl	277.15	3.9	2,050

[a]Reprinted by permission from Ref 38.

addition of solute decreases the mole fraction of solvent and the vapor
pressure of the solution is decreased according to the equation

$$P = P_0 X_1$$

where P is the vapor pressure of the solution, P_0 is the vapor pressure of
the pure solvent and X_1 the mole fraction of solvent. The efficacy of this
approach was shown by Freed and Witt (38) with isopropyl 2,4-D alone and
isopropyl 2,4-D with chlorinated biphenyl added (Table 3).

The interactions between the surface properties of foliage and pesticide
chemicals can be modified by inclusion of a surface active agent in the pesticide formulation. These organic compounds (given the general name of
surfactants) include the compounds which are variously described as adjuvants, spreaders, wetters, humectants, penetrants, dispersants, and
cosolvents. It is accepted that surfactants generally increase the penetration of organic and inorganic compounds, Currier and Dybing (39) have
listed nine ways by which the effect of surfactants might be mediated: (1)
improving coverage, (2) removing the air film between the spray droplet
and the leaf surface, (3) reducing interfacial tension between relatively
polar and apolar submicroscopic regions of the cuticle, (4) inducing stomatal entry, (5) increasing the permeability of the plasma membrane,

TABLE 3

Evaporation of the Isopropyl Ester of 2, 4-D at 40°C[a]

	Percent loss	
	24 hours	48 hours
Isopropyl 2, 4-D, alone	1.83	8.55
Isopropyl 2, 4-D and chlorinated biphenyl	0.41	4.17

[a]Reprinted by permission from (38).

(6) facilitating cell wall penetration in the region of the wall cytoplasm interface, (7) inducing the cosolvent effect, (8) interacting with the herbicide, and (9) acting as a humectant. However, not all organic compounds show increased penetration after the addition of a surfactant as was demonstrated by Darlington and Barry with sucrose (40) and by Goodman with antibiotics (41). Some surfactants decrease the effectiveness of foliar applied pesticides (42).

As the concentration of surfactant is increased, a concentration is reached called the "critical micelle concentration" at which there is an abrupt change in many of the characteristic properties of the surfactant solution (43). The osmotic pressure ceases to increase, and further increase in the surfactant concentration fails to cause a corresponding decrease in the surface tension (liquid-solid interface) as shown in Fig. 2.

It has also been shown that although there is a close relationship between contact angle, surface tension and spreading coefficient in some systems there is not necessarily a good correlation between these parameters and herbicidal activity (44). Further evidence that lowering of the surface tension does not necessarily provide an adequate explanation for enhanced response is supplied by Freed and Montgomery (45) who obtained quite different herbicidal responses after the application of 3-amino-1,2,4-triazole solutions to which different surfactants were added at concentrations such that the surface tension of each aqueous solution was the same.

Jansen (46) points out that although the maximum reduction in surface tension occurs at surfactant concentrations corresponding to the critical micelle concentration of 0.01% to 0.1%, surfactants generally exhibit their

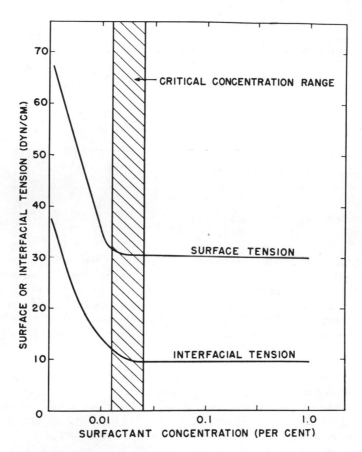

FIG. 2.   Effect of increasing concentration of a typical nonionic surfactant on surface and interfacial tension of aqueous system.   Reprinted
by permission from Ref. 43, p. 87, courtesy of University of Chicago
Press.

greatest biological effects at concentrations exceeding this level, thus
strengthening the idea that the effects of surfactants are due in part to
changes in physical–chemical properties other than surface tension.   The
equivalent conductance, electrophoretic, and osmotic properties have been
shown to exhibit anomalous behavior depending upon the concentration and
surfactant homologue used.   In an attempt to correlate the biological effects
of surfactants with ionogenic class, Jansen et al. (47) investigated 63 surfactants but could find no correlation.   They concluded that surfactants

increase or decrease herbicidal effectiveness according to the species sprayed, the nature and inherent activity of the herbicide, and the concentration of the surfactant. Any attempt to gain an understanding of the mechanism underlying the response to surfactants is further complicated by consideration of some of the other effects on biological systems brought about by the addition of these compounds.

Temple and Hilton (48) increased the solubility of the herbicides ametryne, diuron and atrazine by the addition of surfactants. A number of surfactants have an inherent toxicity. The extent of this toxicity has been investigated (48-51) and found to vary according to ionogenic class (49), concentration, and molecular structure (50). At a more specific level Parr and Norman have cited in their review (43) a number of papers which have reported the effects of surfactants in bacterial systems, on protein synthesis and enzyme activity, on the stimulation of some growth regulator effects, and on the biosynthesis of some essential plant constituents.

The known effects of surfactants are so diverse, it is not surprising that little progress has been made in determining the mode of action for any one pesticide-surfactant combination. Until more is known of the route by which foreign molecules gain access to plant cells and of the chemistry of the components of the pathway, there is little possibility that any clear understanding of the role of surfactants in the absorption process will be obtained.

One approach to the comparative evaluation of surfactants has been suggested by Freed and Witt (38). If absorption can be regarded as a rate process, then physical-chemical principles can be used to describe the process. In order to transfer an organic compound from an external phase to an internal phase, an energy barrier similar to that encountered in reaction kinetics must be overcome and may be thought of as the energy of activation. Then, the reaction rate constant, (k), for the penetration process may be expressed in terms of the activation energy (ΔE) using the Arrhenius equation:

$$\frac{d \ln k}{dt} = \frac{\Delta E}{RT^2}$$

which after integration becomes:

$$\ln \frac{k_2}{k_1} = \frac{\Delta E}{R} \left[\frac{T_2 - T_1}{T_1 T_2} \right]$$

where k_1 and k_2 are the reaction rate constants at absolute temperatures of T_1 and T_2. R is the gas constant. Then

$$\Delta E = R \ln \frac{k_2}{k_1} \left[\frac{T_1 T_2}{T_2 - T_1} \right]$$

By determining the reaction rate constants at two different temperatures, the activation energies of systems with and without surfactant can be calculated. If we assume that the rate of uptake follows a first order rate equation then we can write the integrated rate expression for a first order reaction in the form

$$\ln \frac{a}{a - x} = kt$$

where a is the original concentration of absorbent and (a - x) is the concentration after time t. The rate constants for the reactions with and without surfactants can be determined from the slopes of the plots of $\ln [a/(a - x)]$ versus time t. The values of k_1 and k_2 can then be substituted in the Arrhenius equation and activation energies for the penetration process calculated. The results obtained by Freed and Witt (38) indicate that there was a general trend for the surfactant to lower the activation energy as illustrated in Table 4. In attempting to summarize the advantages of adding a surfactant to pesticide formulations, it may be noted that the general effect of surfactants is to reduce inherent biological variation thus increasing the efficiency of the penetration process.

E. The Cuticle and Cuticular Layers

The cuticle refers to the outer or oldest layer which contains cutin, possibly impregnated with waxes, but with cellulose and hemicelluloses absent. The layers underlying the cuticle, but above the pectinaceous layer if such occurs, are termed the cuticular layers. It should be realized that to consider the various layers as discrete and separate is undoubtedly a simplification, but it offers some advantage in considering the chemical nature of these layers. The pectinaceous layer is an exception in some species in that it appears to be quite discrete with little transition into the cuticular layers.

The ultra-structural studies made of the cuticle and cuticular layers using x-ray analysis and polarized, phase, interference, fluorescence, and electron microscopy have shown the variable nature of the disposition and quantity of the various constituents (52). It is probable that this variation markedly affects the cuticular penetration of pesticides applied to different plant species.

TABLE 4

Effect of Surfactant on Absorption Energy[a]

| Herbicide | pK | Ea (Kcal/mole) | |
		Without surfactant	With surfactant
dicamba	1.94	18.9	
picloram	2.94	12.0	10.4
2,4-D	3.31	6.5	3.0

[a]Reprinted by permission from Ref 38.

The chemical composition of cutin was investigated by Legg and Wheeler in 1929 (53). They reported the presence of two acids, cutic and cutinic acid, in the cutin of Agave americana and Agave rigida. More sophisticated techniques lead to the isolation of an increasing number of fatty acids (54, 55), and in 1965 Crisp (3) used a stepwise degradation of cutin from Agave americana to its monomers, concluding that the degraded cutin of this species consists of hydroxy fatty acids ranging in chain length from tridecanoic to octadecanoic. Crisp identified 19 fractions of which nine were present in excess of 1%. Ester, alkyl peroxide and ether bonds were found in a bond ratio of ester:alkyl peroxide:ether of 7:2:0.12, indicating a high degree of polycondensation with some cross linking.

Based on analytical studies of cutin structure, several oxidative steps are involved in both the biosynthesis and decomposition of cutin. Linskens et al. (56) have proposed a structure for cutin polymerization which shows the types and ratio of bonds to be similar to those found by Crisp (3), with the exception of the ether bonds, as shown in Fig. 3.

The degree of polymerization of fatty acid precursors of cutin probably varies throughout the cuticular layers, with the outermost layers more highly polymerized than the cutin closer to the cell wall (1, 57). Crisp (3) has reported that ultraviolet irradiation promotes the formation of peroxide cross linkages at the outer edges of the cutin polymer. If the hydroxy fatty acids are synthesized in the protoplasm as suggested by Bollinger (58),

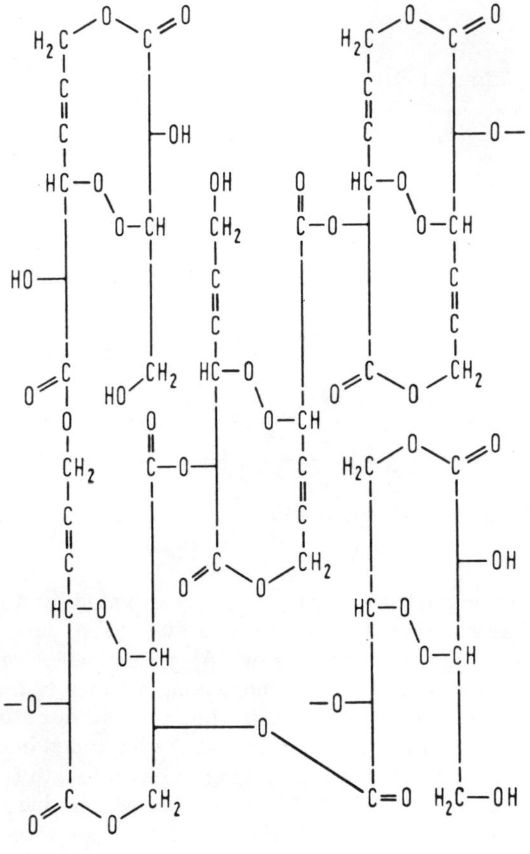

FIG. 3. Proposed structure of cutin from Gasteria verricuosa. Single chains are linked by peroxide groups to form double- and triple-chain units. Esterification of the terminal hydroxyl groups and the carboxyl groups occurs either at both ends (upper right) to form a closed double–chain–unit, or only at one side, the second site being used then to bind other units (lower right). Free hydroxyl groups will be linked to other chains, leading to a three-dimensional polymeric structure. Reprinted by permission from Ref. 56, p. 138, courtesy of Springer-Verlag.

then procutin must penetrate the cell wall and eventually become oxidized and polymerized with an associated loss in hydrophilicity. However, OH and COOH groups are undoubtedly retained on the pure cuticular waxes as the surface of the cuticle carries a negative charge (4). Other apolar compounds are not stained by the frequently used stains for cutin, Sudan III and IV (59).

The chemical composition of the waxes which impregnate the cuticle and cuticular layers is virtually unknown. Kolattukudy (60) considers that the site of synthesis is within the leaf epidermal cells. The hydroxy fatty acid precursors diffuse across the cell wall and then undergo oxidation, reduction and conjugation to form the various constituents of the outer wax layer (10). Where these latter reactions take place is not known but it is probable the embedded waxes are chemically and structurally different from the epicuticular waxes. It is known that within the cuticle are found waxes which are optically negative, soluble in pyridine and able to be stained with lipid dyes (4).

F. The Pectinaceous Layer

A pectin layer is found between the cell wall and the cuticular layers in many species (2, 3, 11, 12) and in some appears to be intercalated with cutin in the cuticular layers. In still other species, the pectin layer would seem to be extremely limited or absent (3). The long chain pectins are formed from glucose monomers in which the CH_2OH side chains have been replaced by COOH groups. There may be some polycondensation or substitution by methoxy groups. This substitution is considered to be without effect on water solubility (3, 4). Polygalacturonic acid is soluble in water although the calcium salt is not (4). Unlike cellulose, the pectinaceous materials are isotropic and amorphous in structure and because of their polar characteristics could impede the penetration of lipophilic compounds.

It would seem to be tacitly assumed that the pectin layers are deposited prior to the laying down of cutin (4, 12). However, Heinen and Brand (61) in studies on recovery after wounding in Gasteria verricuosa reported that cutin is deposited on top of cellulose and pectin impregnation follows later. Pectin synthesis has also been reported to occur during primary wall thickening (62). This information is of interest as epidermal cells are undergoing expansive growth by anticlinal divisions prior to exposure to the external environment, and cell plate formation on periclinal walls does not occur at this time. Thus, the formation of a pectinaceous layer may occur by continued pectin synthesis throughout the thickening of the epidermal cell wall, which may ultimately lead to pectin intercalation of the cell wall and cuticular layers. Perhaps the degree of cutin polymerization and impregnation by waxes determines the extent to which pectin ramifies through these layers.

G. Cuticular Penetration

Isolated cuticle and cuticular layers have been frequently used to investigate penetration and absorption of organic and inorganic molecules.

These membranes can be isolated mechanically by removal of underlying tissues, by chemical treatment, or by various enzymatic methods (2, 63, 64, 65, 66). The isolation method employed is apparently of importance. Norris and Bukovac (17) found that the penetration of NAA through astomatous tomato cuticle was 8 to 10 fold greater when the cuticular layers were removed chemically by refluxing with ammonium oxalate/oxalic acid than when the membranes were removed mechanically or enzymatically. This suggests that chemical isolation modifies cuticular structure and indeed some morphological changes in chemically removed cuticles were reported by Goodman and Addy (64). Data obtained by Norris and Bukovac (7) indicated that when pear leaf cuticles were isolated enzymatically, the morphological structure of the surface waxes was retained, and no change in the position or degree of orientation of the embedded waxes could be detected. However, although morphological changes have not been detected in enzymatically isolated cuticular membranes, Kamimura and Goodman (67) found that pectinase treatment of excised astomatous cuticle increased permeability to leucine. This suggests that, at least in some species, incubation in pectinase solution may result in the removal of pectins from cuticular layers, altering the permeability characteristics. Many of the results of permeability and absorption studies using isolated cuticles are difficult to interpret, and extrapolations to in vivo absorption must be made with caution.

The percentage penetration of ions and organic compounds through isolated cuticles is invariably low. From 0.2% to 2.0% of cations and anions penetrated isolated tomato fruit and onion leaf cuticles (66). Less than 2.0% of the applied leucine penetrated lower isolated cuticles of apple leaves (68), and approximately 1.8% of 1 x 10^{-3}M NAA penetrated the upper cuticle of isolated pear leaf cuticle (17). If in fact the penetration of pesticides in aqueous solution is negligible through stomatal openings without the addition of agents to reduce surface tension (39), then the rate of cuticular penetration in intact leaves is considerably greater than that reported for isolated cuticular systems where the concentration of penetrant in the donor solution remains essentially constant throughout the duration of the experiment. However, by allowing a 10 μl droplet of NAA to evaporate after placement on floating disks of cuticle, Norris and Bukovac (17) found that 72% of the donor NAA had penetrated in 24 hr, a figure of the same order as those observed for absorption of other organic compounds by intact leaves (69, 70).

It has also been reported that the inner or mesophyll side of isolated cuticle has a greater capacity to bind both cations and anions than the outer or atmospheric side but permeation of both classes of ions is greater from the outer to the physiologically inner surface (66, 71). It has been suggested that the preferential binding on the inner side is due to cuticular protrusions and cell wall fragments providing an increased inner surface area (71). Binding is also enhanced by the polarity of the cuticular layers adjacent to the cell wall. A convincing explanation for the greater inward ionic penetration and the failure to eventually reach a steady state with equal penetration in both directions has not been given.

Similar to ions, the organic molecules urea, maleic hydrazide, and N, N-dimethylaminosuccinamic acid penetrated isolated astomatous tomato fruit cuticle to a greater extent from outer to inner surfaces, but showed no differences in binding capacity between surfaces (72). However, the rate of penetration of NAA through isolated astomatous pear leaf cuticle was the same irrespective of the surface to which the donor solution was applied. The permeability characteristics of urea are somewhat different compared with the other organic compounds so far investigated in that movement through astomatous tomato fruit cuticles was 6 times greater than maleic hydrazide and 10 to 20 times greater than inorganic ions (72). Urea also appears to increase the permeability of cuticular membranes to inorganic ions such as rubidium, chloride, and phosphate (1). Yamada et al. (72) have made the suggestion that urea alters cuticular structure by breaking hydrophobic bonds, thus facilitating the penetration of urea and accompanying inorganic ions. However, Kannan (73) found that iron penetration was reduced when accompanied by urea. As Franke (1) points out, the disruption of cuticular bonds would seem to require enzymatic catalysts, and the isolation treatment is not conducive to the maintenance of enzymatic activity if such enzymes exist within the cuticular layers.

It appears that penetration of inorganic ions and some organic molecules across isolated cuticular layers occurs by a diffusion process (74, 75). Yamada et al. (66) were able to fit data for influx and efflux of maleic hydrazide and N, N-dimethylaminosuccinamic acid to a first order rate equation. However, the data for urea penetration does not fit first order rate kinetics, as penetration increased with time. Further, when Yamada et al. (71) investigated calcium binding on boiled and unboiled tomato fruit cuticles, they found that calcium retention on unboiled cuticle was increased by the addition of KCN and decreased by DNP. These chemicals had no effect on calcium retention when boiled cuticles were used nor was there any respiratory activity associated with the nonboiled cuticular layers, thus suggesting that the effect was of a physical-chemical nature.

Little information is available on the effect of molecular structure on penetration through cuticular layers. The cuticular movement of a number of compounds of differing polarity has been expressed in terms of their permeability coefficients (P) by Darlington and Cirulis (74). These values have been calculated from the conventional equation

$$P = \frac{V}{qt} \ln \frac{C_d}{C_d - C_r}$$

where V is the volume of the receiver solution, q is the area through
which the test compound must penetrate, C_d is the concentration of the
donor solution and C_r the concentration of the receiver solution at time
t. No correlation could be found between permeability coefficients corrected for molecular weight ($PM^{1/2}$) and the chloroform:water partition
coefficients for glucose and sucrose. However, a relationship was found
for two homologous series of the α-chloroacetamides such that

$$PM^{1/2} = k(PC)^n$$

where PC is the partition coefficient. The authors concluded that each
type of compound has a characteristic permeation rate which may be altered by substitution but which cannot necessarily be predicted from partitioning between polar and nonpolar phases. Substitution, chlorination,
cyclic substituents, heterocycling of the amide nitrogen, and the inclusion
of an ether oxygen in the substituent increased chloroform:water partitioning of the α-chloroacetamides but not cuticular penetration. The alterations in structure which caused changes in the partition coefficient not
correlated with changes in penetration are those which might be expected
to change the hydrogen bonding characteristics. The correlation between
partitioning and penetration was retained where structural changes between members of the homologous series involved alterations in Van der
Waals bonding. Although it is known that molecular size, charge, partitioning, volatility, solubility and sorption are important factors in cuticular penetration (76), the critical interrelationships between these factors have yet to be evaluated.

H. The Cell Wall

There is generally a gradual transition between the pectinaceous layer
and the underlying primary cell wall during the early stages of leaf development. The cellulose of the primary cell wall is normally intercalated
with substances consisting of homo- and hetero-polymers of hexose (glucose, mannose and galactose), pentoses (xylose and arabinose), and uronic
acids (galacturonic and glucuronic) (77). These substances are known as
the hemicelluloses, a term which includes all the amorphous compounds
comprising the ground matrix of the cell wall. With secondary thickening,
the cellulose of the epidermal wall may become increasingly impregnated
with cutin and wax (5). According to Crisp (3) the chemical analysis for

cork is representative of the epidermal fraction in the mature leaf if one omits the suberin and lignin. The comparison with a quantitative chemical analysis of the primary wall in a young wheat leaf is shown in Table 5.

The cellulose framework in which the hemicelluloses, pectic materials, and later wax and cutin are deposited would appear to be composed of microfibrils, averaging 150 to 250 Å in diameter and with interfibrillar spaces of up to 100 Å. The ultimate structural unit is termed the elementary fibril with a diameter of about 35 Å and made up of approximately 40 glucose molecules (1, 80). These fibrils may be aggregated to form the microfibrils, and variation in microfibril diameter—Northcote and Lewis (81) report microfibril diameters of 57 to 85 Å—may be due to variation in the number of elementary fibrils which aggregate. The interstices between the microfibrils would normally have a high water content, but Roelofsen (82) suggests that the epidermis contains lipid and pectinaceous substances, presumably procutin and prowax in the young leaf and an increasing quantity of cutin and wax as the leaf matures.

Because of the hydroxy content of the cellulose and hemicelluloses and the COOH groups of the polyuronides and wax and cutin precursors, the epidermal cell wall in expanding leaves should be considerably more hydrophilic than the outer cuticular lamellae. Consequently, it is reasonable to

TABLE 5

Quantitative Chemical Analysis of Cork and Primary
Wall of Young Wheat Leaf

	Percentage of dry weight				
	Cellulose	Hemicellulose	Pectin	Lignin	Suberin or Cutin
Cork[a]	2 - 11	Trace	Trace	12 - 20	40 - 60
Wheat leaves[b]	30	11	22	–	–

[a]Data reproduced from Ref. 78.

[b]Data reproduced from Ref. 79.

predict that a gradient of increasing polarity exists from the external leaf surface to the primary wall in the young leaf. As wax and cutin are deposited within the epidermal cell wall with the increasing maturity of the leaf, the polarity gradient would be diminished and the rate of movement of polar compounds across the cell wall would be reduced. Conversely, it could be expected that the rate of movement of apolar compounds would be increased. In fact, Kamimura and Goodman (67) using entire leaves found that the relatively polar compound, leucine, was more readily absorbed by older leaves. However, in this study a metabolic component of absorption cannot be divorced from penetration through the cuticle, cuticular layers, and cell wall.

Irrespective of its polarity, the cell wall constitutes a resistance to diffusing molecules because it represents a static layer and also may act as a molecular sieve restricting the passage of molecules which exceed a size corresponding to the majority of interfibrillar spaces (83). The extent of the reduction in permeation may be seen from Table 6 (83) in which the decrease in the $PM^{1/2}$ values of compounds at the end of the table indicate the reduced rate of permeation once the molecular weight of the compound exceeds 300.

There are numerous references to the existence of extracellular enzymes (84) and it is interesting to consider whether these proteins are quantitatively important as adsorptive surfaces for organic pesticides. Also of interest is whether possibilities exist for the structural modification of pesticide molecules so that lipophilic substituents, added to increase the rate of penetration across the relatively apolar outer layers, may be enzymatically cleaved or transformed thereby forming a molecule which is more readily transported across the plasma membrane.

I. The Absorption Pathway(s)

Considerable controversy surrounds the subject of whether the cell wall and cuticular layers are traversed by pores or canals and whether there are pathways through the extracellular layers along which water and other polar molecules preferentially diffuse as suggested by Crafts (85). The existence of surface pores has been most clearly demonstrated by Hall (86) and Hall and Donaldson (87) after stripping away the surface waxes and employing a carbon replica technique to reveal the surface structure. However, the pores revealed by the electron microscope appear to be associated with epicuticular wax formation and do not pass through the cuticular layers. Also it is improbable that pores, through which surface waxes are secreted, constitute the pathway for the penetration of polar compounds.

TABLE 6

Permeability of the Cell Wall of Nitella mucronata to Nonelectrolytes[a,b]

Substances	M^c	$P \times 10^5$	$PM^{1/2} \times 10^4$
Methanol	32	75	43
Ethylene glycol	62	56	44
Glycerol	92	44	42
Tetraethylene glycol dimethyl ether	222	24	36
Sucrose	342	17	31
"Polyethylene glycol 400"	400	7.6	15
Raffinose	504	8.3	19
"Polyethylene glycol 600"	600	2.7	7
"Polyethylene glycol 1000"	1000	0.8	3

[a]The permeability constants (P) are given in centimeters per second, multiplied by 10^5.

[b]Reprinted from Ref. 83, p. 37, by courtesy of Academic Press.

[c]M = molecular weight.

The existence of plasmodesmata, small protoplasmic extensions of about 0.2 μ diameter through the cell walls of adjacent cells, has long been recognized. Similar structures were reported to occur in outer epidermal walls (59, 88, 89) and were given the name of ectodesmata. Ectodesmata were originally thought to be similar to plasmodesmata in that they were extensions of the protoplasm, enclosed by the plasmalemma, into the outer epidermal wall. However, Franke (1) has proposed that ectodesmata are interfibrillar spaces in the cell wall containing liquid excretion products of the epidermal protoplasts, which include a reducing agent suggested by Crisp (3) to be abscorbic acid, glutathione, or quinone. This reducing agent is thought to cause the reduction of soluble mercuric ions in Gilson Solution fixative (basically FAA fixative stabilized with oxalic acid and mercuric chloride) to the insoluble mercurous form thus resulting in the visualization of ectodesmata in the light

or electron microscope. The quantity and distribution of the reducing substance will then determine the shape and distribution of the ectodesmata.

In his review Franke (1) cites evidence indicating that foliar absorption
of foreign molecules does not occur equally over the exposed leaf surfaces.
The epidermis over veins is more permeable to a number of compounds
than the epidermal cell at the leaf margin (39). Preferential absorption
occurs through trichomes, particularly the basal portion, over anticlinal
walls (90) and through the guard cells rather than through stomatal pores
(1). It has been demonstrated that particularly high concentrations of ectodesmata exist at the base of hair cells or in the surrounding epidermal
cells, in epidermal cells above, beneath or on both sides of the veins,
along the anticlinal walls, and in the walls of guard cells (88, 91). Thus
high concentrations of ectodesmata are found in the regions of the leaf
where preferential absorption appears to occur. To further strengthen
the possibility that ectodesmata are the absorptive pathways for at least
some molecules, Yamada et al. (8) showed that in isolated green onion
cuticle there was particularly intense binding of calcium and chloride ions
above the periclinal walls, on cuticular membranes adjacent to stomatal
apertures, and even on the cuticular membranes lining the stomatal cavity.
Also using green onion leaves, Franke (92) showed that these preferential
binding sites coincided with the major distribution of ectodesmata.

It is of interest to note the reports of Roberts et al. (93) and Palmiter
et al. (94) in which they found that the cutin of the epidermal cells of
McIntosh apple leaves was intercalated with pectinaceous material, and
what appeared to be pectinaceous substances extended from the outer cuticular surface along the bundle sheath extensions to the veins. Pectinaceous
strands extending into the cuticular layers and the primary cell wall have
been found in a number of species (2, 3, 7), and Crafts (85) concludes
that hydrated pectinaceous strands in the cuticle provide an aqueous pathway for the diffusion of ions and polar solutes.

With one exception (93), neither ectodesmata nor pectinaceous strands
have been shown to extend to the surface of the cuticle. However, this may
be a limitation associated with the techniques so far employed for their
visualization in the light and electron microscopes. It is obvious that
physico-chemical characterization of the pathway must await its further
delineation. When this information is available it should be possible to
better define the molecular structures required for maximum rate of penetration via the "aqueous route," if such exists, as opposed to penetration
through the more lipophilic portions of the cuticle and cuticular layers.

J. The Plasmalemma

In entering epidermal mesophyll or phloem cells, externally applied molecules must traverse the plasmalemma, but relatively little is known of the mechanisms by which this is accomplished. The structural organization of the plasmalemma as originally proposed by Danielli and Davson (95) consisted of a bimolecular lipid layer oriented with the polar ends outward and covered on each side with globular protein. A refinement of the original model showed the lipid core stabilized by protein monolayers (96). This concept of membrane structure was later extended by Robertson (97) to a model he claimed was typical of all biological membranes. Thus the idea of the unit membrane originated and was widely accepted for a number of years. Recently, chemical analysis of membranes, electron microscopy, and knowledge of the variation in function of membranes of chloroplasts and mitochondria have caused the concept of a unit membrane to appear most unlikely. However, even the plasmalemma does not appear to entirely conform to the unit membrane as the plasmalemma of the same cell may show variations in thickness (98). Sub-unit patterns have been observed in the synaptic discs where the plasmalemma of two cells become appressed (99) and in the plasmalemma of pea and onion root tips (81, 100). It has been suggested that the globular sub-units might indicate the location of ATPases and other enzymes within the membrane (101, 102).

At this time the structural organization of the plasmalemma, or of any other membrane, is not known with any certainty, and indeed a chemical analysis for plant plasma membranes is conspicuously lacking. Branton (103) in reviewing membrane structure presents evidence indicating that either a protein-lipid-protein or a lipid-protein-lipid model is feasible. A bimolecular lipid with a substantial portion of the protein on the membrane surface provides the best explanation for what is termed "the unspecialized properties" of the membrane in areas devoid of specialized functional sites. Korn (102), after examining the available evidence, has stated that the most reasonable models of membrane structure are those that emphasize protein-protein interactions as the basic forces holding membranes together. These models include appreciable α-helical regions buried in an internal hydrophobic environment and place the polar groups of both protein and lipid at the aqueous interface.

Little is known of the mechanisms involved in the movement of pesticide molecules across the plasmalemma. Diffusion across the membrane probably occurs after dissolution in the lipid components of the membrane according to the lipophilic/hydrophilic balance of the pesticide molecule. In the case of small molecular weight compounds the possibility exists that passage through molecular pores occurs either by diffusion or by virtue of the bulk flow of water. While Stadelmann (104) points out that such pores have never been shown in electron microscopic studies, although reportedly of observable size, the fact remains that small molecules permeate the plasmalemma faster than anticipated on the basis of their relative lipophilicity (83). Wartiovaara (105) suggested that transient pores are formed by thermal agitation of lipid molecules in the membrane. These holes exist only for very short periods of time and larger molecules may only enter the pores if they are of the appropriate dimensions. In the case of long molecules, entry through the pores is a function of the speed at which such molecules become properly oriented at the membrane surface.

As well as a passive movement of molecules across the plasmalemma, a number of studies suggest that a metabolic component of uptake is also involved in the transport of some pesticide molecules across the membrane. Absorption of inorganic ions is usually considered to take place in two phases, the first being relatively rapid and consisting of diffusion into the apparent free space (generally considered the extracytoplasmic tissue volume), and a second slower and more prolonged phase considered metabolically dependent (106). Certain growth regulators and pesticides exhibit uptake characteristics similar to those of inorganic ions in that a period of rapid uptake is followed by a decreased rate of accumulation (107, 108). However, a number of reports indicate that the substituted phenoxyacetic acids, 2, 4-D and 2, 4, 5-T, and the substituted benzoic acids, 2, 4-DCBA and 2, 5-DCBA, had an uptake pattern in stem segments of Pisum and Gossypium where the initial rapid uptake was followed by a period of net loss to the external solution. In Triticum and Avena segments another pattern of uptake was found in which the initial rapid uptake and net loss phases were followed by further accumulation (109, 110, 111).

Saunders et al. (112, 113) suggested that two absorption processes occur in plants which accumulate auxin-like compounds. The first, called Type I accumulation, is unstable and results in release of the accumulated compounds due to changes in binding sites within the tissue. Venis and Blackman (114, 115) have proposed a mechanism to explain the Type I process. They propose that there is an electrostatic binding between the carboxyl group of specific growth regulators such as 2, 4-D and 2, 3, 6-TBA and the ammonium group of the choline moiety of α-lecithin or of phosphatides. Absorption of the solvent buffer or the growth regulator itself is thought to cause the activation of a pH sensitive enzyme, phospholipase D, which is able to break the ester linkage between the nitrogenous base and the phosphoric acid group. This causes binding sites to be destroyed and the growth regulator is released into the external medium. The second accumulation process, referred to as Type II accumulation, is stable and is associated with the metabolism of the compound to forms which are retained within the tissue. It is interesting to note that Saunders et al. (112, 113) found the progressive accumulation of POA by segments of Gossypium hypocotyl was accompanied by conversion to other radioactive metabolites, but there appeared to be no metabolic conversion of 2, 4, 5-T. In segments of Avena mesocotyl all of the substituted phenoxy acids tested underwent some conversion to forms differing from the parent acid. Thus, according to the species and the chemical structure of the growth regulator, a Type I or Type II accumulation may predominate and control the rate of uptake, or there may be simultaneous absorption by both processes.

Both Saunders et al. (112) and Venis and Blackman (111) found in the species investigated that, irrespective of which accumulation process predominated, the concentration of phenoxyacetic acid or benzoic acid within the tissues was greater than the concentration of the compound in the external medium. The question now arises as to whether accumulation within a tissue is the result of simple diffusion or whether an active process, perhaps involving a carrier system, is operative. Certain criteria have been employed in an attempt to make deductions about the mechanism of transport into cells. According to Cirillo (116), if a transport process requires metabolic energy, it is called "active." However, a diffusion process may be the sole uptake mechanism, but because of complexing, compartmentation, or metabolic conversion, a diffusion gradient is maintained across the plasmalemma, even though the intercellular concentration is such as to indicate transport against a concentration difference. It would seem inappropriate to term uptake such as this, although metabolically dependent, an active uptake. Rather, this term should be used only in those cases in which it can be established, as in the case of some inorganic ions, that transport across the plasmalemma is against an electrochemical potential. It is not sufficient to establish movement against a chemical potential, i.e., against a concentration difference, as many pesticides are at least partially ionized at the pH encountered in the cell wall, and consequently diffusion along an electrical potential gradient against a chemical concentration difference could well occur. It would be difficult to establish movement against an electrochemical potential for nonmetabolized organic compounds in higher plants. This would require the measurement of membrane potential and the establishment of flux equilibrium. When the organic compound is metabolized, the task becomes formidable.

Demonstrating the existence of a metabolic component of uptake for a nonmetabolized compound is more easily accomplished as there are a

number of criteria which if satisfied are indicative that such a component
exists: (1) a temperature coefficient (Q_{10}) of greater than 2, (2) uptake inhibited by anaerobic conditions, inhibitors of respiration, and other metabolic inhibitors, (3) adding compounds with similar structures may result
in competitive inhibition, and (4) the intercell concentration is greater than
that of the external solution. (5) The rate of uptake as a function of concentration may be hyperbolic. If the absorption characteristics of a nonmetabolized compound fulfill these criteria then it is possible that the compound is being transported into the cell by an energy-requiring active uptake
process, or, as Donaldson (117) suggests, the metabolic energy may be required to maintain the integrity of membranes and other surfaces on which
adsorption could occur. It is not possible, using the above criteria, to
readily determine whether either one or both of these processes are occurring.

There is considerable evidence that the accumulation of some organic
pesticides depends on metabolic energy (109, 117, 118, 119), but it is not
apparent whether the energy is required to maintain adsorption sites associated with the plasmalemma or whether the energy is required for transport across the membrane. By the development of specific inhibitors and
histoautoradiographic techniques for water soluble compounds at the electron microscope level, answers to these questions may be obtained.

III. TRANSLOCATION IN PLANTS

Translocation as described in this section will involve a discussion of
transport over distances greater than from cell to cell. Local or short
distance symplastic translocation from cell to cell can occur via the plasmodesmata unless toxicity of the herbicide is sufficient to cause severe
enough injury to interfere with this process.

An early report, Maskell and Mason (120), noted a relationship between
the concentration gradient of solutes in the phloem and transport. This
idea was elaborated by Münch (121) in the theory of mass flow in which assimilates move along a concentration gradient from an area of production
to an area of utilization. This theory is generally expressed by the Hagen-
Poiseuille equation to characterize the flow of liquid through tubes,

$$\frac{dP}{dx} = \frac{8\pi\mu v}{A_p}$$

in which μ is the viscosity of flowing liquid, A_p is the cross section area of
sieve tube, v is the velocity of flow, p is the hydrostatic pressure, and π is
the proportionality constant.

Because of the presence of the sieve plates in the sieve elements the equation $dP/dx = -\epsilon v$, where ϵ is a constant and v is the velocity of flow of the fluid in the phloem tube, may be a reasonably accurate expression of the pressure gradient. Horwitz (122) indicates that even the presence of sieve plates would result in a dissipation of pressure that would still be proportional to the first power of the flow velocity.

Attempts to fit translocation into a mathematical model have been unrewarding because of insufficient and conflicting data. A critical evaluation must take into account two major concepts of the translocation mechanism. The first involves independent movement, e.g., activated diffusion, protoplasmic streaming, surface migration, etc., in which the velocity of movement of individual species is different. The second involves mass flow, in which the velocity of movement of individual species is the same. Although water is the solvent for both transport systems, the solutes would be moving independently of the water in the "activated" hypothesis, but the solutes would be moving "en masse" with the water in the mass-flow hypothesis.

A commonly used expression to describe translocation velocities according to the mass flow concept, using phloem exudate data (123, 124) is:

$$\text{Volume transfer} \;=\; \text{area} \;\times\; \text{velocity}$$
$$(\text{cm}^3 \text{ hr}^{-1}) \qquad (\text{cm}^2) \quad (\text{cm hr}^{-1})$$

The mass flow hypothesis considers that all the solutes are moving in the translocation stream "en masse" with the water at the same velocity. An average velocity of an organic substance in this system would be directly related to the overall assimilate transport velocity.

An alternative to the above expression has been proposed by Canny (125):

$$\text{Specific mass transfer} \;=\; \text{velocity} \;\times\; \text{concentration}$$
$$(\text{g cm}^{-2} \text{ hr}^{-1}) \qquad (\text{cm hr}^{-1}) \qquad (\text{g cm}^{-3})$$

From this equation it is possible to calculate the rate of arrival of an organic substance at an area of utilization. According to Crafts and Crisp (126), this expression used by Canny to express transfer of mass through the phloem elements is limited. The expression indicates a "diffusion analogue" in which the organic substances are moving independently of each other and independently of the water in the sieve elements.

An important concept suggested by Crafts and Crisp (126) involves a re-evaluation of the Poiseuille formula, $P = (8 R_1 \eta l)/r^2$, to calculate pressure gradient in terms of velocity of flow, in which P is the pressure in dynes per cm^2, R is the velocity in cm per second, η is the viscosity of the liquid in poises, l is the length of the tube or gradient in cm, and r is the radius of the conducting element in cm.

From the studies of Esau (127), two essential features of the primary sieve plate are apparent. First, the sieve plate pore openings are not angu-lar but are curving, converging at the entrance and diverging at the exit. Laminar flow would be restricted under these conditions and may not occur at all. To account for this, Crafts and Crisp (126) set up a new formula, $P = (4 R_1 \eta l)/r^2$, to cover this nonlaminar flow. Calculations using this formula would indicate the pressures as calculated using the Poiseuille formula above are twice as great as the pressures theoretically needed if the formula of Crafts and Crisp is used.

The second feature, shown in many of the micrographs, is that the diameters of the pores are equal to or greater than the sieve plate thick-ness. If flow is creeping rather than laminar, as has been suggested un-der these conditions, and resistance would be at a minimum, the formula (128) would be $P = (3_\pi R_1 \eta l)/r$. The flow through the lumina of the sieve element would be laminar. Sieve plates in secondary phloem, when con-trasted with those in primary phloem, are generally thicker, and sieve pores are more tubular, indicating that flow may be typically laminar where flow through secondary phloem is involved.

Crafts and Crisp (126) have compiled a table of "Velocities of Trans-location in the Phloem" which indicates a wide range of variation. Veloci-ties of 75 to 100 cm/hr are very common, with extreme ranges of below 10 cm/hr to over 350 cm/hr. Since all plants do not have the same growth potential, and since variation will occur within the same species from day to day, these variations in velocity are understandable as the velocity would depend to a large degree on the growing condition of the plants.

Radioactive tracer techniques have allowed a more detailed and quan-titative evaluation of the translocation process. Much of this work has been of a confirmatory nature but with the added advantage of more pre-cision than was available with the earlier methods. Most data suggest a linear relationship between the logarithm of the radioactivity in the stem at a given point and the distance from the origin. All solutes move togeth-er and, except for preferential distribution along the route, all solutes in the stream will be delivered at the same time.

According to Peterson and Currier (129) bidirectional flow does not occur simultaneously in the same sieve element. However, flow may be in one direction at a given time and in the opposite direction at some later time. The requirement determining the direction of flow would be an osmotic differential favorable to the direction of flow. This movement is a hydrostatic flow along an osmotic gradient created by synthesis of osmotically active solutes at the source and their oxidation and storage in the sink. Since evidence indicates phloem mobile herbicides move with the assimilate stream, accumulation would be expected to occur where assimilates were accumulating or being used.

Translocation of assimilates in plants involves both a metabolically active source (area of synthesis) and a metabolically active sink (area of utilization or storage). Thus assimilates will move "en masse" through the sieve elements only in the direction and rate as determined by the gradient created by the total osmotically active solutes. A unique feature of this system is characterized by the ability of materials in the phloem to move out of the sieve elements at any point along the route or to move into the sieve elements at any point from adjoining cells.

It is important to remember that the phloem is a distribution system serving the entire plant. Since respiration is going on in all the living cells the requirements for these cells must be satisfied. To more easily understand this system, Münch (121) introduced the terms symplast, to describe the total interconnecting living protoplasm of the plant, and apoplast, to include the nonliving continuous cell wall phase around the symplast. Some workers include the xylem stream as part of the apoplastic system.

Studies involving the uptake and distribution of herbicides have contributed to our knowledge of the basic physiology of translocation as well as to an understanding leading to a more effective use of herbicides. Many workers have reported that any factor that affects natural transport in the phloem of a plant will affect the distribution of the herbicide.

It has been well established that formulation, point of application, growth condition of the plant, dosage rate, treatment time, etc., influence the uptake and distribution of a herbicide in the plant. Many compounds are incapable of moving into the symplast from external applications because of physico-chemical factors such as polarity, inability to move across membranes, adsorption to various cellular components, etc. Thus there are structural requirements which if met enable the compound to move from mesophyll to phloem. If these criteria are overcome sufficiently to allow phloem translocation once transport to the phloem is accomplished herbicides should move with the assimilates. Translocation of many herbicides is similar to the translocation of naturally occurring assimilates except when the toxic effects induced by the herbicide prevent such movement. These toxic effects will be of varying degrees depending on the herbicide and how it was used.

Some herbicides including dichlobenil and endothall interfere with assimilate transport by inducing the formation of callose, thus preventing the movement of assimilates. Maestri (130) studied the effects of endothall on translocation and showed that callose formation in leaves of bean and cucumber was especially evident at veinlet endings and as pit callose in vein parenchyma. It was also shown that endothall had a damaging effect on the plasma membrane. If movement of solutes into the phloem is an active transport process as suggested by Bieleski (131), a logical location for the site of endothall action might be in the parenchyma cells associated with the fine veins of the leaf. Since mechanisms of this type must be in close association with membranes it would be logical that movement of solutes into the phloem from areas treated with endothall would be difficult.

Glenn (132), studying the effects of dichlobenil on transport, found heavy deposits of pit callose in mesophyll and vein parenchyma cells and excessive deposits of callose were present on the sieve plates. His conclusion was that dichlobenil blocked transport by plugging the sieve elements, thus preventing movement of assimilates through the phloem.

Many herbicide molecules are mobile in the symplast, while the movement of others is restricted within or on the surface of living cells. These "mobility factors" determine whether a herbicide molecule will have a limited localized distribution or a systemic distribution throughout the entire plant. Although herbicides all move with the photosynthates, differences in the rate of introduction into the phloem might occur. Thus the front of an introduced herbicide may not accompany the assimilate front resulting from $^{14}CO_2$ fixation. Little and Blackman (133), studying the movement of 2,4-D, 2,4,5-T, and IAA in Phaseolus vulgaris, noted movement of 10 to 12 cm/hr for 2,4-D and 2,4,5-T and movement of 20 to 24 cm/hr for IAA. These compounds are absorbed out of the assimilate stream, so these low velocities reflect a combination of retention along the route of export and actual velocity of assimilate flow. Increasing dosage rates will change the pattern of translocation of sublethal concentrations. For example, sublethal concentrations of 2,4-D applied to the roots of plants will tend to remain in the root tips. Increasing concentrations will result in an upward movement into the foliage of the plant

because of injury to the root tissue. Many reports have indicated that herbicidal rates of 2,4-D often induce proliferation of growth and callus forma-
mation in stems and roots, causing excessive crowding and unorganized
growth, eventually resulting in disruption of the phloem and plugging of the
xylem. However, extreme phytotoxic contentrations will limit distribution
by contact killing, thus preventing movement away from the point of application. Phytotoxicity of many herbicides is related to their ability to resist metabolism by the plant (134). Many of these herbicides will cycle in
the plant and eventually accumulate at metabolic sites where they affect a
growth response. There is a direct relationship between the amount of
herbicide absorbed by roots and transported to the foliage and the rate of
transpiration. The greatest accumulation in the foliage will occur when
transpiration is the greatest.

Leonard and Weaver (135) found that grape leaves were importing assimilates when they were less than one-third of their final size. When
they approached one-half their full size the leaves began to export photosynthates. As may be noted from Fig. 4, translocation from leaves close
to the apex is always acropetal, while exporting leaves further from the
apex may export assimilates in both directions or basipetally if still further from the apex. The same phenomena occurs during the growth and
development of many perennial plants. Age of plant and season of growth,
has a strong influence on the distribution of assimilates.

Carbohydrate reserves in perennial plants, depleted during early
growth and development of the plant, are replenished later in the life
cycle. The major direction of assimilate movement is from stored reserves in the underground organs of the plant to the young developing
foliage. Therefore an application of a foliarly applied systemic herbicide would remain in the foliage if applied during the period when underground reserves are being utilized for foliar growth. Very little herbicide would translocate to the underground vegetative system. A later
foliar application of the systemic herbicide applied at or following full
bloom would be translocated to the underground system since the major
flow of assimilates would be from the mature foliage to either the developing seed or the underground storage system.

Herbicides may be grouped into the following four categories depending on their translocation characteristics (137): (1) Herbicides that move
primarily in the apoplast, (2) herbicides that move primarily in the symplast, (3) herbicides that move freely in both symplast and apoplast, and
(4) herbicides that have restricted or no movement in either the symplast
or apoplast.

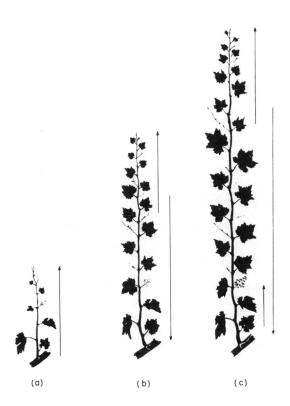

(a)   (b)   (c)

FIG. 4. Diagram of a rapidly growing grape shoot at three different developmental stages showing main direction of movement of photosynthate: (a) movement in a very young shoot or shoot tip is apical; (b) in the pre-bloom or bloom stage export is bidirectional from two or three leaves below the shoot tip, and below this region movement is basal; and (c) after the set of fruit, photosynthate also moves apically into the cluster from leaves below the cluster, and after the rate of shoot growth decreases several weeks following set stage, photosynthate moves basally from the tip. Reprinted by permission from Ref. 136, p. 123, courtesy of California Agricultural Experiment Station.

Detailed translocation information concerning various individual herbicides may be noted in Tables 7 and 8. Additional information has been tabulated by Foy, Coats and Jones (139). Herbicides will differ in the manner in which they are translocated. They may move in either the symplast or apoplast phases of the plant. Those herbicides that are listed as

TABLE 7

Translocation Patterns of Labeled Tracers in Bean Plants[a]

(Results are averaged, hence not specific for time.)

Compound	Leaf application, translocation						Root application, translocation			
	In the leaf, via:			In bud	In stem	In roots	Sorption by roots	Transport to tops		
	Veins	Apoplast	Symplast					Hypocotyl	Epicotyl	Leaves
Alanap	+	++	tr	tr	tr	tr	++	+	+	+
Amiben	++	o	+	+	++	+	++	o	o	o
Amitrole	o	+++	++	+++	++	+	++	+	+	++
Ammonium thiocyanate	++	o	tr	tr	tr	o	++	+	+	++
Arsenate[b]	o	o	++	++	+	+	+++	++	++	+
Atrazine	+	+++	o	o	o	o	++	+	+	++
Barban	+	+++	o	o	o	o	+++	+	+	+
Dacthal	tr	o	tr	tr	tr	tr	++	+	+	tr
Dalapon	+	+	++	++	++	tr	+	++	+	tr
2, 4-D	+	o	++	+++	+++	+++	++	+++	tr	o

TABLE 7 (continued)

Compound	Leaf application, translocation						Root application, translocation			
	In the leaf, via:						Sorption by roots	Transport to tops		
	Veins	Apoplast	Symplast	In bud	In stem	In roots		Hypocotyl	Epicotyl	Leaves
2,4-DB	+	o	o	o	o	o	+++	+	+	+
Duraset	o	++	tr	+	tr	tr	++	+	+	+
Eptam	++	+	o	tr	tr	tr	++	++	++	++
2060	++	o	tr	tr	tr	tr	+++	++	++	+
IAA[c]	+	+	++	-	-	++	+++	-	-	+
Maleic hydrazide	o	++	++	++	++	++	++	+	+	tr
Monuron	o	++	o	o	o	o	++	++	++	+++
PCP	tr	o	o	o	tr	tr	++	tr	tr	o
1607	++	+	tr	tr	tr	tr	++	++	++	+
3-Cp[d]	++	+	++	+++	++	++	++	tr	tr	o
2061	++	o	tr	tr	tr	tr	+++	+++	+++	++
Simazine	o	++	o	o	o	o	++	+	+	++
Sodium acetate	+	o	+	+	+	+	+++	tr	tr	o

Compound										
Sodium benzoate	++	o	+	++	+	+	+++	tr	tr	o
2, 4, 5–T	o	o	++	++	++	+	++	tr	tr	o
2, 4, 5–TB ester	+	o	++	++	+++	+	++	tr	tr	o
TCP[e]	+	+	+	–	+	+	++	tr	tr	o
Ca^{45} [f]	++	+++	o	o	o	o	–	–	–	–
p^{32} [f]	o	o	+++	++	+++	++	–	–	–	–
Zn^{65} [f]	tr	o	++	++	+++	+	–	–	–	–

Key to symbols: tr = a trace; + = light gray; ++ = dark gray; +++ = black; – = no data.

[a]Reprinted from Ref. 137, p. 80, by courtesy of California Agricultural Experiment Station and Extension Service.

[b]Coffee

[c]Barley

[d]3-chlorophenoxy-alpha-propionic acid

[e]Trichloropropionic acid

[f]Leaf treatment only, bean

TABLE 8

Translocation in Phloem[a]

Tracer compound or ion	Chemical name	Translocation
1 Atrazine	2-Chloro-4-ethylamino-6-isopropylamino-s-triazine	In apoplast only; transport limited to nonliving structures—cell walls, including tracheary elements of the xylem
2 Calcium-45	--	
3 CIPC	Isopropyl N-(3-chlorophenyl)carbamate	
4 Diuron	3-(3,4-Dichlorophenyl)-1,1-dimethylurea	
5 Fluometuron	3-(m-Trifluoromethylphenyl)-1,1-dimethylurea	
6 IPC	Isopropyl N-phenylcarbamate	
7 Kinetin	N6-Furfuryladenine	
8 Monuron	3-(p-Chlorophenyl)-1,1-dimethylurea	
9 Simazine	2-Chloro-4,6-bis(ethylamino)-s-triazine	
10 Sodium lauryl sulfate	--	
11 Strontium-89	--	

12	T-1947	Polyoxyethylene polyol	
13	--	Tetramethylenedisulfotetramine	
14	Tween 20	Polyethylene sorbitan monolaurate	
15	Tween 80, Polysorbate 80	Polyoxyethylene (20) sorbitan mono-oleate	
16	Amiben	3-Amino-2,5-dichlorobenzoic acid	In symplast mainly, transport limited to the living system, including protoplasts of parenchymal cells; also, in the open lumina and sieve-plate pores of the sieve tubes
17	2,4-D[b]	2,4-Dichlorophenoxyacetic acid	
18	Dichlorprop, 2,4-DP[b]	2-(2,4-Dichlorophenoxy)propionic acid	
19	Fenac	2,3,6-Trichlorophenylacetic acid	
20	Maleic hydrazide, MH[c]	1,2-Dihydropyridazine-3,6-dione	
21	MCPA	2-Methyl-4-chlorophenoxyacetic acid	
22	Phosphorus-32, as phosphate	--	
23	2,4,5-T[b]	2,4,5-Trichlorophenoxyacetic acid	

TABLE 8 (continued)

Tracer compound or ion	Chemical name	Translocation
24 Amino acids (some)[d]	--	In symplast and apoplast, freely
25 Amitrole	3-Amino-1, 2, 4-triazole	
26 Cesium-134	--	
27 Dalapon[c]	2, 2-Dichloropropionic acid	
28 Dicamba[c]	2-Methoxy-3, 6-dichlorobenzoic acid	
29 Picloram[c]	4-Amino-3, 5, 6-trichloropicolinic acid	
30 Potassium-39		
31 2, 3, 6-TBA[c]	2, 3, 6-Trichlorobenzoic acid	
32 Amino acids (some)[d]	--	Limited to symplast and/ or apoplast; principal limiting factor is uptake by companion and parenchymal cells
33 Alanap, NPA	N-1-Naphthylthalamic acid	
34 Arsenic-77, as arsenate	--	

			Little or no mobility
35	Barban	4–Chloro–2–butynyl m–chlorocarbanilate	
36	Diallate, DATC	S–2, 3–Dichloroallyl N, N–diisopropylthiolcarbamate	
37	Eptam, EPTC	Ethyl N, N–dipropylthiocarbamate	
38	Gibberellin, mixed[e]	– –	
39	IAA	Indoleacetic acid	
40	Ioxynil	3, 5–Diiodo–4–hydroxybenzonitrile	
41	Iron–59	– –	
42	Magnesium–28	– –	
43	Paraquat	1, 1'Dimethyl–4, 4'–bipyridinium salt	
44	Propanil	3', 4'–Dichloropropionanilide	
45	Zinc–65	– –	
46	Dacthal, DCPA	Dimethyl 2, 3, 5, 6–tetrachloroterephthalate	
47	2, 4–DB	4–(2, 4–Dichlorophenoxy)butyric acid	
48	DNBP	4, 6–Dinitro–o–sec–butylphenol	

TABLE 8 (continued)

Tracer compound or ion	Chemical name	Translocation
49 Dinitrocresol	4,6-Dinitro-o-cresol	
50 Diquat	6,7-Dihydrodipyrido[1,2-a:2',1'-c]pyrazidiinium salt	
51 Endothall	7-Oxabicyclo[2.2.1]heptane-2,3-dicarboxylic acid	
52 PCP	Pentachlorophenol	

[a]Reprinted from Ref. 138, p. 737, by courtesy of Federation of American Societies for Experimental Biology.

[b]Mobility limited by uptake by living cells en route.

[c]Leaks from roots. Sodium, rubidium, and cesium ions also leak from roots.

[d]Details on amino acid translocation available in Ref. 137.

[e]Synthesized by a culture of Fusarium moniliforme.

moving only in the apoplast may have limited movement in the symplast but since they have a high affinity for the apoplast they are rapidly partitioned out of the symplast and back into the apoplast.

There is a definite relationship between molecular configuration of organic compounds applied to leaves of plants and enhanced uptake and translocation. Mitchell et al. (140, 141) found that by substituting a lactic acid group for the isopropyl group of isopropyl-N-phenylcarbamate and 3-chloro-isopropyl-N-phenylcarbamate they were able to increase absorption into the plant and change the translocation pattern from apoplastic movement to apparent symplastic movement. It should be noted that structurally, these compounds have a carbon atom which is in an alpha position with respect to a carboxyl group and which is associated with a hydrogen and a methyl or methoxyl group.

Symplastic movement is usually greatest following foliar applications, while apoplastic movement is of greatest importance when herbicides are applied to the soil. Some herbicides may "leak" from roots into the soil where they may be re-absorbed again by the plant. Those herbicides that possess the ability to move in either the symplast or apoplast should have the greatest use potential. It is of utmost importance to understand their translocation potential if they are to be used intelligently as herbicides.

REFERENCES

1. W. Franke, Ann. Rev. Plant Physiol., 18, 281 (1967).

2. W. H. Orgell, Ph.D. Thesis, University of California, Davis, 1954.

3. C. E. Crisp, Ph.D. Thesis, University of California, Davis, 1965.

4. A. S. Crafts and C. L. Foy, Residue Rev., 1, 112 (1962).

5. J. van Overbeek, Ann. Rev. Plant Physiol., 7, 355 (1956).

6. F. M. Scott, Nature, 210, 1015 (1966).

7. R. F. Norris and M. J. Bukovac, Am. J. Botany, 55, 975 (1968).

8. Y. Yamada, H. P. Rasmussen, M. J. Bukovac, and S. H. Wittwer, Am. J. Botany, 53, 170 (1966).

9. G. S. Hartley, 8th British Weed Control Conf. Proc., 3, 794 (1966).

10. D. A. Fisher, C. E. Crisp, and D. E. Bayer, unpublished work, 1969.

11. F. M. Scott, K. C. Hammer, and E. Baker, Science, 125, 399 (1957).

12. R. H. Schieferstein and W. E. Loomis, Am. J. Botany, 46, 625 (1959).

13. L. E. Mueller, P. H. Carr, and W. H. Loomis, Am. J. Botany, 41, 593 (1954).

14. H. F. Linskens, Planta, 41, 40 (1952).

15. J. A. Sargent, Ann. Rev. Plant Physiol., 16, 1 (1965).

16. F. Radler and D. H. S. Horn, Austr. J. Chem., 18, 1059 (1965).

17. R. F. Norris and M. J. Bukovac, Physiol. Plantarum, 22, 701 (1969).

18. G. Egglington and R. J. Hamilton, Science, 156, 1322 (1967).

19. A. DeBary, Botan. Ztg., 14, 128 (1871).

20. B. E. Juniper, Endeavour, 18, 20 (1959).

21. B. E. Juniper, J. Linn. Soc. (Bot.), 56, 413 (1960).

22. D. R. Kreger and C. Schamhart, Biochim. Biophys. Acta, 19, 22 (1956).

23. N. K. Adam and G. E. P. Elliott, J. Chem. Soc. (London), 1962, 2206 (1962).

24. P. J. Holloway, J. Sci. Food Agr., 20, 124 (1969).

25. A. E. Dimond, Mod. Methoden Pflanzenalyse, 5, 368 (1962).

26. G. E. Fogg, Discussions Faraday Soc., 3, 162 (1948).

27. H. F. Linskens, Planta, 38, 591 (1951).

28. C. G. L. Furmidge, J. Sci. Food Agr., 13, 127 (1962).

29.   G. S. Hartley and R. T. Brunskill, in Surface Phenomena in Chemistry and Biology (J. F. Danielli, K. G. A. Pankhurst, and A. C. Riddiford, eds.) Pergamon Press Ltd., 1958, pp. 214-223.

30.   R. T. Brunskill, British Weed Control Conf. Rep., D11, 1956, p. 593.

31.   A. C. Evans and H. Martin, J. Pomol., 13, 261 (1935).

32.   C. C. Thompson, J. Sci. Food Agr., 9, 650 (1958).

33.   N. K. Adam, Nature, 182, 809 (1957).

34.   C. G. L. Furmidge, J. Sc. Food Agr., 16, 134 (1965).

35.   R. N. Wenzel, Ind. Eng. Chem. (Indust.), 28, 988 (1946).

36.   J. W. Mitchell, B. C. Smale, and R. L. Metcalf, Adv. in Pest Cont. Res., 3, 359 (1960).

37.   J. W. Mitchell and P. J. Linder, Residue Rev., 2, 51 (1963).

38.   V. H. Freed and J. M. Witt, Adv. in Chem., No. 86, 1969, A. C. S., Washington, D. C.

39.   H. B. Currier and C. D. Dybing, Weeds, 7, 195 (1959).

40.   W. A. Darlington and J. B. Barry, J. Agr. Food Chem., 13, 76 (1965).

41.   R. N. Goodman, Adv. in Pest Cont. Res., 5, 1 (1962).

42.   L. L. Jansen, Weeds, 9, 381 (1961).

43.   J. F. Parr and A. G. Norman, Botan. Gaz., 126, 86 (1965).

44.   C. L. Foy and L. W. Smith, Weed Soc. Am. Abstr., 1964, p. 77.

45.   V. H. Freed and M. Montgomery, Weeds, 6, 386 (1958).

46.   L. L. Jansen, in Plant Growth Regulation, Fourth Intern. Conf. on Plant Growth Regulation, Iowa State University Press, Ames, Iowa, 1961, pp. 813-816.

47.   L. L. Jansen, W. A. Gentner, and W. C. Shaw, Weeds, 9, 381 (1961).

48.   B. E. Temple and H. W. Hilton, Weeds, 11, 297 (1963).

49.   C. G. L. Furmidge, J. Sci. Food Agr., 10, 274 (1959).

50.   J. F. Parr and A. G. Norman, Plant Physiol., 39, 502 (1964).

51.   B. B. Stowe, Plant Physiol., 35, 262 (1960).

52.   H. M. Hull, in Absorption and Translocation of Organic Substances in Plants (J. Hacskaylo, ed.), 7th Annual Symposium, Southern Sect., Am. Soc. Plant Physiol., Emory University, 1964, pp. 45-93.

53.   V. H. Legg and R. V. Wheeler, J. Chem. Soc. (London), 131, 2444 (1929).

54.   E. A. Baker and J. T. Martin, Nature, 199, 1268 (1963).

55.   M. Matic, South African Ind. Chem., December 1956, pp. 1-2.

56.   H. F. Linskens, W. Heinen, and A. L. Stoffers, Residue Rev., 8, 136 (1965).

57.   D. B. Anderson, Ohio J. Sci., 34, 9 (1934).

58.   R. Bolliger, J. Ultrastruct. Res., 3, 105 (1959).

59.   P. Lambertz, Planta, 44, 147 (1954).

60.   P. E. Kolattukudy, Science, 159, 498 (1968).

61.   W. Heiner and I. V. D. Brand, Z. Naturforsch, 18, 67 (1963).

62.   P. Albersheim and V. Killias, Am. J. Botany, 50, 732 (1963).

63.   P. K. Biswas and M. N. Rogers, The Tuskegee Veterinarian, Fall Ed., 1962, pp. 61-63.

64.   R. N. Goodman and S. K. Addy, Phytopathologische Z., 46, 1 (1962).

65.   M. F. Roberts, R. F. Batt, and J. T. Martin, Ann. Appl. Biol., 47, 573 (1959).

66.   Y. Yamada, S. H. Wittwer, and M. J. Bukovac, Plant Physiol.,
       39, 28 (1964).

67.   S. Kamimura and R. N. Goodman, Physiol. Plantarum, 17, 805
       (1964).

68.   S. Kamimura and R. N. Goodman, Phytopathologische Z., 51, 324
       (1964).

69.   S. C. Fang, Weeds, 6, 179 (1958).

70.   L. W. Smith, D. E. Bayer, and C. L. Foy, Weed Sci., 16, 523
       (1968).

71.   Y. Yamada, M. J. Bukovac, and S. H. Wittwer, Plant Physiol.,
       39, 978 (1964).

72.   Y. Yamada, S. H. Wittwer, and M. J. Bukovac, Plant Physiol.,
       40, 170 (1965).

73.   S. Kannan, Plant Physiol., 44, 517 (1969).

74.   W. A. Darlington and N. Cirulis, Plant Physiol., 38, 462 (1963).

75.   Y. Yamada, Ph. D. Thesis, Kyoto University, Japan, 1962.

76.   W. H. Jyung and S. H. Wittwer, Agr. Sci. Rev., 3, 26 (1965).

77.   A. Frey-Wyssling and K. Muhlethaler, Ultrastructural Plant
       Cytology, Elsevier Publishing Company, Amsterdam, 1965.

78.   P. A. Roelofsen, Handbuch Pflanzenanatomie, 3, 1 (1959).

79.   H. Burström, Kgl. Fysiograf. Sällskap. Lund Förh., 28, 53 (1958).

80.   K. Muhlethaler, Ann. Rev. Plant Physiol., 18, 1 (1967).

81.   D. H. Northcote and D. R. Lewis, J. Cell Sci., 3, 199 (1968).

82.   P. A. Roelofsen, Acta Botan. Neerl., 1, 99 (1952).

83.   R. Collander, in Plant Physiology: a Treatise (F. C. Steward, ed.),
       Vol. 2, Academic Press, New York, 1959, pp. 3-102.

84.   J. A. Sacher, in <u>Absorption and Translocation of Organic Substances</u> <u>in Plants</u> (J. Hacskaylo, ed.), 7th Annual Symposium, Southern Sect., Am. Soc. Plant Physiol., Emory University, 1964, pp. 29-44.

85.   A. S. Crafts, <u>The Chemistry and Mode of Action of Herbicides</u>, Interscience Publishers, New York, 1961.

86.   D. M. Hall, <u>J. Ultrastruct. Res.</u>, <u>17</u>, 34 (1967).

87.   D. M. Hall and L. A. Donaldson, <u>Nature</u>, <u>194</u>, 1196 (1962).

88.   W. Franke, <u>Planta</u>, <u>55</u>, 390 (1960).

89.   W. Schumacher and W. Halbsguth, <u>Jahrb. Wiss. Bot.</u>, <u>87</u>, 324 (1939).

90.   K. Rudolph, <u>Botan. Arch.</u>, <u>9</u>, 49 (1925).

91.   W. Franke, <u>Am. J. Botany</u>, <u>48</u>, 683 (1961).

92.   W. Franke, <u>Am. J. Botany</u>, <u>56</u>, 432 (1969).

93.   E. A. Roberts, M. D. Southwick, and D. H. Palmiter, <u>Plant Physiol.</u>, <u>23</u>, 557 (1948).

94.   D. H. Palmiter, E. A. Roberts, and M. D. Southwick, <u>Phytopath.</u>, <u>36</u>, 681 (1946).

95.   J. F. Danielli and H. Davson, <u>J. Cellular Comp. Physiol.</u>, <u>5</u>, 495 (1935).

96.   J. F. Danielli, <u>Collston Papers</u>, <u>7</u>, 1 (1954).

97.   J. D. Robertson, <u>Biochem. Soc. Symp.</u>, <u>16</u>, 3 (1959).

98.   J. D. Robertson, <u>Cellular Membranes in Development</u> (M. Lock, ed.), Academic Press, New York, 1964.

99.   J. D. Robertson, <u>J. Cell Biol.</u>, <u>19</u>, 201 (1963).

100. D. Branton and H. Moor, <u>J. Ultrastruct. Res.</u>, <u>11</u>, 401 (1964).

101. E. J. DuPraw, <u>Cell and Molecular Biology</u>, Academic Press, New York, 1968.

102.   E. D. Korn, Federation Proc., 28, 6 (1969).

103.   D. Branton, Ann. Rev. Plant Physiol., 20, 209 (1969).

104.   E. J. Stadelmann, Ann. Rev. Plant Physiol., 20, 585 (1969).

105.   V. Wartiovaara, Physiol. Plantarum, 3, 462 (1950).

106.   G. G. Laties, Ann. Rev. Plant Physiol., 10, 87 (1959).

107.   R. J. Poole and K. V. Thimann, Plant Physiol., 39, 98 (1964).

108.   R. Prasad and G. E. Blackman, J. Exptl. Botany, 16, 545 (1965).

109.   C. L. Foy and S. Yamaguchi, in Absorption and Translocation of
       Organic Substances in Plants (J. Hacskaylo, ed.), 7th Annual
       Symposium, Southern Sect., Am. Soc. Plant Physiol., Emory
       Univ., 1964, pp. 3-28.

110.   P. F. Saunders, C. F. Jenner, and G. E. Blackman, J. Exptl.
       Botany, 16, 683 (1965).

111.   M. A. Venis and G. E. Blackman, J. Exptl. Botany, 17, 270
       (1966).

112.   P. F. Saunders, C. F. Jenner, and G. E. Blackman, J. Exptl.
       Botany, 16, 697 (1965).

113.   P. F. Saunders, C. F. Jenner, and G. E. Blackman, J. Exptl.
       Botany, 17, 241 (1966).

114.   M. A. Venis and G. E. Blackman, J. Exptl. Botany, 17, 771 (1966).

115.   M. A. Venis and G. E. Blackman, J. Exptl. Botany, 17, 790 (1966).

116.   V. P. Cirillo, Ann. Rev. of Microbiology, 15, 197 (1961).

117.   T. W. Donaldson, Ph.D. Thesis, University of California, Davis,
       1967.

118.   C. F. Jenner, P. F. Saunders, and G. E. Blackman, J. Exptl.
       Botany, 19, 333 (1968).

119.   S. Yamaguchi, Hilgardia, 36, 349 (1965).

120.   E. G. Maskell and T. G. Mason, Ann. Botany, 43, 615 (1929).

121.   E. Münch, Die Stoffbewegungen in der Pflanze, Jena, Gustav
        Fischer, 1930.

122.   L. Horwitz, Plant Physiol., 33, 81 (1958).

123.   A. S. Crafts, Plant Physiol., 6, 1 (1931).

124.   A. S. Crafts, Plant Physiol., 7, 183 (1932).

125.   M. J. Canny, Biol. Rev. Cambridge Phil. Soc., 35, 507 (1960).

126.   A. S. Crafts and C. E. Crisp, Phloem Transport in Plants,
        W. H. Freeman, San Francisco, 1971.

127.   K. Esau, The Phloem, Gebrüder Borntraeger, Berlin, 1969.

128.   J. Happel and H. Brenner, Low Reynolds Number Hydrodynamics:
        With Special Applications to Particulate Media, Prentice-Hall,
        Englewood Cliffs, N. J., 1965.

129.   C. A. Peterson and H. B. Currier, Physiol. Plantarum, 22,
        1238 (1969).

130.   M. Maestri, Ph. D. Thesis, University of California, Davis, 1967.

131.   R. L. Bieleski, Plant Physiol., 41, 455 (1966).

132.   R. K. Glenn, M. S. Thesis, University of California, Davis, 1971.

133.   E. C. S. Little and G. E. Blackman, New Phytol., 62, 173 (1963).

134.   J. L. Key, Ann. Rev. Plant Physiol., 20, 449 (1969).

135.   O. A. Leonard and R. J. Weaver, Hilgardia, 31, 327 (1961).

136.   C. R. Hale and R. J. Weaver, Hilgardia, 33, 89 (1962).

137.   A. S. Crafts and S. Yamaguchi, The Autoradiography of Plant
        Materials, California Agricultural Experiment Station and Exten-
        sion Service, Manual 35, 1964.

138. A. S. Crafts, in Respiration and Circulation (P. L. Altman and
      D. S. Dittmer, eds.), Federation of American Societies for Ex-
      perimental Biology, Bethesda, Maryland, 1970, pp. 737-738.

139. C. L. Foy, G. E. Coats, and D. W. Jones, in Respiration and
      Circulation (P. L. Altman and D. S. Dittmer, eds.), Federation
      of American Societies for Experimental Biology, Bethesda, Mary-
      land, 1970, pp. 742-791.

140. J. W. Mitchell, P. C. Marth, and W. H. Preston, Jr., Science,
      120, 263 (1954).

141. J. W. Mitchell, I. R. Schneider, and H. G. Gauch, Science, 131,
      1863 (1960).

Chapter 10

ADSORPTION, MOVEMENT, AND DISTRIBUTION
OF PESTICIDES IN SOILS*

V. H. Freed and R. Haque

Department of Agricultural Chemistry and
Environmental Health Science Center
Oregon State University
Corvallis, Oregon

I. INTRODUCTION

When a pesticide is released into the environment for pest control the majority of it comes in contact with the soil surface. The physical–chemical behavior of the pesticide in relation to the environment will play an important role in determining its fate, persistence, and biological activity. The biological availability of a pesticide and hence its effects will depend upon many chemical processes. Among these are adsorption by the surfaces of the solid constituents of the soil, leaching and diffusion in the soil matrix, and the degradation of pesticides. This chapter describes the role of the above processes in relation to the behavior of pesticides in soils. No attempt will be made to review all the papers which have appeared in the literature. Much of the work described here was carried out in our own laboratory, and will be discussed in relation to the important results obtained by other workers.

*Data presented here is based on research supported in part by Grants
ES–00040 and ES–00210 from N. I. E. H. S.

II. ADSORPTION OF PESTICIDES

Soil is a heterogeneous mixture consisting of clays, organic matter, sand and inorganic salts. The extent of the surface area of a soil always depends upon the ratio of the constituents in the soil. For example, a soil high in organic matter usually has a larger surface area. There are always residual forces on the surface of the solid constituents of soil and hence the chemical is adsorbed when it comes into contact with the soil. Although most of the adsorption of the chemical occurs from solution, adsorption to some extent also occurs from the vapor phase. At equilibrium, probably an idealized concept in the field, the adsorption of a pesticide (P) on a soil surface (S) can be represented by Eq. (1). During adsorption from aqueous solution, both the pesticide as well as a soil-matrix is usually hydrated to a certain degree depending

$$P(H_2O)_x + S(H_2O)_y \rightleftharpoons P(H_2O)_z S \qquad (1)$$

on the hydration numbers x and y. The hydration number of the adsorbed species $P(H_2O)_zS$ is designated as z. When a chemical is in direct contact with the soil surface, z usually will approach zero. The equilibrium constant K_E for the above process can be represented:

$$K_E = \frac{[P(H_2O)_z S]}{[P(H_2O)_x][S(H_2O)_4]} \qquad (2)$$

Here the quantities in brackets represents the activities. An exact determination of the K_E is not possible since it is difficult to estimate the exact volume occupied by the adsorbed species.

In general, the most convenient way of representing adsorption data for soil-pesticide interaction is with an isotherm. The most common isotherms are the Freundlich; Langmuir; and Brunauer, Emmett, and Taylor (B. E. T.) isotherms. The different types of isotherms are discussed below.

A Freundlich type isotherm can be represented by Eq. (3) where X is the amount of chemical sorbed, m is the mass of the solid on whose surface the sorption occurs, C the equilibrium concentration of the chemical, and K and n are constants. This isotherm has been found to apply widely to the adsorption of pesticides by soils.

$$\frac{X}{m} = K \cdot C^n \qquad (3)$$

The Langmuir type isotherm, which assumes that the adsorption is of a monolayer type and one molecule of chemical is available for each site on the surface, can be expressed by the following expression:

$$\frac{X}{m} = \frac{abC}{1 + aC} \tag{4}$$

Here a and b are constants. The constant b represents the capacity of the adsorbant for a particular adsorbate. This equation has rarely been used for soil-pesticide adsorption. The basic reasons are the complex nature of the soil and the formation of a multilayer during adsorption. The adsorption of isopropyl N-(3-chlorophenyl) carbamate herbicide, CIPC, on activated carbon has been expressed by a Langmuir isotherm (1).

The B. E. T. isotherm (5) which has found important use in adsorption of gases on surfaces has also rarely been used for soil-pesticide adsorption.

$$\frac{C}{(C_0 - C)X} = \frac{Km - 1}{b \, Km} \, \frac{C}{C_0} + \frac{1}{b \, KM} \tag{5}$$

Here C_0 is the maximum amount of chemical available and Km is a constant related to the molar heat of adsorption for the monolayer, Q_1, and molar heat of crystalization, Q_L, by Eq. (6)

$$Km = \frac{Q_1 - Q_L}{RT} \tag{6}$$

Since the Freundlich type isotherm has been applied in the majority of pesticide-soil adsorption studies, we shall discuss it in some detail. Upon taking the logarithm of Eq. (3) one obtains the expression

$$\log \frac{X}{m} = n \log C + \log K \tag{6a}$$

According to Eq. (6a) one should expect a linear relationship between $\log X/m$ and $\log C$. The slope, n, and intercept, K, throw much light on the nature of the adsorption. For dilute solutions of nonpolar substances the value of n usually ranges in the neighborhood of unity. The magnitude of n usually indicates the nature of the adsorption. For values of n greater than 1 the adsorption is usually referred to as S or concave type, whereas for values less than unity it is L or convex type. The adsorption of isocil and bromacil on several mineral surfaces (2) has recently been discussed in terms of L and S type adsorption. Adsorption on montmorillonite and kaolinite surfaces gave values of n greater than unity indicating an S type isotherm. An S type isotherm may arise from strong absorption of the solvent, strong intermolecular attraction within the adsorbed layers, penetration of solute into the adsorbent, and mono-functional nature of the adsorbate. The tendency of these two surfaces to absorb water solvent molecules may be the major factor responsible for the concave nature of the adsorption isotherm.

Sorption by humic acid gave a value of n which was less than unity indicating a convex or L type isotherm. This kind of isotherm is attributed to surfaces having a lower affinity for the solvent. Since humic acid absorbs very little water the results obtained seem reasonable. On the other hand, the value of n for silica gel and illite surfaces ranges between 0.8 to 1.0 indicating an intermediate behavior.

The intercept, log K, of the Freundlich plot is usually an indication of the extent of the binding. It was noted in the early studies of Sherburne and Freed (3) that organic-matter-rich soil showed higher adsorption of 3(p-chlorophenyl)-1,1 dimethyl urea than clay or silt loam soil. The value of K is usually much higher for high organic matter soil. This is demonstrated for the adsorption of 2,4 dichlorophenoxyacetic acid, 2,4-D, as well as isocil and bromacil on many surfaces (Table 1). Adsorption data have also been correlated with the partition coefficient of a pesticide between water and cyclohexane (5), parachor (6), and Hammett function (7). Lambert (8) has introduced a term, Ω, to relate the sorption equilibria to the organic matter of a soil. This semiempirical equation is shown in Eq. (7), where X represents the amount of indicator

$$K_P = \frac{X}{\Omega C} \left(\frac{\text{grams of soil}}{100} \right) \tag{7}$$

chemical sorbed. C is the equilibrium concentration and K_P is the equilibrium constant for the same chemical using the standard soil. This equation produced a satisfactory correlation between percent organic matter and the Ω values.

The pH of a solution or soil also has a significant effect on the adsorption. Since many organic pesticides are not very stable at either high or low pH values, the pH of the soil or solution will greatly influence the adsorption characteristics. Furthermore a change in pH would also influence the nature of the soil surface. The effect of pH on adsorption of several s triazine herbicides on montmorillonite surface was demonstrated by Weber (9). It was noticed that maximum adsorption of prometone, a methoxy s triazine, occurred in the vicinity of its pK value (9). Formation of protonated species of prometone was suggested.

TABLE 1

Freundlich Isotherm Constants K and n for the Adsorption of
2, 4D, Isocil and Bromacil on some Surfaces

System	pH	Temp, °C	Slope n	Intercept log K
Illite-2, 4D	6-7	0	0.658	1.11
		25	0.719	1.02
Montmorillonite-2, 4D	6-7	0	0.925	- .63
		25	1.004	- .186
Silica gel-2, 4D	6-7	0	0.9	0.58
		25	0.95	0.11
Sand-2, 4D	6-7	0	0.671	- .984
		25	0.827	1.45
Humic acid-2, 4D	6-7	0	0.86	0.931
		25	2.01	1.9
Alumina, 2-4D	6-7	0	0.97	- .06
		25	1.01	- .08
Silica gel-isocil	5.2	0	1.0	0.58
		25	1.0	0.31
Silica gel-isocil	3.5	0	0.8	0.97
		25	1.0	0.31
Silica gel-bromacil	4.6	0	0.8	1.03
		25	0.8	0.68
Illite-isocil	7.4	0	0.8	0.79
		25	0.9	0.3
Illite-bromacil	7.0	0	0.9	0.48
		25	0.9	0.42
Montmorillonite-isocil	6-7	0	1.1	0.49
		25	1.2	0.1
Montmorillonite-bromacil	6-7	0	1.3	-0.03
		25	1.3	-0.14
Kaolinite-isocil	6-7	0	1.1	-0.36
		25	1.2	-1.07
Humic acid-isocil	6-7	0	0.7	2.19
		25	0.7	2.05
Humic acid-bromacil	6-7	0	0.7	2.1
		25	0.7	1

In soil systems adsorption also depends on the nature of the adsorbate. Usually inorganic cations adsorb on the clay portion of the soil through an ion-exchange mechanism. Many organic cations (such as diquat or paraquat) are also adsorbed on a soil surface through an ion-exchange process (10). However, most neutral organic pesticide molecules follow a physical-type adsorption. As a result, solubility parameters of the adsorbate may occasionally relate to the extent and strength of binding. Usually, for neutral organic molecules, there is an inverse relation between a function of the solubility and the extent and strength of binding (11).

Relatively little work has been done on the temperature effects in relation to pesticide adsorption, although processes involving pesticides are exothermic and the amount of pesticide adsorbed usually increases with a lowering of temperature. However, in many systems, temperature has very little effect on the amount of chemical sorbed (12). For most of the pesticides, the heat of adsorption is very low, in the range of only a few kcal/mole. A low value of heat of adsorption (<10 kcal/mole) indicates a physical type of adsorption, whereas, for chemisorption processes, this value is high (>10 kcal/mole). The most common experience with the neutral organic pesticides is physical type adsorption. Chemisorption, indicating formation of a chemical bond in such a system, has rarely been observed except at extremes of pH if an ionic species can be found.

In a soil-pesticide system, there is first a formation of a monolayer on the surface and then the build-up of multilayers. Using the analogy of the adsorption of gases (13) on solids, the heat of adsorption for a pesticide from aqueous solution obtained for monolayer should be in the range of the heat of solution. For the adsorption of 2,4-D and DNSBP pesticides on charcoal, the heat of adsorption was about 4 kcal/mole, which is comparable to their heat of solution values (14). The solubility of thiolcarbamates decreases (15) with an increase in temperature, hence one should expect a decrease in adsorption with the lowering of temperature.

For the adsorption of vapors on solids, an isosteric heat of adsorption (16), which depends upon the amount of chemical sorbed, has been used to describe the strength of binding. The isosteric heat of adsorption, ΔH, can be expressed with Eq. (18), where C is the equilibrium concentration

$$\Delta H = R \left(\frac{\partial \ln C}{\partial \frac{1}{T}} \right)_x \tag{8}$$

of the pesticide at a fixed amount of pesticide sorbed, x, T the absolute temperature, and R the gas constant. This expression requires that the chemical potential in solution, $\mu_1{}^0$, is the same as in the adsorbed state, $\mu_1{}^0$, and where a_1 and a_s are the activities of the adsorbate in free and

$$\mu_1{}^0 + RT \ln a_1 = \mu_s{}^0 + RT \ln a_s \tag{9}$$

adsorbed states respectively as shown in Eq. (9). For the adsorption of 2, 4-D (4), isocil (2), and bromacil (2) on surfaces such as illite, montmorillonite, kaolinite, silica gel, sand and humic acid, the isosteric heat of adsorption was calculated as a function of chemical sorbed. The ΔH values varied exponentially with the amount of chemical sorbed and tended to become more positive with increasing amounts of sorbed chemical. Sand and silica gel surfaces always gave higher value of ΔH indicating a stronger binding.

For the systems where ΔH of absorption is large, a hydrogen bonding is probably involved in the proposed mechanism of adsorption. The positive value of the heat of adsorption has been explained on the basis that, during the adsorption, some solvent molecules are displaced from the surface. This increases the translational degrees of freedom and consequently increases the entropy change in such a manner that $T\Delta S > \Delta H$, thereby resulting in a negative free energy change. Such a mechanism is proposed for the positive ΔH value obtained for the adsorption of polyvinyl acetate on iron powder (17).

Most adsorption processes involving pesticides and soil take only a few hours to attain 70-80% of the adsorption; the total completion of it takes a much longer time, sometimes days to weeks. First order kinetics are not usually applicable to the pesticide-soil adsorption system. Kinetics of adsorption of many pesticides has been explained on the basis of a diffusion controlled process (18). The adsorption of 2, 4-D on clay surfaces has been discussed in light of Eley's theory (19), which takes into account the changes in the free energies of activation and surface coverage. The equation proposed by Fava and Eyring (20), which discusses both the adsorption and desorption process, has also been applied to 2, 4-D adsorption.

A general equation describing the adsorption as well as desorption of pesticide on surfaces has been derived recently (21). This equation can be described as

$$\frac{d\phi}{dt} = k_1 \left\{ (1 - \phi) \left(1 - \frac{\phi}{2} \right) e^{-b\phi} \right\} + k_2 \left\{ \left(1 - \frac{\phi}{2} \right) e^{b(2-\phi)} - \frac{\phi^2}{2} e^{b\phi} \right\} \tag{10}$$

In this equation $k_1{}'$ and $k_2{}'$ are the rate constants for the adsorption and desorption process respectively, ϕ, the fraction of chemical sorbed at a certain time and, b, a constant. The equation takes into account the desorption of chemical as well as the probability factor. This model has the main advantage in that the parameters for both the adsorption as well as the desorption process can be determined by performing only an adsorption experiment. The rate constants and activation energy values for the adsorption and desorption processes for 2,4-D, isocil, and bromacil on many surfaces are given in Table 2. As expected, k_1' is always greater than k_2'. Similarly the energy of activation for desorption is always greater than for the corresponding adsorption process. This is reasonable since a chemical has to cross a higher energy barrier for the desorption process than for the adsorption.

As indicated previously, qualitative information about the mechanism of adsorption can be obtained from the magnitude of the heat of adsorption data. Spectroscopic techniques such as infrared or ultraviolet can give details about the mechanism of adsorption. Infrared spectroscopic techniques have been used extensively in elucidation of the adsorption mechanism for s triazine herbicides. The protonation of triazine on clay surfaces has been demonstrated by the observed changes in the infrared vibrational frequencies of the triazine (22). The mechanism of adsorption of EPTC and aminotriazole herbicides on montmorillonite surfaces has also been studied by infrared techniques (23). Diquat and paraquat, being quaternary pyridinum cations, adsorbed readily on montmorillonite surfaces through an ion-exchange process. Infrared and ultraviolet studies have shown that diquat and paraquat form charge transfer complexes on montmorillonite surfaces (24). The uv spectrum of diquat has two peaks at 309 and 318 mμ, whereas paraquat has a uv maximum at 258 mμ. The addition of montmorillonite to either solution produced a change in the adsorption maxima of these compounds towards longer wave lengths. This shift was about 10 mμ for diquat and 20 mμ for paraquat. These shifts were explained on the basis of a charge transfer complex formation between the montmorillonite and the cation diquat or paraquat. Diquat and paraquat also form charge transfer complexes with halide ions. It may be possible that due to lattice substitutions, montmorillonite donates electron(s) to cause interaction with diquat or paraquat, resulting in a charge transfer complex. Infrared (ir) spectroscopic studies also show such a complex formation. The ir spectrum of diquat or paraquat show changes in the vibrational frequencies when the anion is replaced by Cl^-, Br^-, I^- or montmorillonite. The out-of-plane C-H vibrational frequency (commonly known as an umbrella mode) gives peaks at 854 cm^{-1} for diquat and 793 cm^{-1} for paraquat. However, on montmorillonite surfaces, these peaks shift to 834 and 782 cm^{-1} for diquat and paraquat respectively.

TABLE 2(A)

Kinetic Parameters for the Adsorption Processes

System	Temp, °K	Rate constant k_1 (sec^{-1})	Activation energy ΔE kcal/mole	Enthalpy of activation ΔH kcal/mole	Entropy of activation ΔS esu	Free energy of activation ΔG kcal/mole	Constant b
Silica gel	273	5.9×10^{-2}		5.5	-43.8	17.4	4.2
2, 4-D (10 ppm)	298	1.5×10^{-1}	6.0	5.4	-43.5	18.4	2.2
Silica gel Isocil	273	2.7×10^{-4}		5.5	-54.5	20.4	1.9
(100 ppm)	298	7.0×10^{-4}	6.0	5.4	-54.0	21.5	1.2
Silica gel Bromacil	273	2.0×10^{-5}		7.4	-54.0	22.0	4.0
(300 ppm)	298	6.8×10^{-5}	7.9	7.3	-52.0	22.7	3.8
Illite	273	1.5×10^{-2}		2.8	-56.5	18.2	4.4
2, 4-D (10 ppm)	298	2.5×10^{-2}	3.3	2.7	-56.5	19.5	3.5
Illite Isocil	273	1.1×10^{-2}		-0.54	-69.0	18.3	0.4
(100 ppm)	298	1.1×10^{-2}	0	-0.55	-70.0	20.3	0.4
Illite Bromacil	273	2.0×10^{-2}		0.95	-62.5	18.0	1.7
(300 ppm)	298	2.5×10^{-2}	1.5	0.94	-63.0	19.7	1.4

TABLE 2(B)

Kinetic Parameters for the Desorption Processes

System	Temp, °K	Rate constant k_2 (sec^{-1})	Activation energy ΔE kcal/mole	Enthalpy of activation ΔH kcal/mole	Entropy of activation ΔS esu	Free energy of activation ΔG kcal/mole	Constant b
Silica gel 2,4-D (10 ppm)	273	4.0×10^{-6}	7.0	6.5	-61	23.1	4.2
	298	1.2×10^{-5}		6.4	-62	24.9	2.2
Silica gel Isocil (100 ppm)	273	2.2×10^{-7}	8.5	8.0	-59.4	24.2	1.9
	298	8.2×10^{-7}		7.9	-58.5	25.3	1.2
Silica gel Bromacil (300 ppm)	273	8.6×10^{-8}	15.3	14.8	-36.4	24.7	4.0
	298	9.2×10^{-7}		14.7	-38.4	26.3	3.8
Illite 2,4-D (10 ppm)	273	4.5×10^{-7}	5.6	5.1	-68.7	23.9	4.4
	298	1.1×10^{-6}		5.0	-68.2	25.3	3.5
Illite Isocil (100 ppm)	273	1.0×10^{-3}	0	-0.54	-74.6	19.8	0.4
	298	1.0×10^{-3}		-0.55	-74.8	21.8	0.4
Illite Bromacil (300 ppm)	273	1.0×10^{-4}	4.5	4.0	-62.2	21.0	1.7
	298	2.0×10^{-4}		3.9	-62.0	23.4	1.4

III. MOVEMENT, LEACHING, AND DIFFUSION

The movement of a pesticide in a soil matrix with water, commonly known as leaching, is an important factor in relation to the pesticide's biological activity. In a three dimensional coordinate system leaching may occur in a downward, lateral or upward direction. The downward movement is the one of major interest, although leaching in the other two directions is also sometimes important. In practice the important factors controlling leaching are adsorption, water solubility, soil type, and amounts of moisture and percolation velocity. The downward leaching of a chemical accompanies the movement of water. Water arriving at the soil surface first dissolves the chemical and carries the chemical with it as it passes through the soil. The situation in the soil is anologous to column chromatography. Under rapid water percolation, the bulk movement of a chemical will be in the direction of the water flow. During the leaching process there is always a dynamic equilibrium between the free chemical and the chemical in the adsorbed state. Consequently, a tightly-bound chemical should leach slowly and vice-versa. It was stated earlier that the heat of solution can give a qualitative indication of the strength of binding. The qualitative relation between leaching and the heat of solution of a chemical is indicated by the information in Table 3. A highly soluble chemical, if not tightly adsorbed, will be leached more easily because of its tendency to go into solution. Temperature, which affects the solubility, will also play an important role. The intensity and frequency of moisture or rainfall will markedly affect the distribution of a chemical. Since the adsorption process depends to a great extent on the nature of the soil, the type of soil will also influence the leaching. Under static conditions of soil moisture or where the percolation rate is very slow, diffusion will become an important factor in determining the distribution of the chemical in the soil profile, at least over short distances.

The process of leaching may be considered to be analogous to chromatography, and consequently one should expect a maximum concentration of a chemical at some point of the soil profile. The chromatographic approach has been described for the leaching of many chemicals (25). The mobility of many chemicals using a film of soil as a thin layer chromatography has also been measured (26).

The theoretical treatment of leaching has lead to the development of mathematical models of process (27). The basic equation in building all the models for leaching is the well known Fick's Law. Recently, Lindstrom, et al. (28) have treated the leaching of herbicides theoretically and, using the methods of Lapidus and Amundson (29) and of Brenner (30), have

TABLE 3

Heat of Solution, Adsorption Characteristic, and
Leaching of Some Herbicides

Chemical	Heat of solution kcal/mole	Leaching	Adsorption on soil
2, 3, 6 TBA	~1.6	Readily	Small
Fenac	~6.7	Resistant	Large
Amiben	~2.8	Readily	Intermediate
2, 4-D	~6.1	Intermediate	Large
2, 4, 5-T	~8.9	Resistant	Large
Simazine	~9.0	Resistant	Large
Fenuron	~3.9	Intermediate	Intermediate
Monuron	~6.0	Resistant	Large
CIPC	~4.9	Moderate	Intermediate
Casoron	~2.8	Moderate	Small
Dacthal	~12	Resistant	Large

come up with a model. The model is based on the following diffusional plus
a convective-type

$$C_t + VC_x = DC_{xx} - \frac{1}{\gamma} N_t \qquad (11)$$

differential equation, where C is the concentration of chemical in the voids,
V the velocity of the carrier flowing through the voids, D the diffusion coefficient, and γ the fractional void volume of the packed bed. The flux, J_x,
across the boundary of a plane sheet of soil at a depth x is given as

$$J_x = K_0 (C) C_x + U_0(x,t) C (x,t) \qquad (12)$$

where $K_0(C)$ is the diffusion coefficient of the chemical in soil-water complex, U_0 is the average water velocity in the interparticle voids at depth x

and time t and C the concentrations of chemical in the water. Assuming that the adsorption of a herbicide on soil surface obeys the Freundlich isotherm, the following final leaching model resulted:

$$C\,(x,t) \;=\; \frac{C_0}{2}\left\{ L\,(x,t) \,+\, M\,(x,t) \,+\, N\,(x,t) \right\} \qquad (13)$$

where

$$L\,(x,t) \;=\; \mathrm{erfc}\!\left(\frac{x}{2\sqrt{KT}} \,-\, \frac{U}{2}\sqrt{\frac{t}{K}} \right) \qquad (13a)$$

$$M\,(x,t) \;=\; \left(\frac{4U^2 t}{K\pi} \right)^{1/2} \exp\!\left[-\left(\frac{x}{2\sqrt{Kt}} \,-\, \frac{U}{2}\sqrt{\frac{t}{K}} \right)^{\!2} \right] \qquad (13b)$$

and

$$N\,(x,t) \;=\; -\frac{U}{K}\left(x \,+\, Ut \,+\, \frac{K}{U} \right) \exp\!\left(\frac{xU}{K} \right) \mathrm{erfc}\!\left(\frac{x}{2\sqrt{kt}} \,+\, \frac{U}{2}\sqrt{\frac{t}{K}} \right) \qquad (13c)$$

Here, K, U and R are defined in Eq. (14). Here α is the ratio of the surface area of active sites to the total surface area, k the Boltzman constant, T

$$K \;=\; \frac{K_0}{1+R} \quad ; \quad U \;=\; \frac{U_0}{1+\alpha R} \quad ; \quad R \;=\; \exp\!\left(\frac{\Delta G}{kT} \right) \qquad (14)$$

the temperature, and ΔG the free energy of adsorption. This model describes satisfactorily the effect of free energy of adsorption on the movement of the chemical and predicts how the movement is retarded with increasing adsorption. The leaching of substituted urea herbicides, fluometuron (3-m-trifluoromethyl phenyl 1,1 dimethyl urea) and diuron (3-3,4 dichlorophenyl 1,1 dimethyl urea), in glass beads and norge loam soil columns has been described with a similar theory (31).

Quantitative treatment of leaching data has been used to estimate the diffusion coefficient of pesticides in soils. Lindstrom et al. (32) have calculated the diffusion coefficient of 2,4-D in a variety of soils. They obtained the value of the diffusion coefficient from a modified form of the leaching model. For a larger value of V_0 and smaller value of time the leaching expression reduces to

$$C(x,t) \;=\; C_0 \,\mathrm{erfc}\!\left(\frac{x}{2kT} \right) \qquad (15)$$

where C and C_0 are the concentrations of 2, 4-D at time t at initial time, K
the diffusion coefficient, and x the length. The value of k ranged from
0. 06 x 10^{-6} cm^2/sec for mulky loam soil to 1. 0 x 10^{-6} cm^2/sec for a
Deschute sandy loam. Sand, which is a poor sorber, gave a higher value
of diffusion coefficient than an organic rich hembre soil (3. 9 x 10^{-6} cm^2/
sec).

Ehlers et al. (33) have developed a model for the diffusion of vapors and
nonvapors of lindane insecticides in Gila silt loam. The diffusion coefficient was not significantly different for vapors and nonvapors (~5 mm^2/
week) and was dependent on concentration of lindane. Grahm-Bryce (34)
has described the diffusion characteristics of organophosphate insecticides, disulfoton, and demethoate, in silt loam soil. The observed diffusion coefficients, although independent of concentration of insecticide,
were highly dependent upon the moisture content. For dimethoate the
value changed from 3. 31 x 10^{-8} cm^2/sec to 1. 41 x 10^{-6} cm^2/sec when
the moisture content increased from 10% to 43% volumetric. Disulfoton,
which is less water soluble and more strongly sorbed on soil surface, did
not show any moisture dependence. The diffusion coefficient changed from
2. 83 x 10^{-8} cm^2/sec to 2. 74 x 10^{-8} cm^2/sec when the moisture content
was changed from 43% to 8% volumetric.

IV. DECOMPOSITION OF PESTICIDES IN SOIL

As stated in the beginning, the chemical upon application distributes
or partitions itself in the various compartments of the environment in accord with its physical-chemical properties and interactions with existing
conditions. For most chemicals, applied in a liquid or solid form, the
bulk will ultimately reach the soil surface. There, in either the air or
the water, the chemical is subjected to a variety of forces and processes
that tend to result in an alteration of the chemical. The physical processes, leaching, vaporization, or adsorption, in themselves have but little
effect on the chemical though decomposition may occur during these processes. The three basic processes leading to decomposition of a chemical
are chemical reaction, photochemical reaction, and biochemical or biological reaction (35, 36).

An example of chemical decomposition is that which might be fostered
by excess acidity or alkalinity of the soil. The acid decomposition is illustrated by the transformation of the chloro s triazine herbicides to their
corresponding hydroxy analog (37). Alkaline-fostered decomposition is
found with compounds such as the organophosphates, diazinon (38), dian-
oxon and the chlorinated hydrocarbon, DDT (39). It is probable that the

decomposition of a number of organic chemicals in soil is brought about by variations in pH, reactions catalyzed by inorganic surfaces, and interaction with various ions indigenous to the soil (40).

  Photochemical decomposition is another process by which a chemical may be broken down. A number of studies have been performed on the ultraviolet decomposition of pesticides (41, 42). These studies have been heavily dependent upon laboratory work, since the measurement of photo-decomposition at the soil surface is most difficult experimentally. Diquat and paraquat are good examples of strong ultraviolet absorbers that probably undergo photo-decomposition. They form mono-cation free radicals when exposed to sunlight (43).

  Perhaps the most important method of decomposition of a pesticide in the soil is through the attack by organisms inhabiting the soil (36). A wide variety of organisms present in the soil assures that almost any organic substance will be attacked by one or more of the species of organisms. In so complex a milieu as the soil, with its variable conditions, it would be unreasonable to expect a simple straight-forward behavior in breakdown. Rather, the process is influenced by availability of moisture, nutrient level in the soil, pH, oxygenation of the soil, and such physical processes as diffusion, leaching and adsorption (44, 45). Compounds, by virtue of their structure and physical properties, vary in rate at which they are decomposed (46). Many of the more refractory compounds may show a considerable period of time between application and the first detectable initiation of decomposition (46, 47). The interval between application and initiation of breakdown is called "the lag period." The rate at which it breaks down in the soil has been determined for many pesticides. Various factors influencing this rate have been elucidated, and expressing the quantitative relationship to rate of breakdown has been attempted, at least to a first approximation. For some compounds, the rate appears to follow an order analogous to that of the first order rate law in chemical kinetics. Here C is the concentration of chemical and k the rate constant.

$$\frac{dC}{dt} = k \cdot C \qquad (16)$$

However, the observations on breakdown of pesticides in soil and calculations based on assumption of a first order rate should never be mistaken as representing the actual detailed kinetics. It rather measures the differences between the boundary conditions, without reflecting the intervening mechanisms. Nontheless, the application of this approximation to study of rates of breakdown of pesticides has proven useful. From such data, one derives estimates of an energy value that gives an index of the stability of the compound in the soil and, hence, its probable persistence under specified conditions. It has, in certain instances, even been used to suggest the probable mechanism of breakdown.

Another concept frequently used in relation to decomposition of pesticides in the soil is the half-life expressed in Eq. (17).

$$t_{1/2} = \ln \frac{2}{k} \qquad (17)$$

where k is the estimated reaction or decomposition constant. While the half-life is a useful parameter in estimating the disappearance of a substance from the soil, it does not necessarily tell when the concentration has been sufficiently reduced so as to have little or no further biological activity. Further, because of the various uncertainties and assumptions that must be made in deriving this so-called half-life, it should be recognized only as a useful first approximation and not a precise figure. Some measured half-lives of a few compounds are measured in Table 4. The decomposition of chemicals in soil is a complex phenomenon. It is influenced by a variety of physical factors, concentration of hydrogen ion and other ions, moisture, and the population of soil organisms. It is, however, an important area and one engaging increasing interest.

TABLE 4

Half-life of Some Herbicides in Soils

| Chemical | Half-time (months) | |
	at 15°C	at 30°C
Fenuron	4.5	2.2
Monuron	5.0	4.1
Diuron	7.0	5.5
Tenoron	3.0	1.0
Atrazine	6.0	2.0
Ametryne	6.0	4.5
Bromacil	7.0	4.5
Terbacil	7.5	5.0
IPC	0.4	0.2
CIPC	3.0	1.5
Amitrole	1.5	1.0
2,4-D	---	0.1
2,4,6 TBA	---	8.0

V. SUMMARY

In the foregoing pages, an attempt has been made to give a brief description of some of the more important aspects of pesticide behavior in soils. The object was to demonstrate some of the physical-chemical principles involved in the behavior of chemicals and show how the properties of the compound interact with such things as temperature, moisture, type of colloidal surface, and the organisms of the soil as well. Applying this knowledge may assist in developing the quantitative relationships that can yield a better understanding of this behavior and the persistance of chemicals. From such studies may then emerge the knowledge requisite for both safer and more effective use of such chemicals as may be employed.

REFERENCES

1. H. G. Schwartz, Jr., Environ. Sci. and Tech., 1, 332 (1967).

2. R. Haque and W. R. Coshow, Environ. Sci. and Tech., 5, 139 (1971).

3. H. R. Sherburne and V. H. Freed, J. Agr. Food Chem., 2, 937 (1954).

4. R. Haque and R. Sexton, J. Colloid Int. Sci., 27, 818 (1968).

5. T. M. Ward and K. Holly, J. Colloid Int. Sci., 22, 221 (1966).

6. S. M. Lambert, J. Agr. Food Chem., 15, 572 (1967).

7. G. G. Briggs, Nature, 223, 1288 (1969).

8. S. M. Lambert, J. Agr. Food Chem., 16, 340 (1968).

9. J. B. Weber, Am. Minerologist, 51, 1657 (1966).

10. J. B. Weber, P. W. Perry, and P. R. Upchurch, Soil Sci. Soc. Am. Proc., 29, 678 (1965).

11. A. C. Leopold, P. Van Schaik, and M. Neal, Weeds, 8, 48 (1960).

12. M. M. Allingham, J. M. Cullen, C. H. Giles, G. K. Jain, and J. S. Woods, J. Appl. Chem., 8, 108 (1958).

13. S. J. Gregg and K. S. W. Sing, Adsorption, Surface Area and Porosity, Academic, New York (1967).

14.   W. J. Weber and J. P. Gould, Adv. Chem. Ser., 60, 280 (1966).

15.   V. H. Freed, R. Haque, and J. Vernetti, J. Agr. Food Chem., 15, 1121 (1967).

16.   J. J. Bikerman, Surface Chemistry, Academic, New York, 1958.

17.   J. Koral, R. Ullman, and F. R. Eirich, J. Phys. Chem., 62, 541 (1958).

18.   J. C. Morris and W. J. Weber, Environ. Health Sci., AWTR, 16 (1966).

19.   D. D. Eley, Trans. Faraday Soc., 49, 643 (1953).

20.   A. Fava and H. Eyring, J. Phys. Chem., 60, 890 (1956).

21.   F. T. Lindstrom, R. Haque, and W. R. Coshow, J. Phys. Chem., 74, 495 (1970).

22.   J. D. Russell, M. Cruz, J. L. White, G. W. Bailey, W. R. Payne, J. D. Pope, and J. I. Teasley, Science, 160, 1340 (1968).

23.   M. Mortland, J. Agr. Food Chem., 16, 707 (1968). J. D. Russell, M. I. Cruz, and J. L. White, J. Agr. Food Chem., 16, 21 (1968).

24.   R. Haque, S. Lilley, and W. R. Coshow, J. Colloid Int. Sci., 33, 185 (1970).

25.   H. R. Sherburne, V. H. Freed, and S. C. Fang, Weeds, 4, 50 (1956); R. P. Upchurch and W. C. Pierce, ibid, 6, 24 (1958); R. L. Gantz and F. W. Slife, ibid, 8, 599 (1960); S. M. Lambert, P. E. Porter, and R. H. Schieferstein, ibid, 13, 185 (1965).

26.   C. S. Helling and B. C. Turner, Science, 162, 562 (1968).

27.   W. R. Gardner and R. H. Brooks, Soil Sci., 83, 295 (1957); D. R. Nielsen and J. W. Biggar, Soil Sci. Proc., 25, 1 (1961).

28.   F. T. Lindstrom, R. Haque, V. H. Freed, and L. Boersma, Environ. Sci. Technol., 1, 561 (1967).

29.   L. Lapidus and N. R. Amundson, J. Phys. Chem., 56, 984 (1952); P. R. Kasten, L. Lapidus, and N. R. Amundson, ibid., 56, 683 (1952).

30. H. Brenner, Chem. Eng. Sci., 17, 229 (1962).

31. J. M. Davidson, C. E. Rieck, and P. W. Santelmann, Soil Sci. Soc. Am. Proc., 32, 629 (1968).

32. F. T. Lindstrom, L. Boersma, and H. Gardiner, Soil Sci., 106, 107, (1968).

33. W. Ehlers, J. Letey, W. F. Spencer, and W. J. Farmer, Soil Sci. Soc. Am. Proc., 33, 501 (1969).

34. I. J. Graham-Bryce, J. Sci. Food Agr., 20, 489 (1969).

35. B. Day, ed., Weed Control - Principles of Plant and Animal Pest Control, National Academy of Sciences, 2, 1968.

36. L. J. Audus, ed., Physiology and Biochemistry of Herbicides, Academic, London, 1964. p. 163.

37. T. J. Sheets, Res. Rev., 32, 287 (1970).

38. H. M. Gomaa, I. H. Suffet, and S. D. Faust, Res. Rev., 29, 171 (1969).

39. S. J. Cristol, J. Am. Chem. Soc., 67, 1494 (1945).

40. V. H. Freed, J. Vernetti, and M. Montgomery, Proc. West. Weed Control Conf., 19, 21 (1962).

41. L. S. Jordan, J. D. Mann, and B. E. Day, Weeds, 13, 43 (1965).

42. D. G. Crosby, Res. Rev., 25, 1 (1969).

43. A. E. Smith and J. Grove, J. Agr. Food Chem., 17, 609 (1969).

44. R. L. Zimdahl, Ph. D. Thesis, Oregon State University, 1968.

45. P. C. Kearney and D. D. Kaufman, Degradation of Herbicides, Dekker, New York, 1969.

46. M. Alexander and. M. I. H. Aleem, J. Agr. Food Chem., 9 (1961).

47. C. R. Youngson, C. A. I. Goring, R. W. Meikle, H. H. Scott, and J. D. Griffith, Down to Earth, 23, 3 (1967).

AUTHOR INDEX

Underlined numbers give the page on which the complete reference is listed.

A

Aaron, H.S., 39(88), 62
Ablett, R., 360, 385
Achorn, F.P., 117(18), 135
Ackley, C., 308, 339
Adam, N.K., 344, 384, 392, 393, 396(32), 432, 433
Adams, D.F., 327(166), 329(166), 341
Adams, L.H., 123(69), 138
Addison, C.C., 288, 334, 335
Addy, S.K., 406(64), 434
Agnew, Z., 297, 336
Ainsworth, S.E., 134(122), 141
Akesson, N.B., 275(1), 276(5, 6), 278(29), 289(61), 290(61), 298(61), 299(61), 301(61), 303(61), 304(61), 306(122), 308(133), 309(122, 141), 315(145), 316(146, 154, 164), 317(153, 154, 155, 156, 157), 319(155, 157), 320(154, 155), 326(154, 155), 327(157, 164), 329(167), 331, 333, 335, 338-341
Albersheim, P., 405(62), 434
Aldrich, S.R., 118(35), 135
Aldridge, W.N., 43(96), 62
Aleen, M.I.H., 455(46), 459
Alexander, A.E., 351, 385
Alexander, M., 455(46), 459
Allingham, M.M., 446(12), 457
Amsden, R.C., 294(69), 335
Amstein, E.H., 150(13), 207
Amundson, N.R., 451, 458
Anacker, E.W., 74(9), 91
Andelman, M.B., 16(30), 58
Anderson, D.B., 403(57), 434
Anderson, J.F., 117(18), 135
Anderson, S.M., 55(128), 64

Andreason, A.H.M., 150(12), 207
Anstey, D.G., 278(22), 332
Arai, H., 73(11), 102, 105(14), 105, 111
Argauer, R., 319(158), 340
Aronow, L., 8(10), 57
Ashworth, R. de B., 371, 386
Atkinson, W.R., 297, 336
Audus, L.J., 454(36), 455(36), 459
Auron, M., 46(105), 63
Aycock, R., 131(101), 140

B

Bailey, T.E., 276(15), 332
Bair, F.L., 118(43), 136
Baker, D.F., 26(57), 32(57), 60
Baker, E., 390(11), 403, 405(11), 431, 434
Ball, C., 311(139), 339
Balmbra, R.R., 101, 111
Bamesberger, W.L., 327(166), 329(166), 341
Bancroft, W.D., 107(20), 111
Baratier, J.P., 119(49), 137
Barbe, G.D., 278(34), 333
Barlow, R.B., 25(55), 39(87), 60, 62
Baron, T., 297, 336
Barry, J.B., 399, 433
Batt, R.F., 406(65), 434
Bayer, D.E., 329(167), 341, 390(10), 405(10), 406(70), 432, 435
Becher, P., 66(1), 69(3), 70(3), 73(11), 74(13), 77(16, 22), 78(17, 20, 21), 79(22), 84(13), 85(24), 86(23), 87, 88, 89(29), 90(29), 91, 92
Becker, E.L., 28, 61

473